Exceptional Life Journeys

Exceptional Life Journeys: Stories of Childhood Disorder

Edited by

Jac J.W. Andrews
Division of Applied Psychology, University of Calgary, Calgary, Alberta, Canada

Peter Istvanffy
Calgary Academy, Calgary, Alberta, Canada

AMSTERDAM • BOSTON • HEIDELBERG • LONDON • NEW YORK • OXFORD
PARIS • SAN DIEGO • SAN FRANCISCO • SINGAPORE • SYDNEY • TOKYO

Elsevier
32 Jamestown Road, London NW1 7BY
225 Wyman Street, Waltham, MA 02451, USA

First edition 2012

Notices
Knowledge and best practice in this field are constantly changing. As new research
and experience broaden our understanding, changes in research methods, professional
practices, or medical treatment may become necessary.

Practitioners and researchers must always rely on their own experience and knowledge
in evaluating and using any information, methods, compounds, or experiments described
herein. In using such information or methods, they should be mindful of their own safety
and the safety of others, including parties for whom they have a professional responsibility.

To the fullest extent of the law, neither the Publisher nor the authors, contributors, or
editors assume any liability for any injury and/or damage to persons or property as a
matter of products liability, negligence or otherwise, or from any use or operation of any
methods, products, instructions, or ideas contained in the material herein.

British Library Cataloguing-in-Publication Data
A catalogue record for this book is available from the British Library

Library of Congress Cataloging-in-Publication Data
A catalog record for this book is available from the Library of Congress

ISBN: 978-0-12-385216-8

For information on all Elsevier publications
visit our website at elsevierdirect.com

This book has been manufactured using Print On Demand technology. Each copy is produced
to order and is limited to black ink. The online version of this book will show color figures
where appropriate.

Working together to grow
libraries in developing countries

www.elsevier.com | www.bookaid.org | www.sabre.org

ELSEVIER   BOOK AID
            International   Sabre Foundation

Contents

Invited Foreword

The human journey is often not an easy one. And it is further complicated when the homeostasis or balance in life we all so long for is upset by a quirk of nature or some challenging event from the world we live in. What so many of us take for granted as the "good life" is not shared by all and invariably when someone is impacted by any condition or circumstance that limits their capacity for a full and fulfilling life, others are drawn in whether by kinship or professional need. There is the child who is visually impaired, learning disabled, ADHD, chronically anxious, physically disabled ... the parent or parents who try so hard to meet all their child's psychological, social, physical, and educational needs while also trying to ensure their own psychological health and well-being ... and the professionals who are counted on to help both child and parent in all of these respects but who also is a parent or a person who has all of same needs as everyone else. On the one hand, it is somewhat easier, less stressful, and more objective to have a distance between these three. For example, there is the scientist who studies the causes of conduct disorder, anxiety, or schizophrenia in the laboratory in contrast to those professionals who must apply all of this research knowledge and clinical acumen in supporting the family while working with a child facing such huge challenges. On the other hand, being the parents of a child who is low functioning and autistic with a poor prognosis for independent living requires a full and dedicated daily commitment in doing all you can to maximize whatever abilities and opportunity for well-being that may be possible for the child.

So what is it like to be in the "thick of it"? To be that child diagnosed with a severe learning disability after several years of school failure? To be the parent who dreamed of having a healthy, happy, and loving child who is now challenged every day by their child's needs arising from their diagnosed anxiety condition and eating disorder? Or to be the psychologist or teacher who is seen to be the hope for a better future for the child and parent all the while feeling their distress, worry, and fears.

I initially approached the reading of this book as a psychologist who has spent many years in research, teaching, and clinical practice. After all, psychology as a science and field of clinical practice has contributed much to our understanding of human behavior including its causes and expression. As psychologists, along with our colleagues from related disciplines such as medicine and the neurosciences, we have engaged in research that has both identified key individual differences factors such as intelligence, personality, and motivation and described the causes and effects of human exceptionalities and disorders ranging from giftedness and cognitive disabilities to depression and ADHD. Social work, sociology, and cultural anthropology have further contributed to our understanding of the external conditions such as poverty, broken families, and natural and human-made disasters that impact the expression of all that can be described as human potential. Side by side with education and the applied areas of the various physical and social sciences have come methods and techniques for intervening and assisting individuals who have experienced challenges as a function of biological or social factors. For example, ADHD is a neurobehavioral disorder that can impact a child in every area of their life from home to school to their intrapersonal and interpersonal experiences. And what is required for the child and parents is a multimethod approach including social skills and strength-based training, parenting support, cognitive behavioral therapy, psychopharmacological treatment, and best-fitting educational opportunities. Many years ago I heard the multifactor term "psychoneurobiomedchemeducation" … this is the integration of all those professional resources and their knowledge that is necessary for these children to understand, cope with, manage, and in many cases overcome many of the limitations imposed by the diagnosis of ADHD or Asperger's disorder.

As a psychologist, I immediately thought of the *Diagnostic and Statistical Manual of Mental Disorders* (DSM-IV-TR, 2000) when I saw the table of contents for this book. After all, the DSM-IV is a multiaxial classification system for all identifiable psychiatric disorders according to five dimensions (clinical disorders, personality disorders, acute medical conditions and physical disorders, psychosocial and environmental factors, and global assessment of functioning). As before, these conditions all pose challenges that increase geometrically as a function of the severity and comorbidity of the disorder. Yet, as the research and clinical literature reports that with parent support, effective and caring teachers, informed guidance through counseling, the use of medication when required, and especially creating a sense of hope rather than hopelessness and helplessness that these children can confront their diagnosed "disorder" and in many cases live fulfilling lives.

But the key to reading, understanding, and appreciating this book goes far beyond what textbooks on psychological disorders can ever tell us. Drawing from the writings of important psychologists such as Abraham Maslow and the current positive psychology movement, this book mirrors the struggles and the setbacks together with the successes and achievements of all those impacted by a child presenting with a significant and challenging disorder. Maslow, much like Erik Erikson and other psychologists of the humanistic tradition, described the human journey as one where we seemed compelled to address the challenges and conflicts that we are confronted with as we grow up, become parents and professionals. For Maslow, life is comprised of deficiency needs; those psychological needs that are lacking until fulfilled. For example, we all have a need to be recognized as a person, a real person with feelings and aspirations, with creative abilities, and the need to define ourselves as worthwhile persons. These needs for developing a positive sense of self can also be challenged in so many ways. The child who fails at school may come to see him/ herself as "stupid and no good" when in fact, underlying this is a learning disability so significant that without special programming, the child will not "learn" to read. As Maslow suggests, if we can address these various deficiency needs, we can move forward to what he calls growth needs where the goal is self-actualization.

So as I read the first section and then, without putting the book down, the second and then the rest of the book, I quickly found that I was taking on a different perspective from the psychology I often practice in a busy teaching–clinical environment. There I am engaged in, for example, assessing a child referred for withdrawing behaviors and learning problems whom I hypothesized is quite likely extremely anxious with comorbid dysthymia. Instead I saw in these stories, how I and also the authors are alike all others, some others and no others, a son, a parent, a friend, a psychologist! These are individuals like me who live in a family, community, city; go to school or work; and have desires, goals, and aspirations. These are the children behind the diagnosis of conduct disorder, the parents who are committed to their child in every sense of the word, and the professional practitioner who sees the human value in every client.

The stories are poignant and inspiring. Each in its own way tells of the many factors, relationships, and opportunities that have had a significant impact on the lives of the authors; that define who "I" am, and how "I" have come to know and like (or not like) "me." And in turn, the stories further tell how the author knows and relates to others from parent to teachers and friends. We do not live life in a vacuum but instead as part of a complex social world where everything we do is in turn perceived, felt, and experienced by someone else. Hence, we impact others and in turn, our interactions with others, whether in our role as child, parent, or professional, will impact ourselves as well. These stories so effectively tell the reader how each author attempts to understand themselves and others.

Reading the narratives in this book gives us multiple views of each other. But unlike a book that explores human relationships in general or presents only a self-analysis or autobiography, this book begins each section with a description of an individual who has experienced, lived, and grown up with some special or exceptional "characteristic" that adds further to their uniqueness and can in turn influence the dynamics of their interactions with others. These narratives are not about individuals with the kind of exceptionality that is necessarily admired, desired, or even copied such as physical beauty or the musical talent and skill that comes from a mixture of genetics, opportunity, practice, and passion. Instead the first person to tell us about themselves is someone who has an exceptionality or disorder that has for the most part been deemed deviant, atypical, and abnormal; certainly not what we would wish on ourselves or others that we know and love. What all of these persons share in common is that they manifest a complexity of characteristics that make them stand out in what has traditionally been viewed within the medical model, as having a pathological condition. And what we are asked to read and come to know is the person himself or herself telling others what their life is like, their moments of joy and desperation, their success and aspirations, and their failures but mostly their efforts to become more than their handicap. But even more than that is the insight they have gained about how others see them and the impact this has on them as they also engage in the universal human quest to find themselves in a world that is both caring and potentially cruel. They are, with the support of others, attempting to travel the path described by Maslow as leading toward human growth and self-actualization.

The challenges and successes, the highs and lows, the times when all seem hopeless and then so hopeful are also told to us from a parent's perspective. Who are the parents of these exceptional children? What is so special about these parent narratives is the human capacity and herculean efforts made by them to love, protect, and encourage their children by telling us about their struggles to understand and cope with their exceptional child who is often not understood or even accepted by others. The story does not always end happily ... children do not grow up shedding their ADHD, in spite of the efforts of mothers and fathers ... and sometimes there is the sense that more could have been done or a kind of sadness associated with not "winning the battle" as happened in the movies of old. In these narratives, we also come to know the "parent as a unique person" and how they struggle to define themselves. There are certainly too many documented cases of a parent who can no longer handle the stresses of life that lead to a tragic end for them and their child(ren). We know that psychologically healthy and resilient parents can have the most positive impact on their children and what is reassuring is that so many parents of exceptional children realize this ... that there is a shared caring and that parents need to ensure that they care for themselves as well as their child.

The helping professions are just that ... whether psychologist, psychiatrist, teacher, or therapist, their task and mandate is to assist those who so require the help necessary to achieve all that they are capable of and to live the most fulfilling lives possible. That of course can be the child with a physical or psychological challenging condition, a mother or father or the whole family. It often involves going beyond the direct service role and becoming an advocate for persons or services. But again, we quickly come to see them not only as members of the helping profession but also as individuals, some who even share considerable similarities with the children and parents they "work with." We have often marveled at how capable teachers and doctors and nurses are at handling their emotions, how they are able to put their students or patients first. And they do but is there a cost? Mental health and educational professionals and health-care providers endure considerable occupational stress. But the good they do for others is not in question. In the process they also join the children and parents in their quest for human acceptance and self-actualization.

These stories from all viewpoints are compelling, moving, informing, and above all, are the stories of each writer ... but as a reader, I can see so many similarities to my personal, parental, and professional stories. The trilogy-type narratives provided for each of the exceptionalities provide so many insights into each of these exceptional persons but also those other exceptional persons who are parents and professionals. What stands out in this collection is the profound mixture of "interpersonal and intrapersonal" intelligence that is shown in the narratives of all three writers in each section of the book. And in spite of the difficulties and challenges, the highs and lows, the failures and successes, there is an overwhelming sense of our resiliency to cope with adversity ... and sometimes more than a hint of "I am glad to be me."

Donald H. Saklofske
Professor and Associate Dean Research,
Faculty of Education, University of Calgary,
Alberta, Canada

Preface

Purpose of this Book

Exceptional Life Journeys: Stories of Childhood Disorder is composed of an Invited Foreword and Introductory Stories followed by 46 narratives (personal, parental, and professional stories) distributed across 10 childhood disorders in 4 sections (i.e., Parts I–IV). This text is expected to provide a rich reference for undergraduate and graduate students in faculties of education, social science, social work, and medicine across Canada and the United States (as well as Europe, Australia, New Zealand, and other places), and to provide an insightful and useful reference for professionals in the field of childhood disorders. It is also our belief that it will be a well-received reference source for the general public, especially for those who have an interest in childhood disorders.

The purpose of our book is to offer professionals and students in training within the field of childhood disorders as well as people in the general public with information about the personal perspectives of people who have lived their lives with a childhood disorder, who have parented a child with a disorder, and who have treated or provided services for children with disorders (e.g., teachers, therapists, counselors, psychologists, psychiatrists, and clinicians). In this regard, it is proposed that information that comes from a triangulation of personal perspectives (parent, professional, and self) about parenting a child with a disorder, providing health, educational, social, or psychological services for a child with a disorder, and living the life with a childhood disorder will be appealing and well received by professionals and those in training in the field, as well as people in the general public who have either had similar experiences and/or are interested in the life journeys of those with such experience.

Beyond a clinical, empirical, and theoretical description of childhood disorders or a personal account relative to one particular disorder, this book aims to provide rich narratives of experience from a triangulation of multiple perspectives with respect to numerous childhood disorders.

Rationale for this Book

From our experience, most students in training to become teachers, counselors, therapists, psychologists, psychiatrists, and clinicians, as well as many practicing professionals in health-related disciplines do not always get the opportunity to fully understand and appreciate the circumstances of children, parents, and others who have had to cope with and adapt to childhood disorder. Most professionals in the field of childhood disorders are well trained in assessment and treatment methods and are aware of the clinical, theoretical, and empirical foundations of the work they do. In their training, they get experience in diagnosing the educational, psychological, social, and medical problems of children through their supervised clinical internships. In their training and in their professional practice, they get to interview, diagnose, discuss, consult, collaborate with, and provide services to children and their families regarding developmental and other issues; however, they do not always get an opportunity to fully realize and understand what it is like to have a disorder and what it is like to be a parent of a child with a disorder. Moreover, they sometimes do not know about or fully understand and/or appreciate the experiences of other service providers of children and youth with disorders.

This book offers an opportunity for students in training and professionals in the field as well as the general public to gain some insight and awareness of the life journeys of some exceptional children, their families, and those who have provided professional service to them. This book intends to provide readers with some significant insight into their lives by sharing examples of personal contexts and situations, life issues, challenges and barriers, successes, and recommendations relative to their particular circumstances. Hence, this book is viewed as a very important complement or supplement to the informational resources available to students in training, professionals, and others for developing their expertise, insight, awareness, and empathy with respect to the children and families they assess, treat, and live beside in their communities.

Lastly, the intention of the editors of this book of narratives (stories) was to let the stories primarily "speak for themselves." In other words, we did not want to critically evaluate or appraise the stories in this book. We did not want the stories in this book to be academic; rather, we wanted the stories to be personal and from one's own experience. We wanted the stories to reflect the individual's voice.

We have presented to you what we wanted: real stories from people who felt and thought that sharing their stories might be informative as well as interesting and appreciated by you (the readers). What this book provides is real stories from individuals who have lived their childhood and youth with a disorder, real stories from parents who have raised their children with a childhood disorder, and real stories from professionals who have worked with children and youth with disorders.

Organization of this Book

This book is composed of 46 stories that are distributed among 4 parts. Each section presents from two to four childhood disorders. For each disorder, there is at least one personal, parental, and professional story. Each disorder within each section begins with a prologue in which the nature of the disorder is described. At the end of each disorder section is an epilogue in which some major themes and thoughts expressed throughout the stories are highlighted. The book ends with some concluding comments.

In his foreword, Don Saklofske presents his views related to this book. He attests to the need for the personal views and experiences of professionals, parents, children, and youth associated with childhood disorders to be meaningfully placed within the framework of theory, research, and best practices with respect to childhood disorders. In their introductory stories, Dr. Jac J.W. Andrews presents his guiding principles of parenting and working with children with disorders, and Peter Istvanffy presents some of his values and beliefs for schooling children and youth with disorders.

Part I: Behavior Disorders. In this section, there is one personal story, one parental story, and three professional stories with respect to attention-deficit/hyperactivity disorder. In addition, there is one personal story, one parental story, and three professional stories with respect to conduct and oppositional defiant disorders.

Part II: Emotional Disorders. In this section, there are three personal stories, five parental stories, and one professional story pertaining to childhood anxiety disorder. In addition, there is one personal, one parental, and two professional stories pertaining to mood disorders (childhood depression).

Part III: Developmental and Learning Disorders. In this section, there is one personal, one parental, and three professional stories related to learning disabilities; one personal, one parental, and one professional story with respect to intellectual disabilities (Down syndrome); one personal, one parental, and one professional story with respect to developmental coordination disorder; and one personal story, one parental story, and three professional stories related to autism.

Part IV: Eating and Health-Related Disorders. In this section, there are two personal stories, one parental story, and one professional story pertaining to eating disorders. In addition, there is one personal story, one parental story, and one professional story pertaining to somatoform disorders (conversion disorder).

Contributors

We very much want to thank all of the contributors of the stories within this book. All of them have been committed to the book project and undertook lots of time, effort, and consideration to write their stories for this book. It is not easy to write a story about your personal experiences. This is particularly true when you are writing a story that conveys your innermost thoughts and feelings about yourself and the challenges and struggles you have faced as someone living with a disorder, as someone who has been a parent of a child or youth with a disorder, or who has been professionally involved in the assessment and treatment of a child or youth with a disorder. All of the contributors of this book not only wrote stories that were clearly and cogently told, as well as thought provoking and fluid, but they also wrote stories that were difficult to reflect upon and share. We applaud all of the contributors' willingness to share their stories. We have learned greatly from them and have benefitted from their openness, honesty, and genuineness relative to their exceptional life journeys.

We thank the following contributors to this book who consented to being identified in this book as well other contributors to this book who did not wish to be identified. The stories come from individuals, parents, and professionals from Canada, United States, Europe, Australia, and New Zealand. This book could not have been done without you.

Sherri L. Armstrong
Natalie J. Bisson, Clinician, Children's Mental Health
Caroline Buzanko, Registered Psychologist
Jeff Chang, Ph.D., Registered Psychologist and Assistant Professor of Counseling
    Psychology
Dr. Shawn Crawford, Registered Psychologist
Darlene Desgagne, Parent
Stephanie Durante, Student
Paige Feurer, Registered Psychologist
Tyler and Shellia Flaumitsch, Parents
Dr. Frederick French, Registered Psychologist
Dr. Jenny Godley, Assistant Professor of Sociology
Lauren D. Goegan
Susie Greco
Dr. Peter Greven, Child and Adolescent Psychiatrist
Dr. Nina Gudeon, Child and Adolescent Psychiatrist
Jody Harpell, Ph.D., Professor and Registered Psychologist

Bonnie J. Heath, Middle School Teacher and Applied Psychology Graduate Student

Janet Jensen, Parent

Tara Jensen, Sixth-Grade Student

Dr. Paulene H. Kamps, Parent, Educator, Kinesiologist, and Registered Psychologist

Martina Kanciruk, M.A., Doctoral Candidate and Registered Psychologist

Tabby Lundy

Dr. Brent MacDonald, Registered Psychologist

Angela MacPhee, Registered Psychologist

Joseph Mann

Sandra Mann

Adam McCrimmon, Ph.D., Registered Psychologist

Milena M. Meneghetti, Parent

Tracy Mitchell, M.Ed., Registered Psychologist

Nicole M. Nannen

Dr. David Nylund, Professor

Barbara Olson, Parent

Natalie Olson, Student

Natalie C. Parnell, B.A., B.Ed., M.Ed. Candidate

Cathy P. Robinson, R.N., B.S.N., Nursing Instructor

Dr. Shelly Russell-Mayhew, Associate Professor and Registered Psychologist

Ken Ryba

Tyson Sawchuck, M.Sc., Registered Psychologist and Pediatric Clinical Psychologist

Carol M. Shirley, Parent and Registered Psychologist

Dr. Teeya Scholten, Registered Psychologist

Gloria Singendonk, Parent

T.M. Smith, Teacher

Lydia Unger, B.A., M.Ed. Candidate, Psychology Intern

Annette Vance, Registered Psychologist

Dr. Lilian C.J. Wong, Psychotherapist

Dr. Paul T.P. Wong, Registered Clinical Psychologist

Acknowledgments

We are indebted to the contributing authors of this book for their commitment to the project and for sharing some of their life experiences with respect to childhood disorders. This is truly their book. We are very grateful to Don Saklofske for writing a foreword to this book. His support for this book and his great insight is very much appreciated. The publication of this book was made possible by the efforts of two people at Elsevier: We thank Lisa Tickner, Publishing Director, for her support of this project, and Paul Prasad, Editorial Project Manager, for his guidance.

Finally, for their love and encouragement, we share this book with our families who saw tremendous merit in this project and continually made us feel that it was worthwhile. This book is dedicated to them.

Introductory Stories

Guiding Principles for Parenting and Working with Children with Disorders

Jac J.W. Andrews

"The keys for success are within us; once we find them, all we need to know is when and where to use them"

Andrews, 2011

Introduction

Effective parental and professional interactions with children who have disorders requires an understanding that children who have disorders *are more like than unlike* other children who do not have disorders. In this regard, we need to be passionate and loving in our roles, wise and understanding, positive, flexible, empathic, honest, persistent, accepting, trustworthy, helpful, and empowering. Moreover, sometimes we need to seek help from others, effectively think and solve problems, and make good judgments and decisions.

We Are More the Same than Different

We are not much different from those we separate out as different from ourselves. In our lives, most of us have had times when we have acted out and have felt depressed. Sometimes, we have felt defeated and felt like failures. We have also struggled in life, been rejected by others, and have felt badly about ourselves. The difference between those who have disorders in learning and development and those who do not is with respect to the amount of times they have experienced these things, the intensity in which they have experienced these things, and the length of time they have experienced these things.

Although we are more the same than different from each other, it seems that we most often prefer to be considered as exceptional or different unless our exceptionality or difference is considered by others and perceived by us as being unattractive or unacceptable. For example, I can remember a time when I was a college teacher (and before I was married) when I first thought about this. I was in a school cafeteria when I saw a woman I knew talking to her friends and I could faintly hear that they were talking about me. So, I snuck up behind them and hid myself behind some greenery so I could hear what they were saying about me more clearly and not be seen. They were talking about my date with a woman the night before. During this conversation, one of the women said her friend (who I dated the night before) told her that I was "just an average guy." Well, I jumped from behind my hiding place and yelled at them that "I AM NOT AVERAGE, I AM EXCEPTIONAL!" After everyone recovered from this shocking behavior, I reflected for a moment and thought how silly it was for me to become so upset about being called normal or average and how much I wanted to be considered exceptional or different. I realized at that moment that our views about normalcy and abnormality depend on our situation, our experience, and our self-perception.

During the same time, a friend asked me to help a 10-year-old girl (Natasha) who was a fifth-grade student and struggling very much at school. I found out from Natasha that she was very depressed about her life and that she did not feel that her mom loved or appreciated her. In fact, she thought that her mom believed her to be stupid.

While working with Natasha, I discovered that she had a significant memory problem. Moreover, I assessed her reading and mathematical abilities to be at a first-grade level. During my work with Natasha, she very often said to me that the thing she wanted most was "to be like the other kids" (to be normal) and be more appreciated by her mom. In my discussions with her mom, it became obvious to me that she was very disappointed with Natasha and felt that she was not trying "hard enough."

Over the course of the year, I was able to teach Natasha many learning strategies that resulted in her making many academic gains. More importantly, her mom learned to understand more about her difficulties and to appreciate many of Natasha's remarkable personal qualities as well as her potential in many areas. By the end of my year with Natasha, she was able to achieve at about her grade level, with support, and raise her self-esteem to the point where she no longer felt hopeless and ashamed. Even more important, I helped them to create a more empathic and loving bond with each other, and helped her mom to realize her daughter's strengths and weaknesses. They each became able to see Natasha as both average and exceptional! For example, I successfully taught Natasha a math strategy that allowed her to accurately perform math calculations "in her head" that far exceeded the skill level of her classmates. This not only impressed her teacher but also her mom to such a degree that Natasha was often asked to perform her new skill for others in the class as well as for visitors in her home. This made Natasha feel proud of herself and allowed her to receive positive feedback from those who once gave her little positive regard. She and her mom so deeply felt her success that I received a handmade blanket from her mom that I still have today and a bond with Natasha that remained for many years. In fact, Natasha and I kept in contact until she graduated from high school and entered college.

From these experiences, I began to realize that my mission in the field of childhood disorders had to be about helping others to see how children with a disorder are both the same as and different from children considered average.

Sometimes We Need to Change as Much as Our Children

One of the greatest challenges and accomplishments of mine over the years has been with respect to changing my thinking and behavior for the betterment of my relationship with my child. I was late to recognize the learning disability within my daughter. To me, she was perfect. For me, there was no problem, until it became obvious that the reason she found school increasingly difficult and why she was becoming increasingly defiant was not because of a behavior problem but because of a learning problem. I had to stop seeing her as I wanted her to be and start seeing her as she was. It became important for me to change who I was with my daughter in order for our relationship to change and for me to become more of a positive influence. Until this time, I was who I was and did not change who I was while my daughter was naturally changing who she was in her development. I was very often demanding, but reasonably demanding, I thought. I considered myself humorous and clever with my humor. The problem was that my demands became anxiety provoking for my daughter, and she coped by resisting. Although my humor was funny to others, my daughter sometimes saw my humor as insensitive and insulting to her. Of course, these were not my intentions; they resulted from my inability to be sensitive and empathic relative to my daughter's psychological and educational needs and struggles both in school and at home. Once I became more aware of how my behavior was influencing her behavior, I knew I had to change. But, for me, changing my thinking and my behavior was not easy. I never had to really change how I thought and behaved before. I learned to change the manner in which I spoke to my daughter so she would not misperceive or misunderstand my messages. I no longer could tell jokes with hidden messages. Instead, my messages needed to be clear and transparent.

One of my daughter's biggest problems was making connections in what she heard. She is very intelligent, creative, and spontaneous; however, she did not always get the full context or meaning from what I (or others) said to her. So, I had to learn how to talk to and act with my daughter in ways in which she would not be confused. I learned how to become more explanatory, open, clear, consistent, and "real" with her. In fact, this resulted in me not only being a better communicator with her but a better communicator with others. I benefitted from my changes as much as she benefitted from her changes, both in and out of school.

I learned that we often expect our children to change in ways we think are best for them but that we are not always so willing to change ourselves. Most importantly, I learned that sometimes we parents need to change as much as our children do. I also learned that cognitive and behavioral change is not easy. It takes a lot of time and practice to change what we have typically and naturally done in the past. Change requires patience, diligence, commitment, and time. This realization and experience made me a much better dad, and a much better child psychologist and teacher.

In the end, my daughter taught me how to become a better me while I was trying to make her a better her.

Passion Drives Much of What We Do and How We Do it

In our roles with children, everything we do should demonstrate to them that we love them, care for them, and respect them. We do this through our passion and our values.

Our passion arises from our values as well as drives our values. Through our passion and values, our children and others see who we are and what is most important to us. In other words, *passion drives what we do and how we do it.* I am passionate about the people who are important to me, about helping people, and about discovering and sharing new ideas and experiences. As part of my passion, I value generosity of spirit, courage of conviction, a positive approach to life, cooperativeness, genuineness, honesty, openness, and integrity. I reveal my passion and show my values through my behavior and particularly through my various rituals. For example, one of my rituals is to make sure to give my children lots of praise for the good things they do before criticizing them for the things they do not do so well. This allows them to consistently understand that I am mostly a positive person and that I see them mostly in positive ways. It also allows them to better accept and consider my occasional critical comments.

Through our general behavior (e.g., being honest, genuine, open, integral, generous, positive, cooperative, and inquisitive) and our rituals (e.g., positive reinforcement of behavior and sharing of self and things), our children form a sense of who we are, what we appreciate, and why we act the way we do. The more congruent we are between our values and our behavior, the more sure our children will be about what they can expect from us. This is critically important for our children's development and learning. In this regard, our values not only become more explicit to others as we act in ways that are congruent with them but also provide a model for our children to adopt and benefit from. My children have often characterized me in the ways I have noted myself, which affirms my passion and values as well as validates my sense that *they see me as I am.* However, sometimes our passion and values can be masked from others. In fact, sometimes the people we associate with can perceive something else about us. In this regard, I remember a particular experience a while ago, when my daughter was 16 and a friend slept over at our house.

One thing I need to mention before telling this story is that I am a demonstrative and energetic fellow. I am also passionate about life, which has caused some difficulties for me within my family for years. Many times when I am talking to family members, I am dramatic and loud. Hence, for many years, this has sometimes confused my family about how I am feeling because, when I think I am just being passionate and dramatic, they have often thought I was angry. What I am about to share is an example of this and is somewhat of a personal and family revelation.

Things were fine with her sleepover until she invited some other friends to the house late at night. If her sleepover had been during a weekend, it would not have been a problem. However, even though she did not have school the next day, I had to work and be part of a meeting that was significant for my career. After many gentle requests that my daughter and her friends keep quiet between 2:00 a.m. and 4:00 a.m., I finally marched downstairs to her room and assertively let them know that it was time for her friends to leave and time for my daughter and her sleepover friend to go to bed. As I was walking up the stairs and as everyone was beginning to act on my request, I heard one of my daughter's friends say to her, "Your dad is REALLY MAD." I heard my daughter say back to him, "OH NO, my Dad is not mad, he is just VERY PASSIONATE." In a comical way, I realized that such passion could confuse people about how we are feeling, even if those people seemed to get the message.

Not Understanding the Problem Can Lead to Not Understanding Our Children

We often do not see the major problems of our children. Instead, we see things that irritate us, or make barriers for us, or confuse us. Because we do not like to be irritated or confused, we do not see the problems. We do not like when things interfere with our expectations and plans. However, if we cannot see the problems, we cannot effect positive and meaningful change. For example, how often have we been upset with another person when we want him or her to do something for us, and in response, he or she says to us, "This sucks!" Or, how often have we been frustrated with another person when we ask him or her to become involved in something we are doing and he or she replies, "I can't do this, so what's the use!" When we react to these sayings by becoming upset or frustrated, it is because we think that the person is choosing not to do something for us or be involved in something with us. However, that is often not the problem. Many times, the problem is that these people respectively suffer from a history of defeat and a lack of self-efficacy in what we are asking them to do. If we recognized the problem as a lack of self-regard rather than as an act of disregard, then we are less likely to be upset and more likely to be helpful and supportive. Importantly, we are more likely to be appreciated by the other person because we will be able to act in a way that demonstrates our understanding and appreciation of the other person.

In my early days as a professor, a student came to me after class and introduced herself as Rachel. Specifically, she said, "Hi, my name is Rachel and I am learning disabled." The problem here was not that she was learning disabled but that she identified herself as learning disabled. If I had not had insight regarding this, I would have dealt with her in a way that reinforced this "self-label" rather than deal with her in a way that showed to her that I understood that she saw herself as only "one self" compared to how most of us see ourselves as many selves. Hence, by seeing the real problem, I was able to be more understanding of and more helpful to her.

Our Children's Strengths Are More Important than Their Limitations

My training as a teacher (in the area of special education) and my training as a psychologist were largely based from a diagnostic-prescriptive (medical) model. Hence,

I was trained to look for the problem, make recommendations, and provide interventions (treatment) relative to the deficit functioning of the individual. Not long into my practice as a teacher and psychologist, I changed my focus from a deficit model to a "strengths-based model," where I emphasized considering an individual's strengths as the basis of treatment. That is not to say that problem areas or areas of deficit are overlooked; rather, it means that remediation, accommodations, and compensations for the areas of weakness are addressed along with and often through the individual's strengths.

I learned through experience that children and youth are not disordered in all areas or healthy in all areas. I learned to appreciate that all children have strengths and weaknesses and that we need to focus as much or more of our attention on the things children do well in as we do on the areas in which they do not do so well. In this regard, I will share with you that I always wanted my son to be an athlete like I was. I wanted him to grow up and be actively involved in all sports, be very successful in school, and eventually have a career that was well respected and well paid. I think my vision for my son is similar to many fathers' (and mothers') visions of their sons. As it turned out, my son did not have a proclivity for sports and struggled in school. He has decided to pursue an acting career. Besides being a bright, creative, mature, sensitive, respectful, good-natured, generous, and friendly young man, he has strengths in the area of presentation and projection. In other words, his strength of character and strength of portrayal of other characters as well as his passion for dramatic arts is what distinguishes him and will lead him to success and happiness in the world. I always wanted him to be able to rectify his limitations, but in the end, his strengths and further development of these are most important, and these strengths will serve him the best in his future.

The fact is that children move in and out of healthy and unhealthy states throughout their development, and we cannot just focus on the unhealthy states. Although some problems in childhood are chronic and persist into adulthood, others are variable in terms of length of occurrence. In any event, learning and development are transactional and are very much influenced by the people and situations within their environment. I believe and have seen many remarkable changes in children, primarily due to the influence of people in their lives who have focused on their achievements rather than on their failures, on their competencies rather than on their limitations, on their self-regard rather than on the regard of others. These people have helped develop their child's learning style, facilitated their problem-solving and coping skills, fostered their locus of control and positivity, increased their social competence, and promoted their motivation to succeed and their persistence and vigilance in areas of difficulty.

Seek to Understand, Then Seek to Be Understood

Most people like to be listened to when they talk. However, most often, the person "listening" does not really understand what is being said. In fact, they are typically more interested in replying. Most conversations are "collective monologues." By that, I mean most people are filled with their own sense of wonder and their own autobiography. In conversations, people mostly want to share information about themselves rather than listen and understand the other person. Most people (including me) do not tend to *really* listen actively and reflectively. Instead, we often ignore a lot of what is being said to us; we pretend to listen, or we selectively listen. Accordingly, when we respond to someone, we often question him or her from our frame of reference rather than from his or her frame of reference; give counsel based on our experience rather than on their experience; and/or evaluate what they say to us and make some sort of recommendation (suggestion) to them.

To be effective communicators, we need to be able to empathically listen to others. By this, I mean that we need to listen reflectively and understand not only what they say but also how they feel about what they say. Especially with our children, we need to be able to convey to them that we have understood what they have told us before we try to get them to understand us. When we communicate empathically with others, we will demonstrate to them that we are "in tune" with them. Nothing is more gratifying than talking to someone and feeling that we have really been heard. In this regard, a student of mine a few years ago did a study of women who had breast cancer. What she wanted to find out was: What was most important with respect to their effective treatment? She found that what was most significant in terms of their treatment progress was that they adhered to the treatment plan, they were satisfied with the treatment plan, and they were in a stable psycho-emotional state while in treatment. But, what was most significant relative to their treatment outcome was the personal communication they had with their physicians. The women who were most successful in their treatment were the women who felt they were understood by their physicians. In this regard, one of the best things we can do as parents and as professionals is effectively and empathically communicate with the children we parent or serve.

See it Their Way as well as You See it Your Way

We do not see the world as it is. Rather, we see the world from the context of who we are. So, when people do not see things the way we do, we often think it is because something is wrong with their thinking. To be effective with others, we need to be able to appreciate how they perceive their world and how they make sense out of what they experience. We all have our own mental "maps" or "boxes" from which we make sense of the world and people around us. To be able to relate to others, we need to be able to get "out of our boxes" (our own "life" context) and get into their "boxes" (their "life" contexts).

The more we are aware of our "maps" or "boxes" of thought, the more we can appreciate how our own experiences and ways of thinking have influenced us. This allows us to be more responsible for our actions and be able to test our thoughts against the reality that confronts us. This also allows us to appreciate the views and thoughts of others and be open and flexible to their perceptions and experiences. The ability to do these things will lead to our ability to be more empathetic with and empowering of others.

When our paradigms of thought can change from the influence of others' paradigms of thought, we will be more willing and able to accept others' positions. A number of years ago, a mother of a young child attending day care asked me to find out what was happening in the day care that would lead the people there to think her son Jason was a bully (conduct disordered). From her perspective, her son was a sensitive and caring young boy who could not possibly be a bully, and she was concerned that this was the reputation he was getting within the day care.

I went to the day care to observe her son and see if I could find out what was happening. Within a couple of days of being in the day care, I discovered what was happening to lead the people there to think of the boy as a bully. I found out that the staff had recently decided to try to make the kids more self-responsible for their behaviors, as well as self-reliant. They also wanted to reduce the number of times the children would come to staff for help with their difficulties. Hence, whenever a child would come to ask for help, they were told to try working it out for themselves. During my time at the day care, I observed Jason interacting with others. I saw Jason's behavior was consistent with how his mother had described him. He seemed to be a sensitive and caring young boy. I also observed that others often bullied Jason. Some of the other young children in the day care would call him names, push him around, and threaten him. On these occasions, Jason would react in ways that would be considered by most as appropriate ways to deal with this aggression from others. For example, he would try to ignore these other boys or try to walk away. However, these strategies did not work for him because the other boys would continue to taunt him and follow him wherever he went. At one time, he became very upset with how he was being treated and went to ask for assistance from one of the staff members, only to be told to work it out for himself. Upon return to the play area, he found himself back with the boys who continued to bully him, tease him, and push him around. It got to the point where Jason pushed back and hurt one of the boys. One of the staff members saw this and scolded Jason for his behavior. She also seemed to be very upset with him for acting inappropriately.

It became obvious to me that, instead of being a bully, Jason was a victim. He was only reacting to the situation, not creating it. Unfortunately, the staff only saw Jason's reactions to bullying behavior and not the antecedents. Moreover, their approach toward developing children's self-responsibility and self-reliance was misguided. They did not know one of the basic understandings of child development, which is that young children need adults to help them solve some of their problems because they are not sufficiently equipped to know how to behave in some situations. Moreover, sometimes children need us to intervene.

Two significant things happened from this experience: (1) Jason was no longer thought of as a bully, which not only relieved his mom but validated her opinion of her son, and (2) I was able to provide some guidance to the day care staff about how to better develop children's self-responsibility and self-reliance.

The point of the story is that neither the staff members nor Jason were bad people. In both cases, they had the right idea but not all of the understandings and experiences required to achieve better outcomes for themselves and others. By getting to know their paradigms of thought and the extent to which their experiences had influenced them, I was in a better position to appreciate their circumstances and respond to them in a way that could be understood and more constructive.

Making People Feel Able Leads Them to Being Successful

When you move through life without fully knowing how to do many things you need to know to succeed, you will not feel confident and secure in your life. A simple idea in child development is that children need to know the basics before moving on to more difficult things. I believe that everyone can learn particular things to mastery; that some people need more time than others to master certain tasks; and that some people need more assistance than others in mastering these tasks. From my perspective, there is no greater accomplishment than self-empowerment because we need to be able to develop mastery in some tasks and have some control in our life to be successful. Hence, we need to be able to empower children if we want them to be successful in their lives. I learned from my son and daughter what I think is the essence of achieving self-empowerment. I asked my daughter once how she got to be such a good singer, and I also asked my son how he got to be such a good actor. Their responses were the same. They both said: Keep on trying when you don't get it right, imitate others who can do it well, get some help when you need it, and do it (right) over and over again. Of course, what they said is what we have known for a long time. To master something, you need to be resilient, model others, get good coaching, and practice "smart."

Sometimes We All Need a Little Help

Early in my teaching career and then later in my work as a psychologist, I learned that I did not always have the answers to all the problems. In my early adulthood, I thought I had a personal lack of knowledge and skill if I was not able to deal with any learning or behavioral problem that came my way. At that time, I thought asking for advice or guidance from others was an admission of inadequacy. I changed this viewpoint when a young boy with Down syndrome entered my class. I felt unprepared and unskillful in my approach to teaching this student. For the first time, I needed help to better understand and learn ways to reach and teach him. I reached out to others to help me be a better teacher for him. The first people I went to were his parents who, during my visit and many visits during the year, provided me with much useful information. Through his parents, I was also introduced to and participated in the Association for Mentally Retarded Children and Youth in my region, which also "opened my eyes" and taught me so much more about children with Down syndrome. The people in this association helped me become not only a better teacher but also a better resource for others who might need help with respect to children with Down syndrome in their classrooms.

Through this experience and others like it, I learned and came to appreciate the value of reaching out to others for their knowledge, expertise, and experience. I also learned the value of working collaboratively with health professionals and others when dealing with complex cases (or at least complex to me). I learned how I could provide much better service with more resources and assistance from others.

Not only did I learn the value of getting help from others, but I also learned the benefit of reaching out for help when it came to my own children. At times, I discovered that seeking consultation from others about issues or circumstances in my children's life was to my benefit and my children's benefit, in that it could help me become a better father. From both my personal and professional experiences, I became more appreciative of the fact that, sometimes, we all need a little help from others who cannot only help our children but also help us help our children.

Accepting Comes from Trusting

If you want respect from others as well as compliance and cooperation, you need to be worthy. To be considered worthy by others, you need to build trust. You build trust by being courteous, kind, and honest. Among other things, you need to attend to the little things, keep your commitments, clarify expectations, be loyal to others (even when they are not physically present), and apologize when you have done wrong. So, if you want to build trust, you need to demonstrate that you value the other person and spend enough appreciated time with him or her to establish a reserve of trust.

I experienced the importance of trust and appreciation a number of years ago when I volunteered my parental guardian services for a school that needed parents to help supervise students who were receiving swimming lessons at a public pool. I was in charge of five kindergarten kids. My responsibilities included making sure that they were properly dressed and prepared to leave the school, making sure they all got on the bus that was taking them from the school to the public swimming pool, ensuring they got off the bus and into the recreation center, helping them change into their swimming suits, escorting them to the pool, making sure they were properly attended to during their time at the pool, and making sure that they safely returned to the school after their lessons. One of the boys in my care had difficulties finding his swimsuit in his locker and getting ready to leave the school. After considerable time, we finally got on the bus. On the way to the pool, I noticed just before he did that he did not have shoes on his feet. When he noticed this, he began to cry and was very upset that he did not have his shoes. I did everything to comfort and soothe him but it did not seem to work. I told him that having his shoes really did not matter, that we would find them at the school before he left for home, that I would carry him into the recreation center if he wanted, that no one would care whether or not he had his shoes, and so on. When we arrived at the pool and were in the changing room, he was still crying and upset about his shoes. He left the change room and went to the pool area with a frown on his face and was very unhappy about not having his shoes. While watching him and feeling sorry for him, I decided that I would go back to the school, find his shoes, and bring them back to him. When I returned from the school with his shoes and entered the pool area, I showed him that I had his shoes. Upon seeing me and his shoes, he immediately ran to me and gave me a big hug and thanked me many times. For the rest of the day, he had a big smile on his face and seemed to really enjoy his day.

I learned from this experience that, although I thought it was a "little thing" that he did not have his shoes, it was a big deal for him. I learned that the little thing of getting his shoes for him was a big deal for him. I also realized how understanding others' needs, valuing another person's views (even if they do not correspond with your views), attending to little things for people, and showing courtesy and kindness has a huge impact on people when it comes to their feelings about you.

Most importantly, I realized that, by doing these kinds of things, you can build a reserve of trust with people that can last for a long time (especially if you do not deplete the reserve by being occasionally less considerate). This was demonstrated to me by this little boy's reaction to me that day. But it was even further shown to me by his behavior with me every time I met him after that occasion. In fact, years later, I met him when he was a teenager, and he let me know that he still remembered that occasion and still held me in very positive regard after all of those years. It was a testament to the power and influence you can have with someone if you build trust with him or her.

The Greatest Impact We Can Have Comes through Our Ability to Think Well

"Mary walks into the school staff room for a meeting at 12:30 p.m. Everyone else has been there since 12:00. What do you think about Mary?"

When I tell my university students this during class and ask them to tell me what they think of Mary, this is what I typically hear back from them: "Mary doesn't like meetings," "Mary is tardy," "Mary is irresponsible," "Mary is not committed to the staff and the school," "Mary is disrespectful," and "Mary is inconsiderate."

After hearing these comments, I let the students know that the meeting was not until 1:00 p.m. They then are embarrassed that they described Mary in the ways they did and shocked that they jumped to their conclusions about Mary without considering other possibilities for her coming to the staff room later than others.

I tell my students that their remarks about Mary are very typical and that people usually jump to conclusions about other people and about various situations because of the natural tendency for people to make judgments without sufficient evidence. In fact, those who do not do this have usually been trained either formally or informally to think in another way. Without training, people tend to make immediate (impulsive) judgments about people and situations and then sometimes later think about the factors that may have influenced particular circumstances. People who are trained in making judgments typically do so in a reverse order compared to those without training. In other words, trained decision makers often consider the evidence first, interpret the evidence with respect to possible hypotheses for particular actions and events, and then make a judgment from the evidence. When people do this, they typically do not respond to questions (like the question about Mary) with assertions but rather with further questions: "Is Mary a teacher?" "What time was the staff meeting?" "Why were the other people in the staff room since 12:00?", and "What was Mary doing before she came to the staff room?"

We owe our children our efforts to develop their potential to think, and to *think well*. To think well is one of the most important keys for being happy and successful in life. It is not clear to me that we have a more important task. It requires a deep and long-term commitment. Teaching our children to think well also teaches them to be inquisitive and fair-minded. It teaches them to be open-minded, to be skeptical of quick solutions, and to put forth an effort to know more rather than less before making a decision. It teaches them to be less impulsive, listen more carefully to others, appreciate points of view different from their own, find out what they do not know or understand, and compensate more effectively with respect to their limitations. If nothing else, we owe it to our children to know what it means to think well.

From a professional perspective, I tell my students who are in training to be psychologists that they need to be accountable and wise when working with children and youth who have disorders. First, I tell them that they always have to think about the individual. By this, I mean that they always have to think about a child with a disorder, and *not* to think about a disordered child. A child or youth who has a disorder is someone who has to deal with that aspect of his or her life. The disorder does not define the child or youth; rather, the disorder is just a particular aspect of who they are. Second, I tell my students that, to be accountable in their service to children and youth who have disorders, they need to be guided by a number of things relative to their assessment and treatment of children and youth. They need to be aware of and reference the supported general and specific models (theories) relative to child development and childhood disorders. They need to know, for example, what is understood about such broad things as personality, intelligence, affect, academic achievement, and behavior. They also need to know what is normal and abnormal. They need to know what a *dis*ability is by knowing what *ability* is. They need to know what is typical and what is atypical. They also need to know, for example, what is generally meant by anxiety, depression, attention deficit, and other psychological constructs. They also need to know the specific models (theories) related to these constructs that researchers, scholars, and experts in the field have developed. Hence, they need to know the general and specific understandings related to child adjustment and child maladjustment. They also need to be constantly informed about what researchers are learning about assessment and treatment of children and youth from the research evidence. Moreover, they need to have clinical understanding of what works well and what does not work well for individual children and their families, with objective discernment of their day-to-day, month-to-month, and year-to-year experiences with children, youth, and their families relative to their developmental and experiential issues. From their good thinking relative to psychological theory, psychological research, and psychological clinical experience, they will become wise and provide excellent service for children and youth.

By working toward the betterment of children, youth, and their families with good thinking and accountability, they will "fly with the angels."

Some Values and Beliefs for Schooling Children and Youth with Disorders

Peter Istvanffy

For the past 38 years, I've enjoyed an extremely satisfying career as an educator. For the first 12 years, I was a teacher/administrator in community schools. Since then, I've been the head of school for Calgary Academy, a highly specialized private school for exceptional children, and the executive director of the James E. Chaput Center for Educational Research Applications.

During my tenure at the Calgary Academy, I have strived to lead our school and our educational professionals toward maximizing our students' intelligence, social–emotional development, and talents by employing evidence-based methods of teaching and learning relative to the cognitive and social–emotional development of our children and youth. All of our students are provided personalized curricula, learning approaches, and strategies, and associated materials based on both their strengths and weaknesses. We strive to enable our students to be critical thinkers, problem solvers, independent learners, socially skillful, and confident advocates for their needs in their world.

Our students come to us because they have typically failed to achieve success academically and socially in other places they have been. Our student population includes students who have attention-deficit/hyperactivity disorder, learning disabilities, developmental disorders, and social–emotional disorders. Hence, our school comprises much student diversity. Each child in the school is ensured an environment that reflects and promotes strong interpersonal skills, positive attitudes, and high self-esteem. Our guiding principles and values are as much about the development of character in children and youth as they are about academic growth. For example, embedded within all of our students' studies and activities are the positive character attributes of respect, enthusiasm, altruism, commitment, and honesty. Fundamental to all aspects within the Calgary Academy is the recognition and reward for students with respect to their demonstration of their character attributes because we have learned from experience that these are the strongest incentives for our students' academic and social development. We want our students to strive for establishing "new personal bests." We want our students to not only develop their own skills and positive attitudes but also develop within them "helping behavior" for fellow students who are having difficulty.

For many parents of students who have enrolled their children in our school and for most students who have come to our school, life has been fraught with frustration and failure. Frustration from the parents typically comes from their unsuccessful struggles in having others understand their children and their needs and from their inability to have their children's needs met. It also comes from their disappointment in not being able to foster the development of their child in ways that they expected would naturally occur through the parenting and support of their children. Frustration from the students usually comes from their academic failure in previous schools and the subsequently evolving low self-esteem.

We have been pleased by the recognition and support we have received from the parents of our students over the years. It strengthens our resolve and validates our approach to teaching and learning when we hear testimonials from our parents about their regard for our school and the experience our school has had on them and their children.

Once such testimonial came from a parent of two children in our school who shared the following with us:

> *"I have been very pleased with the experience that my children have had at the Calgary Academy. I do not think it is an overstatement to say that the school has positively changed our lives. Our children have learned how to think 'out of the box.' The school has many activities for our children to engage in. My son and daughter have achieved more in this school than what I ever thought was possible. The school enriches our children as a 'whole people' by not only addressing their academic needs but by also fostering a sense of honesty, integrity, and respect within them and between them. These are the same principles that we foster in our home. There are some children who learn differently than other children and we think the school has tapped into how our children learn and how to best teach them in accordance with their unique learning styles."*

Another such testimonial came from a student who has been with us for a number of years:

> *"Without the years I had at this school, I might not have been as successful as I have been. It is the sense of community in this school that stays with you for life. One of the best things about the school is the relationship I have with my peers as well as my teachers. The teachers have been amazing. They have been caring and supportive of me. I have been able to learn at my own pace. And I have come to learn that I can be successful in life."*

However, it needs to be said that not all of our experiences with parents and students have been perfect. We have had parents and children come to our school that have been resistant to change. We have interacted with parents and children and youth who have opinions and ideas that are very different from what we believe and uphold. We have met parents who are negative about their children and experience frictions within their family life resulting from their children not developing in an expected way or from not being the person they had hoped he or she would be.

For example, regarding two kids with the same significant challenges, I've heard one dad say, "he is just like I was at his age" and "I know the struggles he is going through," whereas I heard another dad say, "all he needs is a good swift kick in the butt." So, one dad appears to have walked a mile in his son's shoes and the other apparently has not accepted that his youngster has significant challenges that need to be addressed in more academically and socially appropriate ways. The same contrast applies to siblings of children and youth who have learning and behavioral problems. In this regard, some are able to acknowledge the challenges of their brothers or sisters as well as empathize with their struggles, whereas others are not able to understand and appreciate their siblings' challenges. One sibling commenting on his brother's challenges stated, "Everything like friends, school, and other things I take for granted. These are so much more difficult for him than me. I'm in awe and I do my best to get my friends to include him in our plans." However, another sibling of another child told me, "I find it embarrassing to be around him; he says things that make me cringe. So, I'm not comfortable bringing friend's home."

Our school has also been about working with children and youth with respect to their academic and social development but also about working (when needed) with their families (parents and siblings) toward fostering a more positive family dynamic within and outside the home. We want to promote a better understanding within parents about their children and demonstrate to them that their children can achieve much more than they might otherwise think when they first come to us.

Friendships are important in helping children develop emotionally and socially. They provide a training ground for trying out different ways of relating to others. Through interactions with friends, children learn how to "give and take," set up rules, and make decisions. Part of the experience we provide includes learning how to properly manage feelings such as fear, anger, and rejection. Children also learn how to understand the viewpoints of different people and that different situations often require different strategies, skills, and behaviors. Friends provide companionship and a relatively safe context in which to compare oneself to others—who is bigger, faster, who can add better, who can catch better. Supportive friends are also a help in tough times—moving to a new school, entering adolescence, dealing with family stresses, and facing disappointments. Most children get invited to play dates, birthday parties, and sleepovers. Children who don't fit, because of learning and social struggles, often miss out on these invitations because their peers and their parents in many cases feel uncomfortable about them. Being perceived and treated as a social misfit is devastating. I've found it heartbreaking, and it has brought tears to my eyes on more than one occasion when children have shared with me that they don't have any friends and are not invited to birthday parties. So, our school has also been about developing positive relationships among and between our students. We want to promote a better, more positive understanding within all our children and youth about each other and foster mutually satisfying relationships among our students.

Although each child is different in some way and every family situation is unique, parents of children who have learning and social problems seem to have some common experiences (e.g., having to work hard to get their children appropriate care, a need to actively promote acceptance of their children within their community, and a concern about how a label might stigmatize their child). In my experience, the more accepting the whole family is of a child, the more aligned the family is with respect to how to provide support. In this regard, it is important for parents to have "helping professionals" who are not only interested in the well-being of their children but who are also knowledgeable and skillful in promoting the children's cognitive and social development.

Many teachers find it difficult to understand and accept children with learning and behavior problems. Many find it too challenging to teach these children. In part, this may be because some of these children respond in atypical ways, refuse to cooperate, may be overtly rude, or may be nonresponsive. Teachers who struggle to teach these children often personalize these atypical responses and find it hard to let go of these feelings, even when it becomes obvious that the student was responding, for example, out of a fear of being embarrassed. Moreover, some teachers subscribe to a deficit viewpoint, which stresses what a student lacks. Focusing on what is lacking hinders a teacher from focusing on how a student is able to learn and from determining how to capitalize on the student's strengths to accomplish learning goals. New teachers often interpret learning problems as a lack of ability and consequently lower their expectations of the student. Unfortunately, I have observed that, when students are perceived as having limited ability, they are often called on less to provide answers, given less time to respond when called upon, praised less, and criticized more.

Throughout my professional career, I have strived to change the opinions of some teachers about children who have developmental problems. I have worked hard to foster within the teachers of the Calgary Academy a much better understanding of the developmental needs of children and youth and how to more effectively address their academic, social, and emotional needs.

I have strived to surround myself with teachers who are passionate about teaching, who are intelligent, who are knowledgeable and skillful at developing the academic abilities of children and youth, who are insightful and appreciative of student diversity, who do not see learning and behavioral problems as barriers but as exciting challenges to resolve, and who are not only sensitive and empathic toward children, youth, and their families but who also are wise stewards of their thinking, feelings, and behaviors.

Each year, I am fortunate enough to see a few hundred children:

- who couldn't initially read but can now read,
- who were afraid to answer a question in class but now put up their hands in class,
- who had never written a paragraph but are now able to write a story,
- who didn't have any friends but now have friends and go to birthday parties,
- who have never played on a team but are now a starter on a school team,
- who have never been in a school play but now have a major part in a school play,
- who have never won a school award but now have won several awards,
- who have never thought they would finish high school but have now graduated and went on to postsecondary education,
- who did not think very positively of themselves but now are brimming with self-confidence,
- who were seen as a problem in their family, but now are thought of as a "star."

My view has always been that what our professionals strive for and what I believe we have accomplished at the Calgary Academy can be strived for and accomplished in all schools. From visits to hundreds of schools in Canada and the United States over the years, I have been energized to learn that we are not alone in our visions, principles, and values associated with the schooling of students with diverse learning and behavior needs. I look forward to the day when all schools are viewed as more alike than different and as we move forward to also appreciate and understand that all children and youth are more alike than different.

Part I

Behavior Disorders

1 Attention–Deficit/Hyperactivity Disorder

Prologue

The primary feature of attention-deficit/hyperactivity disorder (ADHD) is a continual pattern of inattention and/or hyperactivity–impulsivity that is more chronically, frequently, and severely demonstrated than what would be typically seen in children and youth at a comparable level of development. Central problems associated with this disorder include lack of inhibition, self-regulation, analysis and synthesis of behavior, goal-directed behavior, problem solving, and cross-temporal organization. Moreover, this pattern of behavior impairs their social, academic, familial, and occupational activities. Although many individuals are diagnosed in adulthood, for the diagnosis to be made, some of the hyperactivity–impulsivity and/or inattention symptoms must have been present before the age of 7. Problems in sustainable attention, concentration, and persistence relative to assigned tasks and activities and goal attainment as well as problems with behavioral regulation can lead to gaps in acquisition of knowledge and skills. Hyperactivity, impulsivity, and inattention appear early in life, and although they typically attenuate during adulthood, the symptoms can continue to be problems in adulthood.

Typically, children and youth with attention deficits will often have difficulty organizing tasks and activities, completing schoolwork, sustaining attention to tasks and activities, following directions, and remembering some things. Children and youth who are hyperactive will often have difficulty keeping still, remaining seated, and keeping quiet. Children and youth who are impulsive will often have difficulty waiting for their turn, and withholding their responses. Moreover, they often will interrupt others and/or "butt in."

In this section are five stories: one personal story, one parental story, and three professional stories pertaining to ADHD.

The personal story, entitled "You Can't Have ADHD—You're Just Like Me!" is written by a woman who came to realize that she had ADHD from childhood and reflects on her childhood and youth with this disorder and its impact on her life. The parental story, entitled "Signs Appear Early in Life and Significant Ones Will Persist," is written by a mother who struggles to make sense of her son's behaviors as he progresses through school. She shares her journey of adaptation with respect to his development. The first professional story, entitled "Attention to a Child's Strengths: A Lesson in Resiliency," is written by a school psychologist who learned about the importance of considering the strengths and resources that a child and family bring to their situation from her experience with a girl in junior high school who had ADHD. The second professional story, entitled "In the Trenches with ADHD," is written by a family therapist who discusses his balancing approach between viewing and treating ADHD as a biological disorder versus a social disorder. The last professional story, entitled "Identifying the Problems and Working

Exceptional Life Journeys. DOI: 10.1016/B978-0-12-385216-8.00001-1

Collaboratively to Make the Best Decisions," is written by a psychologist who reflects upon his experience in working collaboratively with parents and school personnel to bring about positive change in a young boy with ADHD.

Personal Story

"You Can't Have ADHD—You're Just Like Me!"

I am 50 years old, and became licensed as a registered psychologist about two and a half years ago. I am in private practice, where I work with individual adults, couples, and families. I have a married daughter who is an elementary schoolteacher, and have recently remarried. Three years ago, I was diagnosed with ADHD.

Several years ago, I was counseling families and youth at risk. Many of these young people (and some of their parents) exhibited problems like defiance, aggression, impulsivity, poor school achievement, and criminal behavior. Many had been diagnosed with ADHD, among other disorders. In no way did I connect their problems with my experiences. This changed about 4 years ago, when a girlfriend, a veterinarian by profession, asked me about adult ADHD. She described having difficulty reading and recalling full paragraphs, having to review the same page repeatedly, and being forgetful and disorganized. I laughingly replied, "No, you don't have ADHD, because that sounds just like me." Then the lightbulb in my head went on. The ADHD-diagnosed youth with whom I had worked presented very differently. However, as I thought about my friend's experiences, and mine, I became curious. I promised to look into it. Thus, my journey with ADHD began.

As I reviewed the recent literature, I connected with what I was reading—it was like finding the missing pieces to the puzzle of my life. About that time, I was studying for the psychology licensing examination—the dreaded Examination for Professional Practice in Psychology (EPPP)—a 200-question, 4-h multiple-choice exam, with finely nuanced wording requiring a great deal of concentration to decipher. I had started studying twice before, and this time was much the same—I struggled to sustain my attention to the readings and retain what I did manage to read. I consulted a psychologist, who confirmed my growing belief that I'd struggled with ADHD my whole life, visited my long-time primary care physician, and started a trial of Dexedrine. As the pieces of my puzzle came together, this is how I made sense of them all.

As I reflected on my childhood, I realized that I recalled very little of it. Even though I grew up in a small town, I'd forgotten names and events. As a young child, my fondest memories were of living on a farm we had rented for about a year. I dug in the dirt to make the garden grow, made mud pies, climbed trees, and brought the odd wounded bird or animal home to "fix." Once school started, things began to change. … My first day of school, I remember my mother walking me into the classroom, where I was allowed to pick my own seat. Like a self-conscious, timid mouse, I quickly chose the closest seat. It was behind a girl with beautiful, long, golden blonde hair; I could hardly resist touching it, as she sat in front of me. I was painfully shy and recall quietly watching others to see how they looked, behaved, and spoke. I was mortified about answering questions in class, and would never raise my hand. I avoided eye contact by looking at my paper, or adjusting my clothes, or picking my pencil off the floor, and so on. I learned to say "I don't know" right away so the teacher would move on quickly. I did this mostly because it seemed like I couldn't think fast enough, and didn't want to be humiliated by having the wrong answer.

My mother tells me she knew "something was wrong" with me—I was fidgety and overactive. My mother complained that she couldn't "settle me down" when I would get home from school. Teachers reported that I would often look out the window and daydream, seemed to fidget a lot, and would "talk in class." By the time I was 8, in third grade, the school was concerned enough about me to send me for some kind of assessment. As I recall, I scored in the 98th percentile in spatial and abstract reasoning, but other areas like math and working memory were below average. My mother recalled that she had slipped and fallen when I was 2 years old. I hurtled over her shoulder and struck the right side of my forehead on the floor, resulting in a large "goose egg." She hypothesized that this might be responsible for my difficulties. The assessment results confirmed my mother's suspicions that something was "wrong" with me. She took me to numerous doctors for 8 months, studiously avoiding psychologists or psychiatrists, to figure out what the "problem" was. Looking back, it seems like she was looking for a "magic bullet" (preferably a physiological one) that would confirm the problem was "inside" me, and not something that she was responsible for, or had to do anything about. Eventually, one physician referred me for an electroencephalogram (EEG), which was "abnormal." Hearing the doctor give the results to my mother, I felt shocked and so empty—my brain was abnormal! I was abnormal! I felt sad, bad, and numb. I was emotionally devastated, thinking, "I guess I really am stupid!"

Even though I had never had an obvious seizure, my daydreaming (blank stare) was interpreted as *petit mal* or "absence seizures." I was diagnosed with epilepsy and was given phenobarbitol initially, which resulted in a severe allergic reaction, including a fever. I was then placed on Dilantin thrice a day, which continued for the next 8 years. Dilantin just seemed to slow me down. When I became excited, active, or simply wanted to play, my mother told me it was time to take my medication. I was basically told to "take a pill." I disliked the way it made me feel. It amplified the feeling that I wasn't very smart or worthwhile as a person.

As school and life continued, I was an easy target for being teased. A very painful experience involved a friend who noticed the long men's underwear showing under my pants and socks, pointed, and snickered out loud—in class. My family couldn't afford to buy me snow/ski pants for me to walk to school, so I was forced to wear the "long johns." Growing up poor in a small farming town added to my sense of unworthiness. The only activities I could ever participate in were Brownies and Girl Guides, and it seemed to me that I never fit in anywhere. I almost always felt, and was often reminded, that I wasn't good enough.

In school, I couldn't seem to *get it*, no matter how hard I tried. I would learn what I could in class, but without doing homework, I was scraping by in school with Cs. I was more interested in social interactions. But, I mainly remembered just feeling stupid and like I didn't belong. At times, I found I could retain more if the classroom was physically arranged so I could hear better. However, when teachers assigned me a seat in the front of the class, again I was embarrassed and felt everyone watching me as I forced myself to pay attention. After some time, it occurred to me that, because I had to pay attention (or risk being rudely reminded by the teacher), I was able to absorb and remember more. This strategy became crucial when I tackled tenth-grade math again—some 15 years later! Although my struggle to learn and fit in didn't seem to change much from year to year, my marks declined. Skimming by with average or below-average marks, I had fallen farther and farther behind. I tried to study but couldn't seem to remember it all … there were so many missing pieces … I just kept doing it the wrong way. For example, in math class, I would sit there and do it, *and do it, and do it*, but it just didn't seem to click. For a while, I had a tutor, which helped a bit, but not enough to feel comfortable with math. Although my sedated behavior was easier to manage, school was always a struggle. I gravitated toward learning activities that utilized audio, video, and hands-on activities, and de-emphasized reading and memorization. Subjects requiring memorization of facts and detail were very difficult. Practical information held my interest and was easier for me to absorb. Accordingly, I did very little homework—the minimum necessary to complete assignments. I didn't see the point in studying because it did not help anyway.

My shyness and introversion followed me into my teens. I felt depressed, and continued to feel stupid. When I failed tenth-grade math (the academic math course), my "stupid" narrative was reinforced. Although I had wanted to become a doctor or a veterinarian, failing math, despite having a tutor, precluded attending university, and I abandoned that dream. I took it as evidence of how stupid I really was. With depression taking hold, I gave up on school and my dreams. I focused my attention on a social life and started to "rebel" by hanging out with a "bad" crowd … kids who smoked and skipped school … and later drugs and alcohol. I started to rebel and tried to assert myself—*I wanted to be who I wanted to be*. Knowing now that I had untreated ADHD, I'm surprised I did not become more drawn into drinking or drugs, given that I lived in a small town with little else to do. My friends were my world. Because I was subdued and quiet, I was seen as a good listener. Others would confide in me, and ask for my opinion. Typically, in adolescence, friendships become very important. In my case, however, they gave me what my family could not. Friends became a safe haven where I could express my feelings, and be accepted the way I was.

Home life was even worse. Even in the context of the prevailing ethos of the 1960s—"children should be seen and not heard"—my mother seemed stricter than usual. As a self-employed tradesman, my dad was often absent. Although my mother stayed home to parent full-time, she suffered depression and could not provide the optimal level of support that children struggling with ADHD require. In retrospect, I consider my parents very authoritarian and harsh. Doing something I wasn't supposed to, making a mistake, or "talking back" (in an attempt to negotiate or ask why) resulted in being spanked as a child, and grounded with loss of privileges in my teens. I could not voice my questions, ideas, or feelings for that matter. I was told what to think and what to do. I blamed myself for my misery—after all, I was abnormal and likely deserved it.

When my parents divorced when I was 16—another risk factor for future problems—I went to live with my dad after a huge fight with my mother. But moving away from my mother facilitated another transition. I had been on Dilantin (a sedative) for 8 years, before an annual EEG test result indicated "normal" results—how that happened I have no idea, but after 8 years of feeling "doped up," I finally went off Dilantin. Despite some clear vulnerabilities, I managed to develop some strengths that pulled me through. I developed some effective coping strategies, for example. The rare times I studied, I did it with others. Discussing content, approaching it from different angles, helped a great deal—even if it was something boring like history. It helped to process in different modes—visual, audio, kinesthetic. The subject matter in nonacademic math courses felt practical and relevant.

Ironically, my parents' authoritarian parenting style supported development of some of my strengths. On one hand, I was indecisive—something that plagues me to this day. I thought if I made a wrong decision, I would be accused of lying and be in trouble. On the other hand, I became very adept at reading people, understanding their perspective, mediating and negotiating, and advocating for others. My predisposition to be a peacemaker was an adaptation to my parents' authoritarian parenting style. My parents' authoritarian parenting also drummed great organizational skills into my head. Everything went into its place immediately after use, was clean, tidy (uncluttered), and organized. This helped me avoid the disorganization that plagues many young people with ADHD. Another strength I developed (through 19 moves in 13 years) was to be able to pack an entire house in order of each room. I can remember the box in which specific items can be found. This helped to keep me organized and able to manage many tasks at once—multitasking was the positive buzzword at the time.

My story does have a happy ending. Fast forward to age 30. I started university as a single parent, conquered my math nemesis, and realized I wasn't stupid! I just needed different tools and methods for me to learn. It took me longer. I needed multimodal strategies (verbal, visual, auditory) and immediate correction as I proceeded through a formula, so I would learn the right way to use it. This was an astounding epiphany in my life. I eventually completed a master's degree in counseling psychology. Being diagnosed with ADHD at age 48 was another revelation. The pieces of my life puzzle came flying into place. I also cried for all those years in school and at home that I will not ever know what I could have been or done.

Based on my experience as a child and youth with ADHD, I would say to kids: "You're not stupid, or worthless, or lazy, or good-for-nothings. You are valuable. You do count. You might feel like you don't belong, can't keep up, or that you aren't as good as other kids. It could be that you have ADHD: a different way of thinking and learning that others may not understand." I would support young people to advocate for them, and ask them, "What are the things that you like? What do you think you do well?" And I'd support them to discuss this with an adult—whoever is available for them. Especially for boys, I would try to help them understand and manage their behaviors, because once they are labeled as "aggressive," or "bullies," adults find it much more difficult to be sympathetic, support them, or understand their behaviors in context.

I would say to parents "stay open and curious." Our children don't come with instruction manuals. They are unique. Parenting and teaching strategies that work for other children may not necessarily work for the child with ADHD. If you have concerns or questions, take account of different opinions (day care, schools, other family members, and so on) and the most recent information possible. This may involve obtaining a competent psycho-educational assessment of your child. In my work, all parents want the best for their child. Many parents may be influenced by their values or culture, and hesitate to have their child on daily medication, especially related to a mental health issue. Or, they may be concerned about side effects or the stigma of being labeled. Recent research finding is that medication, properly administered by a competent physician, with informed collaboration with parents and other supports, gives a child the greatest opportunity for success. ADHD does not mean that you're a horrible parent, or your child is a horrible kid. Medication and other supports may create a context for your child's creativity, intelligence, energy, and dynamism to emerge. I would also urge parents to take action to support your child, to obtain the necessary services and information, in a timely way, before your child starts to believe he/she is "stupid," not worthwhile, or develops an aversion to school. If a child develops a pattern of aggression, anger, and rage, others will be less sympathetic and less interested in supporting him/her. It's especially important to maintain your relationship with your child. Although you may need to set firm limits, provide strong structure, and be the "bad guy" at times; it's important to sustain a caring, connected relationship.

Parental Story

Signs Appear Early in Life and Significant Ones Will Persist

Our son was first diagnosed with ADHD, combined type, in the spring of 2007, at the age of 9, confirming a suspicion we had had for a very long time; we were relieved to finally understand the cause of such a complex set of symptoms. Our son was often lost in his thoughts. He was very bright, and yet could not follow a simple set of two or three instructions without getting distracted by Lego, his sisters, or another similar disruption. He often would become so involved in an activity that interested him, that he could not make a transition to something else. We were often late getting places because of distractions and behavioral problems. Getting him to come off of the playground equipment after school was difficult nearly every day. At school, our son would start assignments but rarely finish them. He talked endlessly, often intruding on other people's conversations, or bursting out with answers before questions were finished being asked. My husband and I were frustrated that our son had so many wonderful qualities, such as kindness, compassion, thoughtfulness, and intelligence, but that his behaviors were getting in the way of his success, his education, and his friendships and negatively impacting his self-esteem, as he received so much more negative than positive feedback.

Our son was a precocious infant and toddler. At 12 months old, he would build towers of seven or eight bottles of spices from my spice rack; he would play in centers I had set up for him for long periods of time. At 16 months, he would say sentences like "moon's flying" or "baby's naked." He could sing his ABCs before his second birthday. At 2 and 3, he would read books with me for long periods. He had an uncanny ability to "hyperfocus" when things interested him, and we were impressed that he could play so long on his own. In retrospect, this was probably the precursor to the "hyperfocus" he shows now when he is really interested in a book, a project, or video games.

When our son was 3, he went to formal day care. Before that, I had been working part-time as a high school teacher, and he was in informal day homes with one or two other children—care that was very child centered. In contrast, day care had schedules; every half an hour, they had to change activities, and making transitions was a struggle. He was the last one to get his jacket on to go outside and the last to come in. He spaced out during mealtimes and often didn't finish eating, or even eat anything during the allotted time. He danced around in front of the TV during the half hour they watched before lunch, while the other children sat quietly. In essence, my son was spaced out when it was time to do things and hyperactive when it was time to focus. This was the first opportunity that I had to observe him interacting with a large group of his peers, and it was clear he was different. Fortunately, he wasn't in day care for very long, because when he was three and a half years old, his twin sisters were born. At this point, our son's ability to keep himself occupied became a real asset. He played on his own for long periods of time as I was trying to cope with having new babies. I remember the day the twin towers were hit on 9/11. I was glued to the TV and, because my twins were 2 months old, they were glued to me nursing. I think my son played alone all day. He was great at entertaining himself. On the other hand, when we needed to get places on time, it was difficult, if not impossible. I remember getting the twins dressed in their snowsuits and putting them into their car seats to carry out into the van, and then calling my son to get ready to go. He wouldn't come—I am not sure if he could hear me, he was so involved with his toys. When I did find him, he hid under the bed, and I couldn't get him out. I was enraged, the twins were crying and overheated and my son was laughing because he thought it was a game. He didn't seem to be able to read social cues. In essence, as long as we did something my son liked to do, such as read, play cars or trains, or entertain himself, life was good. When we needed to be somewhere on time or do something on someone else's schedule, things did not work out so well.

Once we entered formal education, problems escalated. Although our son was happy to go to preschool, he never did any of the crafts, he wouldn't sit in the circle with the other children and sing action songs or play games; he would only come to the circle if they were reading. I remember asking the preschool teacher to make sure that David did some crafts, but she reassured me that children were all different and that if he preferred to play with toys, that it was fine in this setting. At the Christmas concert, our child stared off into space as the other children were singing and doing actions. I thought maybe he was just tired, but this sort of thing was happening more often than not. By this time, as a newly stay-at-home mom, I was trying to make connections with other parents and to help my son make friendships, but I didn't like the friends he was choosing; they also were hyperactive and unlikely to pay attention. Instead of focusing on the friendships David might have wanted to make, which I now realize was a mistake, I made friends with the mother of a boy I thought would be a good role model; this boy sat at the front of the class, did crafts and activities at designated times, and was alert to social cues. As I write this, I am aware that all of this social engineering seems ridiculous. I agree that some kids want to do crafts and others don't, but I think I knew that there was something much deeper going on, something I couldn't put my finger on. I just thought that if he became friends with good students, he would pick up some of their good habits. Although this boy and my son had some similar interests, such as Lego and Rescue Heroes, after a while, this friend didn't want to play with my child anymore. He even refused to invite him to his birthday party, causing me to be so hurt on my son's behalf, that I severed my friendship with the child's mother.

Kindergarten was worse. David was unable to settle in. He would either bother other kids or wander around. He often spaced out during lessons and developed a routine of picking his nose while being spaced out, a habit that drove me wild. He spoke out of turn all the time, making it hard for the teacher to maintain classroom management. This was all horrifying for me as a teacher. I thought that kids who behaved like this had bad parents or ate too much junk food and didn't get enough sleep, and I suspected the teacher must have thought badly of me and my son. I felt ashamed, and I also felt that there was no one I could talk to. My husband thought I was overreacting and I didn't want to share my feelings with my friends who had children the same age, for fear they would think badly of him and not want their children to play with him. I, who had always been gregarious, who had historically had many friends, was becoming socially isolated. This lack of ability to see the positive things about my son and to overcome my embarrassment about some of his behaviors, at least in the early days of elementary school, has had long-lasting implications. Although I am friends with many other women and their families who have children my twins' ages, my son and I have not yet been able to overcome this early social deficit, something that gets harder to rectify the older he gets.

One of the most frustrating things during the kindergarten year was that my son had good days and bad days. I was forever trying to find out if eggs, toast, or cereal in the morning were able to produce a more focused school day, or if a different nighttime or morning ritual would be more beneficial, but my son's behavior seemed independent of whether we'd had a good morning or evening at home ... and by the way, for the most part, mornings were horrendous. I had two toddlers and a 5-year-old who was unable to follow two-step directions, a 5-year-old who I knew was very bright. On a typical morning, we would have breakfast before I sent my son upstairs to change or to brush his teeth. I took to putting the TV on for breakfast, because then at least David would stay at the table to eat. When I sent him upstairs to change, he would invariably get distracted by toys or games, by the dog or his imagination. Often, I would hear the Lego bin dump out, or the fish bowl tip over, and I would have to go upstairs, clean up whatever mess had been made and get him changed myself. A boy who has twin 2-year-old sisters needs to be responsible; this boy wasn't. Another scenario would involve being sent upstairs numerous times to do individual tasks–saying go upstairs to get ready for the day was never enough. I could never understand why he couldn't figure out that he had to get dressed and brush his teeth, but the truth is that he completely forgot his tasks even before he was up the stairs. Our son always seemed stunned that I was upset. I felt that every day the same thing was happening—why was I so helpless to change it, and why couldn't he just get it?

One thing that shocks me about this period in time was how often I would get angry and how easily my sweet son forgave me. He was kind, thoughtful, engaging, and fun to be around. So long as there were no places to go and things that had to be done on a schedule, there were no problems. Mornings did improve once I gave him a handheld checklist that had pictures of his responsibilities, but things at school did not improve. By early October, I had to go into the kindergarten classroom every day to settle my son for learning, no easy task when it involves finding care for one's twin toddlers first. The kindergarten teacher was positive and warm and did what she could to make David's learning experience a good one. David certainly knew how to read and had no trouble understanding new information.

We hobbled through kindergarten this way, but first grade was yet another challenge. David had two old-school teachers who job-shared and who believed that children needed to tow the line. My son's unusual behavior became more pronounced. During lesson time one day, he left his seat and began to crawl around the perimeter of the room looking for stray pencils. After this had been brought to my attention and I asked him what he was doing, he said that it occurred to him that if the pencils were not picked up before the caretaker arrived, they would all go in the garbage, and he could not stand the idea of pencils being wasted. One of his teachers told me my son was "bazaar," not a helpful statement for a parent who already felt she was in a state of crisis. It is true, though, that our son was incredibly easily distracted, not just by the things around him, but also by his own thoughts. In this classroom, the teachers instituted a rewards program to encourage good behavior and learning habits. It was in the form of red slips of paper that the kids could accumulate and then trade in for rewards. My child almost never earned the tickets, which should have worked well in improving specific behaviors for him. Ironically, the kids who earned most of these rewards were the kids who would likely have behaved well regardless. When I approached one of the teachers to tell her that my child was excited about this program and to suggest that it could be used to improve specific learning behaviors for him, she told me that no students got special treatment. It was not a very helpful stance for any of us! Any excitement that our son had about this program was dashed when all of his tokens were ripped up after he made a paper airplane out of a language arts lesson. Obviously, this is not desirable behavior, but he wasn't doing it to be disrespectful; he had seen another student do it and thought it looked like fun. Impulse control was a major problem for my son, a problem that I would later find out is a hallmark for children with ADHD. When I suggested that David might be engaging in these types of behaviors because he was so bright, and I knew that bright students sometimes lacked the ability to read social cues, the teachers said they saw no evidence that he was particularly bright. The event that pushed me into finding outside help was when my son won a schoolwide jelly bean in a jar guessing contest, one in which over 650 students participated, and the type of contest he wins whenever there is one. One of his teachers told me that it was strange that he could do so well at these tasks when there was no particular genius about him otherwise. I had him tested for giftedness out of spite. I can still remember how stunned they both were when the results came back.

Being gifted seemed to solve some of the puzzle surrounding my son's behavior. Gifted kids get hyperfocused on activities they are interested in, they appear bored and uninterested when information seems irrelevant, they often have social skills that lag behind their intellectual development, they talk a lot and interrupt, they question authority and rules, and they can often seem careless and disorganized. Although it was an answer that made us feel better and that confirmed our beliefs about his intellect, it didn't make anything better on the day-to-day front.

Second grade was better, not because my child's performance improved, but because his teacher was so flexible and creative at solving problems. She was able to see past his problems to his potential. She understood how bright, kind, and creative he was. She had a sense of humor, and she appreciated my son's. There were still hard times, such as the time he pulled on the overhead projector, and we had to pay for it, but the teacher and I were not adversaries; we worked together. I understood that my son was tough to teach, and I appreciated the effort and kindness that came his way. This teacher understood that David was not trying to be bad, and she could see the good in him—it made an enormous difference. This teacher also employed strategies; she sat my son close to her desk and away from distractions, and she sat him near kids who were good role models. She would sit with him sometimes to get him started on an assignment, because once he started, there was a greater chance he would finish. This teacher and I communicated often through the school agenda or at the end of the day—just for a minute or two. This made a difference for all of us. The teacher knew she could come to me for help, my son was happy to go to school, and I felt like I was looked at as a resource and a help, rather than as a problem.

Third grade was an enormously important year. The second-grade teacher carefully placed my son with a teacher who had a sense of humor and the intellect and resources to deal with such a child. This new teacher promptly referred my son to a community resource center that helps students who have difficulties in school settings. One day, the resource worker, who also came to our home, asked me if I had ever considered that my son might have ADHD. She said that he was very similar to other students she worked with who had been diagnosed. A referral to a psychologist followed. There, we met with a psychologist who took a detailed history of my son's medical and family history, who observed him in the classroom and who had both the teacher and my husband and I fill out questionnaires about our son's behaviors. After meeting alone with the psychologist, we met with both her and the doctor, a specialist in ADHD. We met for probably 2 h and were able to ask in-depth questions and receive answers that helped us to understand our child. They sent us away with books and brochures to help us understand and with photocopies of learning and teaching strategies to help his teachers. Although my husband and I were initially dead set against the idea of using medication to treat ADHD, our increasing frustration led us to agree that if I were to research the medications and their side effects, and if they were found to be safe and effective, we would let our son try them. At the time, I was earning a degree in psycho-educational skills and used the resources that were available to me to research information about ADHD, treatment plans, and success stories. What I have learned in medical journals and case studies, as well as from firsthand experience with our son, has caused me to become an advocate for the use of medication as part of a comprehensive treatment plan for ADHD.

The process of figuring out the right medication and dose is not straightforward. The first medication we tried, a long-acting Adderall, worked well for our son. We started at the lowest dose, 5 mg, and had significant success. The first thing I noticed was that when I called him to come off of the playground, he said OK and immediately came with me. His teacher noticed an immediate improvement on his ability to get work done and to transition from one activity to another. His report card went from being all over the map to almost all 5s (on a scale of 1–5). It was a startling improvement! Because he was on a more even emotional keel, his friendships improved. However, my son had not acquired many skills when he had not been paying attention, and those skills, such as waiting his turn in conversations, and developing more sophisticated writing skills, are taking longer to improve, but we will get there. He is also much less impulsive, which improves his relationships with adults. When we increased our son's dosage to see if we could get more improvement, he began complaining of a headache and stomachache. We persisted for 2 days, at which point I talked to the teacher, who was concerned that our son didn't seem himself. I called the doctor and we reduced his medication to the previous dosage. The medication should not change the child's personality—if it does, the dosage is too much, or the type of medication is inappropriate. We were very lucky—we found the right medication right away, and our son doesn't have many side effects. In the beginning, he was not hungry during the day, and thus didn't eat much of his lunch. In response, we made sure breakfast was hot and nutritious, and we offered him quite a large bedtime snack. He also has difficulty getting to sleep, but that was also the case before he went on any medication.

We are very happy with our son's increased success at school, with his friends, and in our relationships with him. I try not to get upset about small stuff and I realize that if he doesn't answer me right away, or if he has difficulty transitioning from one activity to the next, or if he needs several reminders to clean his room or do his homework, it is just the way he is—I need to allow a bit more time for him. As for my son, he realizes that he has a disorder that requires medication to help his brain chemicals be like other kids'. The medication allows him to act the way he wants to and to be himself, rather than to be controlled by his ADHD. We are proud of our son—he has overcome a lot.

There are a number of important things that we have learned regarding helping our son and our family cope with ADHD. Our son knows we love him and that we view him as a valuable person, the importance of which cannot be overstated. We are successful at reacting with patience and creativity most of the time, and we try to see the humor in things. We have to take him as he is and not be too frustrated by the gap between his potential and the reality—most kids have a gap there anyway. We keep introducing him to new activities to help him become well rounded. He would prefer to play video games and read books nonstop, and we have to make a big effort to get him interested in sports and other activities. We have also made relationships within our community a priority. Our son knows that if he runs into trouble, there are a number of adults who could help him. As a family, we camp as much as we can every summer, and family activities and traditions take precedence over other aspects of our lives. My husband spends a lot of one-on-one time with our son and I truly believe that the strong relationships in our family will help us weather the storms to come.

Teachers make a huge difference! We try to foster strong relationships by volunteering in class and by having open communication with the school. We make it a priority to be supportive of teachers and when we have been frustrated, we have paid for a psychologist to come to meetings with the teacher to help her understand our son's challenges. Not a lot of people outside of the medical and psychological community seem to understand very much about ADHD, and therefore I have to be an advocate for my son, helping teachers and coaches to understand the best way to deal with him and helping them to see the good in him. I also talk with a lot of other parents whose children have been diagnosed with ADHD and it makes me feel angry when I realize how many parents are unwilling to try medication, knowing what I know about how much it has helped my son. ADHD is as biological a condition as is diabetes, and who would deny a diabetic his insulin or tell him that he just needs to try harder to produce his own insulin? Well-meaning friends and relatives, and even professionals who are not trained in ADHD, will tell you many things about how to improve your child, from dissecting your parenting skills, to offering your child special diets, or sending him to military school. Although it is true that a small minority of kids who have food allergies can have symptoms that mimic ADHD, kids who actually have ADHD are not going to be significantly improved by restrictive diets. Seek help from people with knowledge and experience, and ignore the people who will tell you that ADHD isn't real—they are wrong and they do harm to people who are trying to do the best for their kids.

Our son has just started seventh grade, and we have many obstacles to face. At the first set of interviews, we heard that some assignments haven't been handed in and that our son interrupts the teacher during lesson time—a concern both because it affects classroom discipline and because it will affect the way his classmates perceive my son. We will soon be making an appointment with our pediatrician to adjust our son's medication to an optimal level, something we have had to do a few times now. We will also be registering our son in a study skills class to prepare him for the increased expectations of junior high school and beyond. Junior high is a difficult transition because there are so many teachers to deal with and there is so much freedom. This school has done away with agendas this year, a disaster for kids who need to have effective communication between home and school. My son is also in a class where he can choose his seat every day—also, not helpful for his learning. We will soon be meeting with his teachers and the resource team. For him to be successful academically, it takes a lot of teamwork, but he is worth it and we are up to the task. For us, however, our biggest priority is our relationship with our son, for it will persist long after junior high, high school, and university, and the respectful and loving way that we treat him will be the model for the way he treats others and the expectations he has of other relationships.

Professional Stories

Attention to a Child's Strengths: A Lesson in Resiliency

I am now a school psychologist, but I began my career as a teacher. In that role, I worked with small groups of children who experienced challenges with learning, attention, or behavior. My greatest rewards in teaching were in getting to work with students over an extended period of time, to form real connections, to provide support, and to witness growth. When I made the decision to embark on a new career in psychology, these were aspects of my work that I wanted to retain. I was lucky enough to find a setting that allowed me to do just that. As a school psychologist, I work with children and adolescents and their families in a multidisciplinary setting (including psychologists, pediatricians, educational consultants, speech–language therapists, and occupational therapists) over an extended period of time. I have the opportunity to get to know these families well and to be a part of their journey. This is the story of my work with Kenzi.

Kenzi was referred to our team by the teachers and administrators at her junior high school. She had previously been identified as having a learning disability and was in a specialized school program to address her learning needs. However, her teachers raised concerns with her attention and concentration. They referred her to our clinic for an evaluation of these concerns. Before I met Kenzi, I met with her mother, Rose. An initial parent interview typically takes about an hour. Rose and I met for two and a half hours. I came out of our meeting feeling invested in this family (before I'd even met Kenzi), but also, if I'm honest, a little overwhelmed. Kenzi had been through a lot in her 12 years, and so had her family. Both of Kenzi's parents had battled addictions, and Kenzi had briefly been in foster care during this time. After her parents' recovery, the family went through the major transition of moving to a new province, away from family and friends. Shortly after moving, the family suffered the loss of Kenzi's grandmother, who had lived with them since Kenzi's birth and had been an enormous part of their lives. Both Kenzi and Rose were still coping with the grief of this loss. When Kenzi was in sixth grade, a psycho-educational assessment revealed that she had a learning disability. This assessment shed light on her ongoing academic struggles, but it also produced a sense of guilt for her parents, Rose and John, who had battled with her for years over homework and school. Now that Kenzi was getting support at school for her learning disability, another challenge was becoming evident: attention.

The diagnosis of childhood disorders is often done from a deficit-based perspective. Areas of difficulty, or symptoms, are evaluated for their severity, frequency, and duration. In the case of ADHD, these symptoms may include difficulties with organizing, sustaining mental effort, listening, paying attention to detail, starting and completing tasks, planning, and managing time. They may also include difficulties with restlessness, fidgeting, excessive or inappropriate motor activity, excessive talking, and impulsivity. For a disorder of attention to be present, symptoms must be severe enough, frequent enough, and pervasive enough to cause significant impairment for the client in multiple settings (e.g., home and school). Kenzi's symptoms met these criteria, and a diagnosis of ADHD, predominantly inattentive type, was made through our clinic. But all of that is just the background to the story that I really want to tell, which is the story of what I have learned from working with Kenzi and her family over the past year and a half. Working with this family has provided me with a powerful reminder that operating from the deficit-based, diagnostic perspective is not enough. If the story ended here, one could take the deficits identified, propose some generic strategies and supports to remediate or accommodate them, and call it a day. Thankfully, the story does not end here. A picture of symptoms, difficulties, and challenges is never the complete picture and, although we can't ignore it, it is essential that we move beyond a basic diagnosis to find the strengths that are always present. These strengths form the foundation for moving forward.

In Kenzi's case, the strengths are plentiful. To begin with, she has an extremely supportive and resilient family. Watching the dynamic between Kenzi and her parents in our clinic, it was clear that they are a close family that values openness and honesty. There is an abundance of humor and caring in the way that they interact. Her parents communicate a deep respect for and acceptance of Kenzi. They are up front with her about problems that they have struggled with in the past, and they are there for her when she wants to talk about her own troubles. They support her ambitions and take a genuine interest in her day-to-day life. Kenzi's parents are also open and honest about their current challenges. From our first meeting, Rose acknowledged that she and John were not very organized or structured at home. This needed to be considered in developing strategies for Kenzi. One common approach for dealing with the difficulties with organization, planning, and time management that are often a key feature of ADHD is to have significant others (e.g., parents and teachers) provide external organizational structure. In Kenzi's case, Rose and John were up front about the fact that it would be difficult for them to provide this structure. However, by looking at the strengths of the family, we were able to identify that Kenzi's grandfather, who lives with the family and is very organized and structured, would likely be an excellent candidate to provide some of the external structure needed. We were also able to draw on Rose and John's strengths at providing emotional support for their daughter. We discussed ways that Kenzi could be more independent in getting organized (e.g., using checklists, developing routines), so that she could take on a primary role, with her parents providing support. Because of the close relationship of the family, we could also suggest that they work together to support each other in using tools like calendars and schedules. In this way, the strategies could become a family endeavor.

Kenzi also exhibits many personal strengths. The same psycho-educational assessment that identified her learning disability (or "distortion" as she likes to think of it) identified that she is highly intelligent. Her verbal and nonverbal reasoning abilities are in the superior range for her age. She is an articulate and insightful young woman. She also displays a strong sense of self. Kenzi is not afraid to be herself and to be different from her peers (her hair is a gorgeous shade of blue at the moment). Kenzi's difficulties with learning and attention clearly needed to be supported in the classroom. However, her strengths also needed to be addressed. To that end, in addition to suggesting supports for deficits, we recommended providing enrichment and opportunities for intellectual challenge. An interesting conundrum often presents itself in the classroom for students who have attention difficulties but are also highly intelligent. Educators appreciate the need for providing additional challenging and enriching learning experiences for students with strong cognitive ability. However, when these students also have attention deficits, they tend to have difficulties starting and completing their basic class work. Enrichment can become a "reward" to be earned once regular assignments are completed rather than an integral part of the educational program. Unfortunately, students with high intelligence and attention deficits may have such difficulty completing the basic class work that they rarely get to experience the enrichment activities that would stretch their intellect. The irony in this is that attention deficits are almost always at their worst during mundane, routine tasks. In this way, the highly intelligent student with ADHD may unintentionally be given an educational program that actually exaggerates the disorder by limiting opportunities for truly challenging and engaging work. In Kenzi's case, we recommended that enrichment opportunities be viewed as a key component of her educational experience and that they be provided whether or not she finished her regular class work. Other consequences for incomplete work could be implemented and other strategies to support task completion were suggested, but we recommended that enrichment be viewed as a right rather than as a privilege.

Kenzi's strengths as an intelligent and insightful young woman also came into play in designing strategies and supports. She was a key part of the discussions of what would work well for her and provided her own ideas of organizational approaches that she thought would be a good fit. Kenzi's personal strengths and the respectful, open relationship that she has with her parents were both important in the discussion of medication as a possible treatment option. Kenzi's parents listened to information and options presented and asked key questions. However, they left the ultimate decision of whether to do a medication trial up to Kenzi. Her decision to try medication was therefore her own undertaking. During the trial, she observed a positive effect from the medication and decided to continue on with it. She also developed her own method for remembering to take it each day.

It is not my intention to paint an unrealistic picture. Kenzi still has challenges ahead of her. She is entering adolescence, with all of the social and emotional hurdles that entails. She is also in the process of transitioning to a new school with fewer learning supports in place. Finally, she will soon be transitioning to high school, where the academic, organizational, and attention demands will be greater. She also needs to continue to develop her own understanding of her strengths. She does not fully identify her own intelligence, insightfulness, and creativity. I don't know the end of this story, but I do feel hopeful about it. The experience of working with Kenzi and her family has been a lesson in resiliency. This is a family that has encountered and overcome a multitude of obstacles. They have reminded me of the importance of looking beyond the basic framework of diagnostic symptoms, to the pattern of strengths and resources that a client and family bring to their situation. They have reminded me that we are not defined by our challenges, but by our responses to them.

In the Trenches with ADHD

Sam, an 8-year-old white male, was brought in to see me by his single-parent mother, Joyce, due to academic problems. Sam was not performing to his abilities in his third-grade class. He attended a prestigious parochial school in an upper-middle-class suburban neighborhood. Sam seemed pretty unhappy with his C grades. It wasn't failing but his mother and school expected much more from Sam. It didn't help that Joyce had completed a graduate degree and Sam had an IQ of 130. Sam's mood in my office ranged from thoughtful to gloomy and reflected, disappointing himself, his mother, and his teacher. As the interview continued, I was impressed with Sam's knowledge on a variety of subjects including US history, the batting statistics of the San Francisco Giants starting lineup, and computers. He also had read every *Harry Potter* book cover to cover. But Sam had trouble getting his homework done. When sitting at his kitchen table to do his coursework, his mind would drift away to subjects more interesting to him than completing monotonous Math and English worksheets.

Joyce was at her wit's end. She had tried everything to motivate Sam and get him on task. But the constant verbal reminders to get his homework done, sitting next to him at the kitchen table, and various rewards and punishments had no effect. It would take Sam all evening to finish his homework. Joyce became increasingly frustrated, losing her temper and yelling at Sam. After losing her temper, she would feel guilty, thinking that she wasn't a good mother. At a loss, Joyce wondered if Sam had ADHD. Sam clearly wasn't hyperactive, Joyce acknowledged, but wondered whether his problems of organization and attention were beyond his control. Perhaps Sam's problems were the result of a brain problem, she speculated. Maybe he needed Ritalin or some other stimulant medication such as Adderall or Concerta. Joyce wanted me to confirm a diagnosis of ADHD and refer him to a child psychiatrist in my department (a larger managed care health facility).

Sam hardly struck me as a young man with a brain disease or a psychiatric disorder. He was brainy, interesting, and affable. Yes, he was underachieving in school but only according to traditional prescriptions for homework completion and course participation. In addition, I was (and still am) concerned with the increasing biological fundamentalism gripping the mental health field. As a strength-based family therapist who situates problems in context, seeing problems as a brain disease or chemical imbalance is reductionistic and deficit based. I suggested to Joyce that I work with her and Sam on nondrug interventions. Yet, Joyce was adamant to see the psychiatrist for a medication evaluation as first treatment option. Being client centered, I honored Joyce's request.

This case example is all too commonplace in my practice. ADHD was virtually unknown 20 years ago, but now is the most widespread behavioral disorder in American children. As the ADHD diagnosis has gained popularity, so has medication treatment as a first and only treatment option been uncritically accepted by parents, teachers, and mental health professionals. This recent surge in the "popularity" has given many mental health clinicians, including myself, cause for alarm. Is ADHD the latest "fad" or is it a truly pervasive affliction whose scope is just now being discovered?

Views and arguments of the legitimacy of the ADHD diagnosis tend to take one of two major arguments. The more prevalent view is that ADHD is an actual identifiable disorder in the brain whose true extent among the American public is just now being discovered. A second view, which is gaining due to the massive surge is ADHD diagnoses, is that ADHD is not a discrete biological condition but a social invention to explain certain behavioral problems. The professionals who hold this perspective posit that ADHD has no pathology, no biological marker in the brain that clearly demonstrates its existence. Thus, its diagnosis is always subjective. These two viewpoints seem incompatible with each other. There is great tension between the researchers and practitioners who subscribe to the opposing viewpoints. Yet, as a frontline therapist, I have found a way to navigate these seemingly opposing perspectives. The following describes how I work with children with ADHD, holding these contradictory views alongside each other.

Diagnosing ADHD has the consequences of tending to cement a totalizing deficit-based description of the child. I have always been suspicious of the medical model and its pathologizing tendencies. In the past, I would impose my view on the parents and children who came to consult me. Many parents would feel invalidated and blamed. My critique of the medical model was of no interest to them; they wanted help for themselves and their kid. The diagnosis and Ritalin offered that relief and hope. So, with time, I learned to listen and privilege the parents' and child's meaning(s) of ADHD. By honoring their views and understanding, my therapeutic alliance was enhanced. Once this therapeutic relationship was developed, the parents I work with are more open to new perspectives. I might ask a question, "Now that Johnny is diagnosed with ADHD, what's possible that wasn't before?" Asking such a question links a psychiatric diagnosis with hope and possibility rather than deficit and settling for less in life. I then can invite a parent to see that the ADHD label paints an only partial picture of a child by asking a question such as, "Who is Johnny apart from ADHD?" This question can give me rich information on the child's skills and abilities that I can harness in therapy to bring forth solutions and improvement in self-esteem.

I also feel an obligation to alert families about the economic, social, and cultural factors that are involved in the ADHD diagnosis and Ritalin use in North America. To not raise alarm would make me complicit with values and factors I feel are harmful to children and their families. How I might raise these sociocultural factors is by asking parents and children their reasons why ADHD has become such a popular diagnosis. Sure, some parents say that, historically, ADHD is underdiagnosed, and we now have more information and better assessment tools. Yet, others share their concerns of having to rely on medication, their distrust of pharmaceutical companies, and the potential long-term side effects of stimulant medication. Some parents tell me about the stresses of parenting and trying to balance both work and home. Many of the ADHD youth disclose to me the pressures of school performance, hours of homework, and needless, mind-numbing worksheets.

Once ADHD is located in a broader context, creative solutions become available, ones that help parents realize that their child has unique abilities that are not valued by dominant culture. The child is seen as more than just Johnny the ADHD kid.

And the parents are more open to nonmedication interventions. They come to see that I'm against the use of medication in children, that I'm critical of Ritalin as a first and only choice for a wide variety of children's performance and behavior problems. Ritalin works but it is not a substitute for, or equivalent to, better parenting and schools for children. Critiquing the larger cultural forces that create that context for ADHD is key to a more holistic treatment approach. Yet, I don't want to ignore the brain and biological explanations for ADHD. It is not an either/or (culture/brain) but a both/and. Instead of brain disorders, I think it is more effective to talk about brain differences. Instead of regarding large portions of the public as suffering from deficit, disease, or dysfunction in their mental processing, neurological diversity suggests that we instead speak about *differences* in cognitive functioning. Just as we talk about cultural diversity, one can use that same kind of thinking in talking about brain differences. A calla lily is not pathologized for not having petals (e.g., petal-deficit disorder), nor are individuals with brown skin diagnosed with suffering from a "pigmentation dysfunction." Similarly, we ought not to pathologize individuals who have different ways of thinking, relating, attending, and learning.

By using the concept of neurological diversity to account for individual neurological differences, a discourse is created whereby labeled people may be seen in terms of their strengths as well as their weaknesses. People with dyslexia, for example, can be seen in terms of their visual thinking ability and entrepreneurial strengths. People with ADHD can be regarded as possessing a penchant for novel learning situations. Individuals along the autistic spectrum can be looked at in terms of their facility with systems such as computer programming or mathematical computation. Those with bipolar disorder can be appreciated for their creative pursuits in the arts. Although proponents of the concept of neurological diversity do not shirk from the realization that people with dyslexia, ADHD, autism, bipolar disorder, and other psychiatric conditions often suffer great hardships, and that those hardships require a lot of hard work to overcome, they realize that, until an individual's strengths have been recognized, celebrated, and worked with, nothing substantial can be accomplished with regard to their difficulties.

The neurological diversity concept also gives a template for modifying the surrounding environment to fit the needs of one's unique brain. ADHD brings with it special abilities as well as difficulties, and appropriate career selection can be an important part of determining whether one will be successful or unsuccessful in a particular job. The notion of neurological diversity helps combat "ableism" and helps to challenge what our society sees as "normal" and "abnormal." Neurological diversity brings with it a sense of hope rather than the cynicism of biological fundamentalism.

In summary, I have learned to take a modest, balanced approach to ADHD. I can be concerned about the overmedicating of kids (and "Big Pharma") and still recommend stimulant medication as one of many therapeutic options. I've learned to not ignore biology but talk about brain differences in a nonreductionistic manner through the discourse of neurological diversity. I have learned to decry the excessive labeling, the unreasonable pressures from schools and parents, the aggressive advertising by "Big Pharma" to parents and physicians while still honoring the ADHD label as a potential meaningful construct for some parents and their children with behavioral and attention problems. As my grandmother said "moderation in life is the key to contentment!"

Identifying the Problems and Working Collaboratively to Make the Best Decisions

I am writing a story about hard work, persistence, and collaboration between a family, a school, and a psychologist in the efforts to support a wonderful child. This story is told from the perspective of the psychologist, who received a referral for assessment and support for an 8-year-old boy named Stefan. Both Stefan's parents and school personnel were concerned about his behaviors, which included significant attention challenges, as well as acting-out behaviors such as yelling, swearing, and hitting. In a large rural school division, many specialized services, such as psychological assessments, often involve a long wait until the school board psychologist can see a child, so families sometime seek support privately when possible. That is how Stefan and his parents, Nancy and David, came to see me in the spring of his second-grade year.

Actually, my first meeting was only with his parents, as a visit to the school for an observation seemed appropriate given the nature of the concerns, and I did not want Stefan to know who I was prior to that time. As is my practice, my initial questions to his parents focused on Stefan's strengths. Nancy and David described their son as loving, full of energy, and possessing a great sense of humor. They also described him as an avid reader who always had a book in his hand, and had a strong interest in cars. One of his challenges was putting down his book or car toys, in lieu of a less preferred activity. At home, this was a challenge, and at school, it was a greater concern. In addition, Stefan's parents described him as highly distractible and disorganized, which made following daily routines, such as morning and bedtime preparation, a challenge to complete. Stefan also frequently joked around when people were trying to accomplish other tasks, such as getting ready to leave for school or work. But besides the distractibility, disorganization, and difficulty with transitions, Stefan demonstrated some other behaviors that were also concerning to everyone who knew him. These involved his impulsivity and high emotionality in the face of challenges, reprimands from adults, problems with peers, or requirements to do tasks that he did not want to do. He tended to quickly become frustrated and either yell out loud, lash out physically, or make self-deprecating remarks like "I'm stupid." Nancy and David told me that these behavioral challenges began to emerge for him when he entered preschool, where difficulties following routines, high distractibility, low frustration tolerance, and rough play with his peers, were not uncommon. Stefan had even hit other children and left the school grounds on occasion. Stefan's parents had already taken him to their family doctor before seeing me, and were told that he likely met the criteria for an attention-deficit disorder and oppositional defiant disorder (ODD). A brief trial of stimulant medication had been conducted, with limited results in terms of reducing his symptoms. The parents were interested in a more comprehensive evaluation, as well as support for developing and implementing a broader range of strategies to assist him.

In my conversations with school personnel, they mirrored many of the concerns parents expressed. They described Stefan as sociable, creative, energetic, and enthusiastic. They commented on his excellent reading and math skills. They also mentioned his considerable impulsivity, distractibility, and low frustration tolerance. Stefan apparently did not like writing, and often avoided activities where he had to put pencil to paper. School personnel also mentioned Stefan's negative self-talk, including self-deprecating remarks and comments about his dislike for school. Stefan himself mentioned that his favorite thing about school was recess, but that he did not like writing. It was interesting to see that his writing seemed to be generally adequate for his age and grade. School staff had already been trying many strategies to support Stefan, including adding fidget tools and sit cushions, teaching calming strategies, using a daily self-rating scale, and using a daily concrete reward system.

I first saw Stefan in his second-grade classroom. Several aspects of his behavior were of interest to me. One was his tendency to want to follow his own agenda. For example, he brought a toy to circle time, which he quietly played with while the class was engaged in their morning routine. In addition, as the students later worked on individual tasks, Stefan often engaged them in joking or talking about topics of interest, which tended to distract everyone from their work. A concrete reinforcement system was being used with Stefan to encourage on-task behavior. Even with this system in place, Stefan required considerable teacher redirection to stay on task. Stefan also tended to be quite loud in his talking and interactions with peers. Finally, Stefan's tendency to frustrate quickly became apparent. He was quick to yell at peers if he was challenged, which he occasionally was, either because he was not following routines or due to typical dynamics between children at this age. On two occasions he yelled at his classmates when they did or said something that he did not approve of, and on one occasion, when his initial yelling did not produce the desired results (he wanted a peer to stop playing with his toy), he struck him with a pencil.

When we met in my office, Stefan's behavior was consistent with reports from the adults in his life, as well as what I had seen in class. Stefan was very quiet at first, not responding to my questions and barely raising his nose from the book he had brought in. Although he did eventually converse with me and participated in tasks I had for him to do, he needed considerable redirection and several breaks to maintain his attention. Stefan also started to become quite talkative as we got to know one another, particular when it was about something he was interested in, requiring frequent redirection to the task at hand. He also often resisted completing tasks, frequently negotiating with me in terms of how much longer we would have to work before he could take a break. Because the reasons for referral were behavioral and there were no academic concerns, I did not conduct extensive assessment of his cognitive or academic abilities. I did conduct some brief assessments, however, and, to no one's surprise, he performed well on all cognitive and academic tasks that I gave him. Behavioral rating scales clearly reflected information that I obtained from parents, school personnel, and through my own observations. I concluded in my assessment report that Stefan's inattention, impulsivity, and activity level met the criteria for a diagnosis of ADHD: combined type. Further, his pattern of irritability, low frustration tolerance, and aggression met the criteria for a diagnosis of ODD. For me, so many times this is the end of my interactions with a child and family, as I provide recommendations, and students, parents, and school personnel are left with the task of trying to implement them. In this situation however, there was a distinct interest from everyone working with Stefan that I continue in my involvement with him, his parents, and the school to assist in the implementation of support strategies.

I met with Stefan and his family on several occasions over the 2-year period that followed the initial assessment to provide consultation. His parents were very willing and open to exploring avenues for helping their son to cope with challenges in school and at home. They were also interested in support for themselves in terms of reducing the stress involved in parenting their child, as well as increasing their own skills for coping with times of high emotionality and noncompliance. Strategies for Stefan included using analogies to help him to visualize and rate his anger, as well as to develop calming strategies. We used an "anger mountain" as our main analogy, and incorporated the use of calming strategies such as taking some time away from a situation or using "turtle" (a deep-breathing technique). For parents, similar calming strategies, as well as using meditational language to assist Stefan to recognize his emotions and implement possible strategies, were employed. Nancy and David reported that the manner in which they handled high-stress situations allowed Stefan the opportunity to also handle things differently. Routines became easier at home, there was less conflict, and Stefan seemed to have more self-awareness regarding his feelings.

In addition to behavioral strategies, another avenue the family explored was medical management of Stefan's attention challenges. I explained to Nancy and David that research supported this avenue for treatment as a powerful part of an overall intervention plan, but only a part. I also mentioned the high emotionality and strong opinions that often surround the area of using medication with children. Many people have strong opinions about this topic, but not all opinions are equally well informed by solid research. We discussed their thoughts regarding the previous medication trial, which had not produced significant results, but had not been closely monitored and was short lived. As with any type of support, the bottom line for the parents was to be well educated on the pros and cons of the treatment so they could make informed decisions for their child. I provided the parents with some resources on the subject and left the decision in their hands. In the end, a referral was made to a pediatrician who specialized in ADHD. A systematic, closely monitored trial of medication was conducted, and parents found that medication made a huge difference for Stefan, not only in his attention and ability to follow through on requests, but also on his oppositional behavior. This seems to be fairly common among children with these types of challenges because, although the stimulant medication often prescribed for managing ADHD symptoms does not directly impact on compliance, oppositional behavior, or high emotionality, it does impact on impulsivity. As a result, children who are prone to lashing out verbally or physically may think first before acting, more often deciding not to act on their initial thoughts and feelings. I mentioned to his parents that the success of this aspect of intervention provided an opportunity for everyone to be more proactive in terms of developing other strategies. Most importantly, with Stefan's increased attention and reduced impulsivity, he may be in a better place for learning other ways to manage his own behavior.

Another crucial aspect of support was collaboration between home and school. An individual education plan (IEP) was developed for Stefan, and an educational assistant (EA) was made available to support him for some of his day. This support was very important in the initial stages of the intervention process, as the EA increased the teachers' abilities to be proactive rather than reactive in helping Stefan to manage challenging behaviors, as well as to provide Stefan with more timely feedback and reinforcement for appropriate behaviors. This is a crucial aspect when dealing with oppositional behavior, because appropriate behavior may actually be the absence of overt inappropriate behavior (e.g., yelling and hitting), and without close monitoring, may go unnoticed and unreinforced in a busy classroom. Stefan's parents were very appreciative of the school's efforts to support Stefan, and were eager to integrate and coordinate this support with what they were doing at home. Through regular collaborative meetings, Stefan, his parents, and school staff worked closely together on a reinforcement system that was used both at home and at school. In this regard, we focused on Stefan's strengths and what was working, as well as challenges that needed to be addressed. This provided regular feedback and reinforcement for Stefan, as well as information for everyone regarding his progress. As time progressed, his initial outbursts lessened considerably. Other issues then became more evident, including the need to assist him in managing transitions between activities, getting back to work in a timely manner after taking a break, and handling conflict with peers during less structured times or when there was not an adult readily available to mediate. As these challenges arose, they were each addressed and dealt with in a positive and proactive manner.

Instead of being adversarial, which can sometimes be the case when people are dealing with challenging behaviors in children, as well as different perspectives from school and home, these meetings were very collaborative and focused on how to improve Stefan's ability to succeed. This was in no small part due to the parents' support and understanding that the school personnel had their child's best interests at heart and were working toward improving his ability to succeed in the classroom. My involvement in these meetings was part of the parents' effort to coordinate school and home strategies and interventions. This appeared to be appreciated by school personnel, who welcomed the opportunity to have another professional opinion in this situation.

Over time, the family developed skills and abilities to address various challenges that they encountered, and our meetings, which were initially quite regular, decreased in frequency. The parents realized, and I reinforced, that our lessening contact was a signal that they were growing in their independence in terms of addressing issues that arose. Instead of regular meetings every other week, we moved to monthly meetings, then to checkup meetings once every school term. Even with the lessening contact, the parents maintained their proactive stance, often calling me after some time had passed to indicate that they felt it was time to touch base or to address some new concerns that were coming up. I feel very fortunate to have had the opportunity to work with these wonderful individuals, to help support a young person through some challenges, and see him succeed.

Epilogue

From our perspective, some themes within and across many of the stories in this section include the following.

- The lack of expected social and classroom behaviors of many children with ADHD often have a negative impact on each individual's social–emotional success and their self-esteem. In general, the challenges of these children include their difficulty in making and keeping friends and their inability to effectively deal with the consequences of not meeting the expectations that are required of them (e.g., following directions, organizing, planning, and managing their time). Each child's self-esteem is typically quite low and becomes an issue as they often receive much more negative than positive feedback.
- Adults (professionals and parents) responsible for these individuals also seem to struggle with how to manage their own emotions. In this regard, their emotions are often influenced by their perceptions that "others" are attributing the child's undesirable behaviors to their inadequacy as a parent, teacher, therapist, or psychologist.
- Strong relationships between a child with ADHD and his or her family and associated "helping professionals" (e.g., teachers, therapists, and psychologists) are "keys" to a child's social and academic development.
- It is important to always view the child as a child with ADHD rather than to view the child as an ADHD child.
- Every child with ADHD has personal strengths as well as appealing and positive characteristics, and these strengths and character will be the foundation for the child's future success and fulfillment.
- Medication can very often be very helpful for children and youth who have ADHD. Medication can create and maintain attention and behavioral balance within many children and youth, but cognitive and behavioral interventions and strategies provided by parents and helping professionals and effectively utilized by the children are also very important in the treatment and developmental plan.

Specific thoughts expressed in the stories that we thought were insightful and that we have heard children and youth express in similar ways over the years include the following.

- Every time I became excited, active, or simply wanted to play, I was told to take a pill. I disliked the way the pill made me feel, and it amplified the feeling that I wasn't very smart or a worthwhile person.
- I mainly remember just feeling stupid and like I didn't belong. Friends sometimes became a safe haven where I could express my feelings, and be accepted the way I was.

Specific thoughts expressed in the stories that we thought were insightful and that we have heard parents express in similar ways over the years include the following.

- After a while, even kids we considered his friends refused to invite him to their birthday parties and I was so hurt on my son's behalf that I severed my friendship with their families.
- The lack of ability to see the positive things about my child and to overcome my embarrassment about some of his behaviors has had long-lasting implications.
- When I approached the teacher to tell her that my child was excited about the rewards program that she was implementing for her students and asked if it could be tweaked slightly to provide an opportunity for him to succeed, we were told that she couldn't give anyone special considerations as that wouldn't be fair to the other kids; thus, the teacher and I became adversaries.

2 Conduct and Oppositional Defiant Disorders

Prologue

Two common disorders of children and youth who demonstrate antisocial and aggressive behaviors significantly beyond what would be considered within the normal realm of disruptive and acting-out behavior exhibited by peers of similar age and development are oppositional defiant disorder (ODD) and conduct disorder (CD).

The primary feature of ODD is a pattern of defiant, negativistic, and hostile behavior such as arguing with authority figures, conning others, annoying people, refusing to comply with adults' requests, defying rules of behavior, and being angry and resentful to such a degree as to result in significant problems in social, academic, or occupational functioning.

The primary feature of conduct disorder is a pattern of behavior that persistently violates the basic rights of others and/or the age-appropriate norms of society. This pattern of behavior includes serious violations of rules (e.g., school truancy, staying out at night without the permission of parents), destruction of property (e.g., vandalism, arson), theft, and aggression to people and animals (e.g., threatening and bullying others, being physically cruel to people and animals, fighting with others).

In this section are five stories: one personal story, one parental story, and three professional stories pertaining to ODD and conduct disorders.

The personal story, entitled "Different Roads with Different Ends," is written by a young man who recalls some of his childhood journey and how it led to being diagnosed as conduct disordered. The parental story, entitled "A Matter of Soul," is written by a mother who reveals her disillusionment and disappointment about raising her son. In this regard, she lets us know that having a son with ODD is, for her, having "death symbolic syndrome" (not being able as a parent to "live out" the hopes and dreams you have for your relationship with your child due to your child's disorder). The first professional story, entitled "Sometimes Persistence and Hope Is All We Have," is written by a school psychologist who presents his views from his experience as a teacher of a student with ODD. He highlights the human dynamics in working with this young girl and the lessons he learned. The second professional story, entitled "Seeking to Understand Before Trying to Be Understood," is written by a behavioral caseworker who recalls her experiences working with a teenage girl and how important building a relationship was for each other's growth. The final professional story, entitled "Sometimes We Gain More than We Think We Will in Our Work with Children and Youth," is written by a psychiatrist who, while working with a troubled youth, found himself changing his thinking paradigms and behaviors regarding conduct disorder while trying to change the thinking paradigms and behaviors of this troubled youth.

Exceptional Life Journeys. DOI: 10.1016/B978-0-12-385216-8.00002-3

Personal Story

Different Roads with Different Ends

When I was 11, the most remarkable thing that happened to me was when my sister and I were sent across the country to live with my father's brother and his family. I did not know these people and it felt very strange to be sent to them. The reason for going was understood a little bit at this time and much better understood during the later years. My mother became very sick and had to be hospitalized. So, we were sent away because my father felt unable to care for us while she was being hospitalized. There is not much I remember about the day that my sister and I were told that we would be leaving our home to go live with my father's brother and family. We were going to travel by train to his home. We went to the train station in my father's brand-new convertible. It was a "hot" car and a real "pride and joy" of my father. The thing I remember most about this car ride was the panic I felt. My father was under the influence of alcohol and he seemed to enjoy driving fast and scaring us. My sister did not seem to be scared but I am sure she was. I was on the floor in the backseat and hoping we were not going to get in an accident and all get killed. I remember my father looking back and down on me as I was screaming for him to slow down, and him just looking at me with a smile on his face as we were going well over a hundred miles an hour on a city street. We made it to the train station, but this experience was so strange to me. The only person who seemed to be having fun was my father who did not seem to "give a ****" about his son or daughter at that time. I am sure this had a lot to do with the alcohol during that time.

At the train station, I remember telling my sister that I was looking forward to our journey and to our change in life. She said that she wanted to stay so she could be with her friends, something that I was not going to miss. My mother had left a few weeks before we left. I cannot remember her saying "good-bye" or my father talking to us about what was happening. Why was she leaving? Why were we leaving? Why was he staying? Everything just seemed to happen without any explanation.

A few days before we left to go live with my father's brother, we took a trip to a city a few hours from where we lived to visit our mother. We were not told where we were going, or where my mother was staying, or what was happening to our family. We just got in the car and headed off to a place to see my mother. I did not know that this would be the last time I would see her for a while. The place she was staying was old. It looked like a prison. It was very creepy and seemed very weird to me. Once in the building, we were led down a long and dreary hallway until we got to the place where my mother was staying. It was like a prison cell. It had bars across the door, which needed to be unlocked by a key from the outside. My mother obviously did not have the freedom to come and go as she pleased. Her living space was very small. She had a single bed and a toilet and that was all. She was dressed in a long nightgown and I could not see any other clothes in sight. She seemed very distant to me. She seemed to be very tired and unhappy. We did not stay long and I do not remember anything that was said (if anything was). I remember leaving and the terrible sound the door made when it closed behind me. There were no hugs or kisses good-bye. We just seemed to walk in, take a look at each other, and then leave. I did not know until much later that she was undergoing shock treatment for depression. I never wrote any letters to her and never received any letters from her. Although we received occasional letters from my father, we did not seem to know much about what was happening to him or my mother.

We boarded the train in quick fashion. We said our good-byes to our father and got on the train like we were going on some day excursion rather than on some long journey. Again, our departure occurred without emotion. My sister and I just diligently got on the train and headed for our compartments, which would be where we stayed most of the time on our trip to the west. Once we arrived, we were taken to our uncle and aunt's place. They were never very welcoming, and it seemed to me that they felt obligated in having my sister and I stay with them. The time we stayed with them was mostly uneventful. I never felt as if I was part of a family, and neither did my sister. We were treated relatively well but there wasn't any bond between us. I always felt like we were in a "holding place" until the time came for us to go.

Coming back home a couple of summers later was an anxious time because I did not know what to expect. In some ways, coming home was different because, although I was ready to come back, I was not really excited about it. However, this was different for my sister because I think she would have rather stayed. In any case, we boarded the train again and headed back east. I do not remember much about the trip back home because not much happened and I did not meet any interesting people. What was most remarkable about the trip was when we arrived and we first met my parents who were waiting at the train station. Upon leaving the train, my sister and I walked out into the station where my father and mother were waiting for us. Once they saw us, my father rushed to give my sister and me a hug. This was a bit strange because I do not remember getting hugs from him in the past, and it seemed a little exaggerated. I was mostly reserved during this time and did not receive the welcoming in a warm and loving way. It was more like getting a hug from someone you do not know very well and wondering why they seemed to be so interested in you. After my hug, my father went to my sister and gave her a hug, which I know was received the same way. As my sister and father were engaged in "the hug," I saw my mother looking at me from a distance. She was about 10 feet away and although smiling at me, she seemed nervous and shy. My father, sister, and I walked toward her with smiles, but there were no hugs or kisses from her. There was just a warm and distant greeting that seemed as puzzling to me as my father's hug.

We left the train station and went back home. I do not remember any conversation on the way back home and still do not remember anything much that was significantly different from how I felt about being with my family before. We just seemed to settle back into routines and live our lives as we did before we had separated. Although my mother did not seem as depressed as she did before, she also did not appear happier with her life.

The summer of that year was unremarkable. Nothing much had changed, and my neighborhood was the same as I had remembered it and had experienced it before I had left. The same kids were in my neighborhood, only now they seemed much more attached to each other and more detached from me. No one seemed to be excited to see me back, and I seemed to be back to the same lifestyle. I always felt that I was an outcast and unappreciated. I spent a lot of the summer sort of hanging around but not really being a part of anything. Life was boring at this time as I spent most of the time alone and around my house. In fact, I found myself untypically looking forward for school to start so I would have something to do.

My first day back at school, I was really looking forward to something different and for some involvement in things that might make life much less boring and much more worthwhile. Although I did not have any friends and was somewhat worried about how the day might go, it had to be better than what summertime had been. I was back to walking to school alone and seeing all the other kids with their friends and enjoying each other's company. I felt lonely and a little nervous as I walked toward the school, but I hoped that things would start to come together and that this might be the start of much better times for me.

Once at school, all the kids had to hang around the schoolyard and wait for the bell to ring that would be the signal that the teachers were ready to have us come into the school and go to our classes. I felt most lonely during this time because I was by myself. Although I knew some of the kids in the schoolyard, no one seemed much interested in me. They all seemed to be having fun with their friends and quite comfortable with their situation. I still remained hopeful that something would happen that would lead me to having more fun maybe meeting new friends. When the bell rang, everyone rushed to the school doors where teachers stood waiting for us. They called out our names and let us know which teacher we should follow to their classrooms. My name was called and I was going to be in the principal's class. Once in the classroom, we were able to choose our seats and await class to begin. I just took the first seat that I saw and that was closest to me. I was happy with my seat selection because I got to sit beside a girl who I thought was the best looking girl I had ever seen and she seemed to be nice and friendly.

The morning went fairly quickly. The principal introduced himself and let us know what to expect in his seventh-grade class. He distributed books, as well as paper and pencils, to us. He talked to us for most of the first hour of school and made many notes on the blackboard, which we were told to copy into our books because they were the rules and expectations he had for us in his classroom. I spent most of this time doing what I was told and looking at the girl next to me in awe. Then, the recess bell sounded. Everybody quickly got out of their seats to get their coats and get outside as quickly as they could. As I was fumbling around to get my stuff, I bumped into the girl next to me. At that exact moment, I felt passion like I had never felt before and looked at her to see what her reaction might be. Although I felt awkward and a little ashamed, I also felt wonderful that I had made this connection with the girl I had been dreaming about all morning. I hoped I would see a smile like the one that must have been on my face. I hoped she felt about me as I did about her. But, to my dismay, she did not look at me; she did not even look like she knew that I was there and had the feelings I did for her. Instead, she continued to talk to a girlfriend and quickly left me to go outside and join the others. What a disappointment!

As I left the class for recess, I wondered what I could do to get her attention and "win her heart."

Once out in the schoolyard, I was standing by myself, uncertain what to do and where to go. It sure did feel uncomfortable to be the only one standing around by himself while looking at others enjoying each other's company. As I was in this feeling, I noticed some guys playing "touch football" with others, including the girl of my dreams, watching them play. So, I moved closer to where they were playing and to pretend that I was part of the group. All of a sudden, I heard, "Hey you! Can you get the ball for us?" I looked down and slowly rolling past me was the football the guys were playing with. So, I bent down, picked it up, and threw the ball to the boy who had asked for my help. It was a great throw and it seemed to impress the boy that I threw it to.

"Do you want to play?" asked the boy who caught my great pass. "You can be on our team!"

"OK," I said.

At the time of my arrival on this boy's team, we were playing defense. I noticed that the other team had some players who seemed to be the biggest boys in school. They also seemed to be the oldest compared to us. In fact, one of the players appeared to be an adult because he had hair on his face and seemed to act in ways that seemed much more mature than the rest of us. I had noticed him before recess in my class, and he presented himself as a bit of a rebel. He appeared to be someone who should be graduating from high school and not someone still in elementary/junior high school. I also noticed that he liked to swear a lot, and it seemed to me that he thought himself to be special and better than all the rest of us. His name was Paul and, along with looking older, he also looked "rough and tough." He had tattoos on his arms that were readily seen because of the short-sleeve shirt he wore and had a pack of smokes in his back pocket. This was somebody I was not used to seeing, and he scared me a little bit.

During our defensive play, it was obvious that the other team was much better than the team I was playing for. The quarterback was good and was always able to throw the ball to players who seemed able to catch the ball. Once the ball was received, the player was able to run for quite a while before being touched by one of our players. The action was on the other side of the field that I was playing, so I spent most of my time watching what was happening. Within about 10 min of me joining my team, the other team was able to move the ball about 50 yards and was very close to our end zone. It was just a matter of a short amount of time before they would score. As they were getting ready to play their next down, I noticed that Paul had moved from the other side of the field to my side of the field. I think he did this because he thought that by moving to where I was, it would be easy for him to catch the ball from his quarterback and score the big touchdown. I think this occurred to the other players on my team as well because as the other team was ready to go with their play, a number of my players were yelling at me to watch out for Paul. Although I looked awkward and clumsy, what I knew was that I was not as bad at playing sports as they thought I was. So, as the play started, I was ready and knew that the ball was going to go to Paul. I thought this was going to be my time to shine!

I was going to stop them from scoring and I would be considered somebody more special because of it. I was especially looking forward to how impressed my "special" girl was going to be with me. This was going to be the start of new friendships and the beginning of respect from others. Everything was going to be OK now!

I saw the quarterback get the ball from the center and, as he was moving into the pocket, I saw him look toward my side of the field and where Paul was. Everything happened for me as if it was in slow motion. As the ball was released from the quarterback's hand, I saw where it was going and I moved to that place, just ahead of Paul. When it came, I was ready and in the best position to intercept it. It finally came and rested in my hands. As I caught the ball, I noticed Paul slip and fall. He fell right into a muddy patch of the field, where I noticed him slipping and falling while trying to get up to get the ball back and to get me. By the time he got up, I was already 10 yards down the field. To my surprise, not only had I caught the ball and avoided being tackled by Paul, but it also seemed that everyone else on his team was on the other side of the field and had given up on the play. It occurred to me while I was running that they must have thought that the game was over because it was Paul against me and that there was no need to pay much attention to what was happening on my side of the field because Paul was going to score. It would be impossible for anything to happen!

I ran as fast as I could, and within seconds, I had not only intercepted the pass but scored a touchdown for our team as well! Whoa! I could not have dreamed anything better for myself. This was going to be the best day of my life. Right after scoring the touchdown, I saw members of my team rush toward me with smiles on their faces, eager to congratulate me and celebrate our win. As this was happening, the bell sounded for us to go back into school. So, before they could reach me, they began to head the other way in order that they could head back to class and not be late. However, while they were doing this, they were still yelling positive comments to me like, "Way to go!" "Great catch!" "We won, we won!"

I wished the bell had not rung when it did because I wanted to engage in all the celebration and get all the positive regard I could. This was my moment to shine and I wanted it to last as long as it could. I wanted to get all the slaps on the back and all compliments. I wanted all of the glory. I especially wanted to get the recognition from my "special" girl that I just knew I was going to get. But, she had gone inside the school before I had a chance to see her. However, I knew she was going to be proud of me and want to get to know me. It was just going to be a matter of time.

Not everyone was happy with me though. As I was returning to class, I saw the faces of the players on the other team just ahead of me. They were looking back at me and shaking their head with both frustration and disbelief. I understood this because it is never fun to lose, especially when you think there is no way you can lose. As I was getting close to the door entry into the school, there was only me and Paul left outside. He was waiting for me at the doorway.

"You think you are pretty ******* good—eh," Paul said to me as he stared at me.

I looked sheepishly at him and decided to walk past him into the doorway and get to class as fast as I could.

"I'm going to get you today!" He shouted to me as he pushed me forward and I walked past him.

I was really scared as I walked past him and walked toward the classroom. Paul followed me and just kept saying, "four o'clock, four o'clock, four o'clock."

This was scary because school ended at 4:00 p.m., and he obviously had something planned for me and it wasn't going to be good.

I got back to the classroom and went right to my seat. The teacher was just beginning the afternoon class in math when I arrived and sat down in my seat. As he was telling the class to get the math textbook from inside our desks, turn to a page for instruction, and get ready to follow along to his instruction, I sat motionless in my seat in a panicky state and unable to focus on anything that was happening around me, except on Paul. As I cautiously turned to look at him where he sat just to the left and down from me a few seats, he was staring at me and punching his right fist into the palm of his left hand. He was staring at me meanly with a slight grin while gritting his teeth and slowly nodding his head up and down.

I turned away from these threatening gestures and began to feel myself sweating and aching. I tried to put Paul out of my mind and concentrate on the instruction, but I was just able to follow along without much understanding of what was really being presented and taught to us. Everything right now was about Paul and my uncomfortable feelings. I forced myself to look away from Paul as much as I could, but it seemed like every few minutes I would glance over to him, hoping he had enough of me and that he would forget about everything over time. But, every time I glanced at Paul, he seemed to be always still staring at me and continuing to punch his right hand into the palm of his left hand. It was obvious to me that his left palm was going to be my head at the end of school.

I did not know what to do. It occupied my thinking the whole afternoon, with Paul's overtures to me and the clock on the wall that constantly reminded me of my doomsday ahead. It became my routine throughout the afternoon to look at Paul and then the clock and then stare down at my desk. Two o'clock, 2:15, 2:30, 2:45, 3:00, 3:15, 3:30.

The day was coming to an end and I was about to be facing Paul outside. As the end of school approached, I also noticed some of Paul's other friends in the class occasionally look at me with grins on their faces as they looked at me, then Paul, and then back to me. It was beginning to be clear to me that not only was I aware of Paul's intentions, intimidations, threats, and goal but many others were aware of them too and seemed to be increasingly interested in what was going to happen at the end of the school day. They seemed to be eagerly waiting whatever Paul had in store for me. As the clock neared 4:00 p.m., I became resigned to the fact that I was going to have a fight with Paul. I also understood that I was going to get hurt because I had never been in a fight before, and Paul looked like he had fights every day. And, I was sure he won all of those fights.

Then it happened. The bell rang and signaled the end of the school day and time for us to go home. I sat motionless in my seat as everyone else rushed by me to get their coats and leave the class. I waited until almost everyone had left and was hoping that Paul had his fun with me during the afternoon and this might all be over. I got up from my seat, and I was the last one to get my coat, walk out of class, walk down the hall, and leave the school. It was a lonely walk because everyone by this time was outside, and I was the only one left to leave the school and find out my fate.

I walked out the school door and noticed only a few kids ahead of me as they were turning around the corner of the school to get to the sidewalks that would lead them home. As I got closer to the corner, I was hoping that Paul had found something better to do and I would be able to escape the doom he indicated was going to happen to me all afternoon. I turned the corner and immediately my hopes were dashed. In the big courtyard were at least a hundred kids all in a semicircle, blocking any way for someone to get past them. In the middle of this semicircle was Paul punching his fists in the air as if he were preparing for a professional boxing match. I slowly walked toward the crowd and was being greeted by comments from a bunch of kids who were saying, "He's going to get it now," "That kid is dead meat," and "He's really ****** man."

I do not remember much that happened within the next 5–10 min. I remember approaching the crowd and being within a few feet of Paul when he came at me. I remember him saying, "You are going to get it now you ***********." He kicked me in the stomach, knocked the books out of my hands, and started throwing punches at my face. I remember the first two as I was falling to the ground. After that, I remember him kicking me over and over and over again. He repeatedly kicked my stomach and my head. It seemed like it went on for a long time but was likely just a matter of a few minutes before he stopped. As he was punching me and kicking me and yelling at me, the kids yelled for him to do it more. It finally ended, and I was lying on the ground in a fetal position. I can remember looking up at the kids as they walked past me and looked at me with disgust and pity. The look I remembered the most was the one from the girl of my dreams who looked at me as if I was the biggest loser she had ever known.

I laid by myself on the pavement for at least 10–15 min before I managed to get up. I was bleeding; I was sore; I was hurt; I was ashamed; and I was disgusted with myself. I said to myself two things that would eventually shape my future as I saw it that moment. I said to myself I was going to get even with Paul and that I was never going to let this happen to me again.

The next few years defined me for many years to come. Within this time period, I met some guys who were going to be a big part of my life. In fact, they were my life from eighth to twelfth grade. Mike was a chubby kind of guy who seemed to be able to do as he wanted when he wanted. He had a really good sense of humor, and we seemed to like each other from the first time we met. By the time he was 18, he was the biggest dope dealer in the region. At 25, he was brutally beaten by some guys with baseball bats who broke into his apartment and killed him while he was in his bed. They were later found and dealt with. I met Ed by way of Mike. Ed was a "James Dean" type (a rebel, a movie star). He was very good-looking and was well regarded by every girl he met. He was also very strong and tough. When I met Ed, he had already formed a reputation around the city as someone you do not "mess around with." Ed and I seemed to like each other a lot from the very first time we met. He was not only very attractive but also very appealing. I liked his "nerve," and I liked his self-confidence and his bravado. Ed became one of my very best friends. Bruce was a bit like Mike. He was also a bit chubby but much more fit. Whereas Mike's chubbiness was mostly fat, Bruce's chubbiness was almost all muscle. He was from a small town outside our city and hooked up with Ed and I after a night of drinking at one of our mutual acquaintances. Bruce was very bright and very much a charmer. Although he was not as good-looking as Ed, he was well liked by every girl who met him. He had lots of personality. He also won every fight I ever saw him in. Bruce became my best friend. He became my lifetime friend. He became the most successful one of us all. John was "the guy!" Just as Ed was like James Dean, so was John. He was street-smart, tough, and a leader. He was a little older than the rest of us, and he had a car! He became our leader.

By the time we were in tenth grade, we had become known as the "Colors." We became a gang that was revered by all gangs in the region (bike gangs and street gangs). When we walked into a bar, people gave us their seats. When we walked into a poolroom, people would give up their pool table for us. By the time we were in twelfth grade, we had collected over 150 criminal charges among us (assault charges, assault and battery charges, disturbing the peace charges, illegal use of firearm charges, break and enter charges, theft charges). We were lucky because, up until this point, we had spent many nights in jail but had not been sentenced to a penitentiary. Instead, we continuously got extended probation. By the time I was 17, I had accumulated 15 years of probation, had got even with Paul, and had got my name and picture in a number of papers in my city and in other regions. My probation order was given to me during the summer in which I turned 17 and stipulated that one more criminal charge could result in me serving out the rest of my probation time (15 years) in prison. Among other things, I was ordered to report to a probation officer once a month, stay away from my friends, and be off the street (be in my home) by 9:00 p.m. each night until I was 32. By the following fall, I had broken all of these probation orders.

It was near the end of twelfth grade when the most significant thing happened in my life. During my twelfth-grade year, I was amazingly able to get passing grades in all my subjects. By this time, only Bruce and I had stayed in school. Even though we still got in fights and created problems in many situations and with many different people, we were able to avoid more criminal charges during my twelfth-grade year, until April of that year.

In April, there was a beer strike in our region. One night (during the middle of April), while we were at John's house, Ed and I decided to take a walk. During our walk (at around 2:00 a.m.), we came across a beer truck. Naturally, we decided to take a look inside while the driver slept in his hotel room. We broke into the truck and found hundreds of beer cases that we began to remove and hide across the street. Instead of just removing a few cases for our personal consumption, we decided to take many more for the upcoming days and weeks. An hour after we had started, we had removed about a hundred cases of beer. While taking our last beer cases across the street we saw a police car approach the beer truck. We now knew that our mission had ended and it was time to leave. However, instead of leaving the scene by foot, Ed decided to steal a car to make our travel easier for us. It was not long until a police car pulled up behind us with his siren blazing. Instead of pulling over, Ed decided to try and speed away. After a few minutes of the chase, I looked out the passenger side window to see how close the police were to us. When I looked back toward Ed, he had already left the car. To my astonishment, I was now going down a street in a car without a driver. So, I bailed to. As I was tumbling out of the car, I saw the car continue down the street with nobody in it, followed by the police car.

Ed ran to the lake where he submerged himself (up to his head) for 2 h before leaving and heading home. Upon his exit from the lake, he was surrounded by police officers who took him to jail. Meantime, I had gone home. About an hour and a half after I had arrived home and gone to bed, eight police cars had circled my home. There were police all around my home, and four of them came to the door of our house. I was handcuffed and removed from my house in front of my mother and father who were still "half asleep" while this was occurring.

My lawyer told me that it was very likely that my friend and I were going to penitentiary to serve out the time given for the current offenses as well as the time left in the probation. He made arrangements for me to see a psychiatrist. During the second visit with the psychiatrist, I first heard the term *conduct disorder*. According to this psychiatrist, I was considered to be conduct disordered. Based on what he told me of this disorder, it seemed to make sense to me. In this regard, I certainly had broken a lot of rules, had been involved in a lot of fights, caused harm to some people, damaged property, and had been truant from school. If having this disorder was going to help me in court, I was fine with having this disorder. The psychiatrist seemed to like me and thought I had lots of potential.

My trial was in June. In the courtroom were Ed and I and our lawyers, along with the prosecutor and judge. A few people were in the seating area of the court but no one was there that I knew. Ed was first to be sentenced. As predicted, he got prison time. Now it was my turn. As I stood up to face the judge's order, there was an interruption in the court. Someone passed a note to the prosecutor. After reading the note, the prosecutor asked for a meeting with the judge and my lawyer. I did not know what was happening, nor did my lawyer. About an hour later, they all emerged from the judge's chambers and resumed their positions in the court. My lawyer stood beside me, and the prosecutor and the judge went back to their places. What took place next happened quickly and without any discussion. The prosecutor recommended that I be given some more probation to which my lawyer and the judge agreed. Within a few minutes, my life changed from being in prison to being free.

A few years later, I was told what happened that day. I do not know if it is a true story, but it's an interesting story. During the later part of my adolescence, a friend of mine got into a car accident. He was the driver of the car and the reason why one other teenager was killed and another had to live the rest of his life in a wheelchair. For at least a year after the accident, my friend was taunted and bullied by other kids because of the accident. He was called, among other things, a "killer." During this time, I did my best to protect him and stand up for him. This was very much appreciated by his father who was a well-regarded and wealthy individual in the community. My friend told me years later that the reason I did not go to prison was that his father was owed a lot of money by the prosecutor, with whom he made a deal. The deal was that if he let me off with probation he no longer would owe the money. I do not know if this is what really happened, or if it was because the judge found out that I was accepted to a university for the upcoming year, or whether it was another reason for me getting a break. Whatever the reason, I have not been in any trouble for about 10 years. At least I have not been in serious trouble. Although I still sit in bars with my "back to the wall" and still have some rage within me that I have to control, things are looking up and looking good for me, for now.

Parental Story

A Matter of Soul

As a research psychologist who works in the field of developmental and mental disorders, I know it is a truism that most people would prefer to have any physical ailment, even cancer, rather than mental illness. What is it that terrifies people so much about mental problems? There are the obvious factors: no easy cures, not always a good long-term prognosis, social stigma, and so on. But I think there is another factor that is not often mentioned: the fear of lacking rational thought processes. It is probably comparable to the fear that we all have of "losing our minds" to dementia in later life.

In a similar fashion, but speaking as the adoptive mother of a child who at various times has met diagnostic criteria for many disorders (ADHD, Asperger's syndrome, obsessive-compulsive disorder (OCD), attachment disorder (AD), generalized anxiety disorder (GAD), and ODD), I also have preferences; in this case, for parenting of different disorders. The fact is that I would rather deal with anything other than ODD. Why? The answer is similar to the previous one: again, it is the lack of the ability to reason. In this case, the child with ODD lacks the ability to reason in certain situations. And any parent can affirm that dealing with irrationality is an insurmountable challenge. In our own case, our son is totally rational about other topics, but he loses his ability to reason when exposed to the expectations, requests, or demands of other people. This "blind spot" has cost him friends, jobs, and roommates, in addition to posing a huge challenge to his family.

I do not believe that it is only because I am a scientist by training and profession that I feel so strongly about the importance of rational thought processes. I often call ODD the "soul-destroying disorder." How else to explain the feeling of trying to parent a child who simply refuses to accept parental guidance? There is nothing rational in our son's rejection of basic parental teaching and modeling, as he is the one who suffers many losses from the lack of training being offered.

In our case, we first attributed our son's oppositionality to the fact that he had gone through 14 months of foster care at the age of 3, where he had no understanding of the events that happened to him, and certainly no control. When we adopted him as he was turning 5, it seemed logical that he would need to exert control whenever possible. Attachment problems, insecurity, lack of control—one can attribute oppositional behavior to all of these challenges that we were dealing with as a family. It took about 4 years, until he reached approximately 9, for us to acknowledge and accept that our son's oppositionality was not going to go away. In spite of his becoming (apparently) firmly attached to us and secure in our family, he still rejected almost all attempts to guide his behavior. It is difficult to convey how profound this rejection was. Imagine asking your son to pick up something he dropped, and being told no. Imagine asking him to do something like come to dinner and almost never getting a positive response. Imagine offering to show your child how to do something (use a tool, wash his face, hold a toothbrush, organize a notebook, prepare a sandwich) and meeting with resistance and the announcement "I know how," when clearly he does not.

I once asked our son's pediatrician to try to explain the basis of oppositionality. He replied that very little was understood about the phenomenon, but that to him, it resembled the "impulsive negativity" that is found in 2-year-olds. We have all seen these little children respond "no" even before the request has even been completed: Yes, we could see that there was a similarity to the impulsively negative rejection of parental information and guidance that we were experiencing. But why would it extend to the age of 9 (and higher!)? The only way I can understand it is to think of his oppositionality as being the product of nonrational thought processes. It is difficult to comprehend nonrational thought in an individual with a high IQ who is clearly capable of rational thinking in other areas. But it is what it is.

At about age 9, it was clear that our son's other problems (ADHD and so on) were not going to disappear either, but it was the ODD that would prevent us from functioning in the manner we believed to be the role of parents. In retrospect, I believe I went through a period of mourning around then. Naturally, I accepted my son and loved him for who he was. But I also mourned for the loss of the parent–child relationship that I had expected to experience. I think I also mourned for the death of my "normal" expectations that I had developed for a very bright, sociable child (which he always was): academic success, good mental health, happiness in life.

I'm told that the parents of children tragically affected by diseases that will result in an early death go through a mourning period, trying to accept that their child will not have the opportunity to fulfill his or her potential. With the neurodevelopmental problems afflicting our son, imminent death is not the obstacle—ODD is. But the reality is the same: He cannot fulfill his potential, as conventionally defined.

The truth is that it is a singularly joyless experience to try to shape and direct your child toward the better experiences in life and to constantly experience negative reactions. At a certain point, the atmosphere in our home became too negative, and when he was about 23, we purchased a condo that he planned to live in with his friend. When that friendship collapsed and our son wanted to move back home, we did not permit him to do so. The reality is that our relationship with him had improved immensely when he was no longer in our home, and we did not want to lose that improvement. Coming together occasionally could be accompanied with moments of positive interactions, as opposed to constantly being confronted with his determination not to learn anything from us about life. So we insisted that he remain in the condo, and he has gone through many roommates since then, with no lasting situation yet in sight.

So here we are, after 20 years of "parenting," in which our son refused ever to learn how to make a bed, clean his room, prepare for bedtime, keep regular hours, eat properly, or attend to his health or personal hygiene in a systematic way. He has, in fact, made good progress in other areas, in his own good time and in his own way (fits and starts, three steps forward and two back—these are all accurate descriptors). So he continues on his own pathway, and we try to reinforce his strengths.

Does this sound like a grim story? As I said, we experienced little joy during his early years. But here is the amazing thing: A while ago, in frustration, I referred to his lack of cooperation as having prevented him from having a happy childhood. He looked at me, slightly stunned, and said, "I did have a happy childhood." I'm glad of that.

Professional Stories

Sometimes Persistence and Hope Is All We Have

By way of introduction, I am a professional within the education system. I have worn several hats in this field, most notably are those of a school psychologist and a sixth-grade elementary teacher. I do not know anyone else who followed this career path, starting as a school psychologist and then entering a career as a teacher, although I do know others who taught first and then became school psychologists. As such, I see my perspective as one that is unique and the experiences accumulated while making the transition to teaching have been characterized with a spectrum of emotion.

As a school psychologist, one is expected to get to know, correspond, and team with numerous teachers on a daily basis. It is impossible to serve in such a role without formulating an array of opinions about these many personalities, some favorable while others less favorable. For the most part, it really does not matter. With time, one develops a style that allows for functional interaction and consultation with almost anyone, regardless of their personality. When developing programming suggestions for students, there is almost always a strategy that can be pulled from a repertoire of "tricks" that can help level out the emotions and get everyone moving forward. If, for example, a school psychologist feels an administrator might challenge her assessment data, the response is simple. Principals often seem to appreciate open acknowledgment before numerous professionals and peers of their school's previous and current efforts to accommodate student exceptionalities, especially the student who is the focus of consultation. If even a small portion of the standardized assessment data could be drawn upon as evidence of these efforts, this often has a generalizing effect, validating all of the assessment data for the administrator. It is wise to start consultations with such a strategy to achieve "buy-in" early from the administrators. This same approach can be applied to teachers and parents, with early acknowledgment of their efforts, evidenced through the assessment data. Another strategy that can be used to overcome teacher resistance is that of opening the table to suggestions. After analyzing all of the psychological assessment data, often the school psychologist is left with some very tough but limited number of decisions. Often, these decisions have been arrived at after very careful consideration of numerous variables. If a teacher demonstrates an early indication that they might resist your programming decision, one approach is to carefully lay all of the variables out before them and ask them to carefully consider all of the information and arrive at their own decision. Most of the time, this teacher will propose a decision that is consistent with yours. When this works, it has the added advantage of automatic "buy-in" on the part of the teacher because it came from them instead of the psychologist.

You will notice that, in the previous descriptions, words and phrases were carefully chosen to convey that none of these strategies is 100% effective. There are exceptions. If there ever was a condition, syndrome, or exceptionality that consistently challenged even the most skilled of practitioners, it is ODD; striving to find support, acceptance, and cooperation among administrators, teachers, and parents of children with ODD marks lonely times for school psychologists. This is the front line, a place that has eaten, chewed up, and spit out many a practitioner of school psychology. And I am no exception.

As a school psychologist, there has always been a number of assessment goals for students with ODD, including assessment of cognitive ability, school achievement, behavior, and emotional well-being. There has also been a need to determine how the disorder is perceived from parents, teachers, as well as the child. Often, the perceptions vary quite significantly. Typically, the teachers and administrators of children with ODD have a litany of examples of the student's maladaptive responses to various situations. "Johnny continuously refuses to walk single file in the hall," "Johnny argues with adults whenever he is asked to comply with school rules," "Johnny never takes responsibility for his behavior, always blaming others," "Johnny is quick to explode when being asked to follow school rules." It is not uncommon for the parents of children with ODD to also be oppositional by nature. As such, home–school conflict is very common. In this regard, efforts on the part of teachers and school administrators to address maladaptive behaviors through correspondence with parents are often characterized by more finger pointing, denial, and general conflict. The school psychologist is often plugged in the middle of this conflict and forced to call upon all muses responsible for inspiring diplomacy and mediation skills to somehow carve a line down the middle and move things forward for the child. Not a job for the light of heart.

There is something about the behavior of a child with ODD that makes adults in education, administrators and teachers included, realize that their responsibilities, duties, moral obligations, and legal obligations have reached the threshold of that which is reasonable. Daily, educational professionals are faced with numerous special educational challenges that contribute to a stewing sense of being overwhelmed, frustrated, and even disempowered. The vast majority of children with exceptionalities that contribute to these feelings evoke a sense of empathy among teachers. Sometimes, however, empathy is not sufficient. When this is the case, teachers can often rise above their inner feelings and realize that efforts must be made to support a child simply because there is a legal and moral responsibility to do so. Drawing upon empathy, moral duty, and legal obligation, it is clear that one must extend efforts toward supporting children with exceptionalities, regardless of the challenges, frustration, hopelessness, and even disempowerment.

It is the condition of ODD, however, that can become the clear "out" for teachers and administrators. And who can argue when a teacher is regularly being told to "**** ***" by both the student with ODD and that student's parent? At some point, the home–school dynamic becomes exceptionally ugly. Numerous efforts have been made on the part of school, and a cycle of frustration and negative energy orbits the child, the nucleus of this conflict. Soon, the child is frustrated with school, the parents are frustrated with school, the school is frustrated, and emotions have escalated to what seems to be the point of no return. School personnel suddenly feel that all of their obligations to the child and family have been fulfilled well beyond expectation and it is now unreasonable to expect anything further. This issue, lumped among a myriad of other challenging special education issues, suddenly becomes the clear boundary between what a school should and should not do for children with exceptionalities. This boundary is drawn confidently by teachers and administrators, and sometimes the efforts to help the child further are met with steadfast resistance.

This is the front line for school psychologists, a lonely and difficult place, but clearly the duty is to somehow help the child, parents, teachers, and administrators move forward. What, however, can a school psychologist contribute? There are traditional strategies such as behavior contracts, token economies, and reward systems for positive behavior. There are newer holistic approaches that target the dynamic of the school by recognition and reward of virtuous and value-based behaviors. This relies on the fact that teens are most influenced by peers, and the hope is that positive peer/group influences will minimize negative behaviors and maximize positive behaviors among individuals. There is always the latest, empirically validated, and efficacious intervention that helps teachers reduce escalating situations and emotional explosions among children with ODD. Dr. Ross Greene's Collaborative Problem Solving approach is a favorite. This approach discourages "control" strategies that typically serve to escalate tensions between teachers and students, especially students with ODD. Rather, Collaborative Problem Solving outlines how a teacher and student can work together to find solutions that serve some benefit to both parties. Such strategies are so user-friendly that it would be unreasonable on the part of the school to refuse to try, as well as unreasonable on the part of parents and students to refuse to follow through on their parts (e.g., e-mail correspondence between parents and teachers to reduce conflict, acceptance of certain school rules on the part of the child). Suddenly, there is a plan, and things are moving in a forward direction. As a school psychologist, I would walk out of these consultations feeling smug and accomplished. In the back of my mind, however, I would secretly and even subconsciously hope that I just saw the last of that case, for I knew that the next time there was no "bag of tricks" and that the school would have a strong case for removal of the child from the system. Parents would be more enraged than ever, and the child would be the victim. I did not want to carry that responsibility. Please let this plan work.

I guess it was written in the stars. After 10 years serving as a school psychologist, I started looking for chances to broaden my career experiences. On the horizon, I saw myself teaching an education program in university. I was able to serve as instructor for some master's level teacher education courses, but I was not really a teacher. Unlike most of the school psychologists, I did have an education degree and was, therefore, licensed to teach. In my school board, there is no difference between a teacher's contract and a school psychologist's contract, same salary and vacation. I realized that if I became a teacher, I would extend my experiences within the school system, setting myself up well for a position within an education department in a local university. So, although I worked very hard to complete my Ph.D. in school psychology, and worked 10 years as a school psychologist, I found myself teaching sixth grade, an opportunity I embraced with excitement and enthusiasm. This has been not only rewarding and fulfilling but also incredibly challenging. I was suddenly experiencing, firsthand, the life of a teacher in all its glory. And yes, my first class was blessed with a student with a diagnosis of ODD. For purposes of anonymity, I will call her Angel, a name that, on the one hand, hints of sarcasm, but on the other, reflects her role as guide from darkness to light to which I was delivered.

Angel was a 12-year-old girl with a bright smile and plenty of spirit. Her name was not unknown to the school psychology team, but I had no previous interaction with Angel. I was determined to make this year a good year for her, as well as myself. Angel was clearly interested in the same, demonstrated by her hard work and determination during the first week of school. It seemed as though this year would be fine and that my experiences and honed skills as a school psychologist were not so foreign in this world of teaching. For a time, it seemed that Angel's behavioral Individualized Program Plan (IPP), developed the previous year of school, was going to be unnecessary. If anything, there were some significant learning challenges that called for forthright support, but this could be addressed with a more academic-based IPP, and so the process ensued. The Program Planning Team considered her case, and everyone agreed that an academic-based IPP was necessary, particularly in language arts-based activities. Although a full academic IPP was originally suggested, there was very significant resistance on the part of Angel's parents toward a full IPP. A compromise was struck to start with a partial IPP, focusing on language arts, and comprehensive accommodations were developed for math. Everyone worked hard to ensure that the IPP goals neatly paralleled those of the regular curriculum. Extra efforts were made to ensure that the instruction and materials were normalized. It seemed as though a solid plan was underway.

The first sign of ODD came as Angel grew to resent all activities related to math. Her attitude in school went from positive and cheery to stubborn and resentful within a couple of months. Angel was shutting down. She refused to open a math book, pick up a pencil during math, or answer any questions. If she was asked a question that was well within her skill level to successfully answer, she felt as though she was being treated as "a special student," prompting her to become saucy and even insulting to others in the class. If she was asked a question that was not within her skill level, she felt like she was "being set up to look stupid on purpose." If there was a thread of respect for me, there was no hint of respect for the school principal, as there was a long and difficult friction between home and school, dating back to when Angel's father was taught in junior high by this same principal. There was a history of hate and tension between these two parties, often expressed openly in front of Angel by her parents. Angel felt justified in expressing her opinions of the principal within earshot of me, although I did my best to pretend I did not hear the comments.

Angel began to demonstrate more behaviors that were more in tune with her reputation. I started hearing her use swear words in the classroom, often followed by a glance from the side of her eyes to determine whether she was able to evoke a response on my part. Again, I chose to ignore most of these behaviors. Angel was given plenty of latitude for her behavior, despite the fact that these actions often served as catalysts for other students with behavior challenges. For one, the school principal implored me to give her a wide berth, as "dealing with her problems through her parents usually led to more problems." It was clear that Angel was on a fast track to graduating to junior high, regardless of her academic performance. There was no way that the principal was going to spend another year dealing with Angel and especially Angel's parents. Angel's father had been known to threaten a previous principal, necessitating a legal order that he should not come near the school grounds. Now that there was another principal in the school, the tension did not subside. There were threats, followed by meetings, followed by physical altercations, followed by more meetings and finally another legal intervention. The principal decided not to file assault charges. It was clear that he just wanted Angel to move on through the system. He tried on numerous occasions to find alternative placements for Angel outside the school system, but these efforts were all futile. No program was open to accept her, and her parents were never supportive of any such suggestions. My instructions were clear—make sure Angel graduates and ignore her behavior to the greatest extent possible.

So, now I was on the inside. Finally, I understood the resistance from teachers and school administrators that I experienced as a school psychologist. This resistance, I figured, came from "group think," when a group of school personnel are convinced that the only strategy left to employ is that of removal of the student to another school, program, or through outright suspension. When this happens, suggestions for supportive recommendations are met with great resistance. I also understood the alternative response, that which made me feel smug and accomplished following a seemingly successful meeting as a school psychologist. I now realized that maybe teachers and administrators were not so enthusiastic to receive Dr. Ross Greene's Collaborative Problem Solving approach to conflict resolution. Maybe, just maybe, they would have accepted anything I said with open arms just to play the game out until the student moves through the system and is out of their hair for good. This was a sad realization, and I was beginning to feel like a foreigner in a strange land, a land that was uncomfortable to get to know. Psychologically, I knew I was experiencing cognitive dissonance, the condition whereby a mismatch exists between one's values and behaviors. I was programmed ethically to serve the needs of the child, but I felt that my behavior was not achieving this purpose.

I also liked Angel. I would be lying if I said she never frustrated me, but at the end of the day, she was a likeable kid. She did not always like me, but there was the occasional smile directed my way. I was never personally offended by her jeers, nor her under-the-breath "**** ****" that typically followed requests of compliance. She was a victim of her circumstances, and I found it easy to empathize with Angel. Furthermore, she had a real potential to shine in areas such as singing and art. I took advantage of every presenting opportunity to highlight these strengths (e.g., by having her draw diagrams on the chalkboard and use her visual depictions as examples for the class during projects). Despite all of this, Angel was not going to succeed in math, and I was faced with a dilemma. Failing math would mean that Angel would be considered for retention, another year in sixth grade. Passing math would be unfair because Angel would not be prepared for seventh grade.

As my discomfort with the situation grew, I knew that I had to, once again, play the mediator, just as I had done regularly as a school psychologist. After all, there were options that, in my opinion, represented the "happy medium." For example, Angel could be reconsidered for an IPP and move along to seventh grade with her peers. There were also nonacademic, practical life-focused, math program options at the junior high level that would be well suited to her needs. Resistance for any change from the academic stream came from the school principal and the parents. The principal simply wanted nothing to do with any changes, as they would require the involvement of Angel's parents. Angel's parents simply denied any need for alternative math approaches. Suddenly, I realized that there was a complete loss of focus on the most important member of this team, Angel. Avoidance of conflict was characterizing and dictating all decisions made for Angel's learning supports.

Pondering this situation, I thought it would be helpful to understand why Angel's parents were so resistant to programming changes. As it turned out, previous teachers were not entirely honest with Angel's parents, presenting math as a strength area. Although this was true, it represented a relative strength. In other words, she was good at math compared to language arts, but she still struggled with math compared to same-age peers. In the past, this strength was highlighted and utilized as a way to reduce conflict between home and school. It seemed as though the strength in math, then, was somewhat exaggerated, thereby giving Angel's parents the perception that all was well in math. This was not really the case. I would have to do some backtracking.

I met with Angel's parents and expressed my concerns with regard to her preparedness for seventh grade. Although somewhat surprised, they could not deny Angel's frustration at home with all school subjects, including math. I approached them with my genuine concern as Angel's math teacher and made no mention of ODD-style behavior, which characterized most of the home–school meetings in the past. I made an effort to highlight the positive features that Angel brought into the classroom, something they were not used to hearing. This style of correspondence continued for several months. The drawback was that Angel's parents viewed me and the school principal as "good guy versus bad guy." I would have preferred that we were perceived as a cohesive team. The advantage, however, was that I was able to present a programming alternative for seventh grade that everyone was willing to accept. More importantly, it served Angel's needs.

This process, although challenging, made me realize that the human dynamics that underlie education, particularly special education, are incredibly complex and sometimes serve to benefit the child, whereas other times serve to direct attention away from the child. Focus on a child's special educational needs can be lost in the mud of stakeholder conflict. This is especially true when particular psychological disorders render parties, such as parents and students, particularly susceptible to patterns of escalating conflict during face-to-face problem solving. Such is the case when students, and sometimes parents, suffer from ODD. In this special case, success was not found by looking through the repertoire of "tricks" or researching some empirically validated consultative approach and applying it to meetings with Angel's parents. I was able to find success in relying on simple traditional virtues such as honesty, respect, and positive focus, which had been lost early in Angel's elementary education. Hardly a day goes by without thinking about how Angel might be doing. By coincidence, I saw her very recently at a community function. She made a special effort to get my attention and say "Hello," a smile from ear to ear directed my way. To me, this is evidence that she's OK and, even more importantly, she knows I care.

Seeking to Understand Before Trying to Be Understood

I am currently finishing my master's degree in education in applied school psychology, and I am writing this story about my experience working with an adolescent with ODD. My involvement with Angie has spanned almost a decade and has been one of the most rewarding and "eye-opening" experiences of my life. When I first met Angie 7 years ago, I had very little experience working with children and youth and had only limited education in the field. I started working with Angie in the context of a residential care facility and, more recently, have been a part of her life as a one-to-one worker.

I want to talk about the resiliency that has been an integral part of Angie's development and the transformation and growth that I have witnessed in her throughout the years. I also want to talk about how my expectations have changed from being focused on gaining compliance from Angie to building a relationship with her and how this has enabled me to provide a strong source of support for her.

It is difficult for me to describe Angie, as she has changed so much since I first became involved with her. In the months after Angie first arrived at the program where I worked, she was involved with the police, missed curfew many nights per week, experimented with drugs and alcohol, and was in and out of the justice system. At 13 years of age, her file was already littered with stories of arson, truancy, car theft, and ongoing physical assault directed at other youth and various staff members who had previously worked with her. Multiple diagnoses of ODD, posttraumatic stress disorder (PTSD), and AD were assigned to her. Angie was taking a cocktail of prescription drugs including antipsychotics, antianxieties, and sleeping aids. Angie was constantly involved in some sort of crisis, and there was not a day that went by that was free from some kind of trouble. In those first 3 years, I remember referring to my job as one of damage control, and it constantly felt like mayhem was just around the corner. I focused all my energy on containing Angie's behaviors and picking up the pieces of one crisis after another.

The struggles that Angie faced were enormous, her diagnoses were severe, and her behaviors were difficult to manage. Her behaviors fit neatly into a diagnostic category, and it was easy to quickly endorse the label of ODD that she had been given. At that time I saw Angie as a textbook case, and the way I approached her was in terms of her diagnosis, the meds she was on, and the strategies needed to gain compliance from her. Yes, I was able to manage her behaviors (just barely); yes, I was able to do damage control; and yes, I was able to enforce every trite and trivial rule (or at least I tried). But at what cost?

There was a particular situation that occurred that was pivotal in helping me to re-evaluate my beliefs of who Angie was and my role in her life. Although it was a "low point" for me, it allowed me to open up my eyes to the underlying resilience and strength that may have driven some of Angie's "defiant and oppositional" behaviors. It made me question the way that I had approached this young lady for so many years. The experience occurred at the residential facility where Angie was living and where I was working 24-h shifts. At night, I was able to sleep for a period of 8 h and I had gone to bed after an exhausting day. I had just fallen into a deep sleep when I heard a loud knock at my door. I woke up, startled, and stumbled half asleep to the door. Angie and my coworker were standing there and Angie was outraged because she did not want my coworker looking through her purse, which was normal protocol whenever a youth came in from the community. We called it a self-search and we were given permission by each individual youth's social worker to enforce this rule. Immediately, I demanded Angie allow me to look in her purse so that we could all go to bed. She would not budge. Right away, I thought to myself, "Oh, here we go, out come the oppositional and defiant behaviors." We went back and forth for what seemed like an eternity; all the while, all I was thinking about was how I could make her comply with me so I could get back to bed. I finally lost it and, in a sharp tone, begged for Angie to "just let me look through your ******* purse." The minute the words came out of my mouth, I felt completely ashamed and out of control. I never "lost it" with Angie. I always kept my cool and set limits in a firm and emotionless way. Losing my cool made me realize that something I was doing was not working. I knew this was not the way I wanted to treat this young girl that I cared deeply about. I went to bed feeling ashamed and defeated because I felt so far away from the relationship that I wanted to establish with Angie.

It was my loss of "cool" and my shameful behavior that led me to my critical reflection. I had not taken into consideration where Angie was coming from and why she was resistant to accommodating my demands. I had made an assumption that she was being "oppositional" and "trying to make my life hard," and I had made these assumptions many times, not just that night. I had been pushing my agenda on Angie, and I had given absolutely no thought to where she was coming from or why she was refusing to comply with my requests. As I reflected further on what had happened that night, I realized that Angie was acting from a place of self-protection and she was advocating for herself. She was sick and tired of being forced to do the things that she did not want to do. She had been forced her whole life. She was experimenting with her own sense of autonomy. She was advocating for herself. When I tried to push a rigid agenda on her and force her to do something that would make her feel degraded and demeaned, she dug in her heels.

So I had to ask myself: Was her behavior defiant? Was she intentionally trying to make my life hard and watch me suffer? Absolutely not! She was thinking about her own needs and feelings, and there was a complete mismatch between what she needed and what I needed. She needed to feel respected and like she had rights, and I attempted to take that away from her. I wish I would have asked her how she felt instead of wondering what she was hiding in her purse that night. I wish I would have taken the time to let go of my preconceived ideas and ask her if she was okay and what she needed that night. I wish I would have done that in so many situations with her. I had dealt with Angie's behavior as if it was a direct and defiant attack against me or the rules of the house. I had lost sight of who Angie was and how I could connect with her.

What is important to note is that Angie was so much more than her "oppositional" behaviors. As soon as I allowed myself to try to understand where some of those behaviors were coming from, I started to challenge myself to put aside my agenda and to see what would happen if I put Angie's agenda first. Now this did not mean acting like a pushover or never setting limits. What it meant to me was supporting her and doing everything I could to understand what it was like to be her and what she might need to feel safe, respected, and supported. I want to tell you that, when I started to put my agenda away and to seek to understand hers, my relationship with Angie began to change and it seemed like her behavior changed too. In those moments where I was able to see Angie as a normal adolescent who needed nurturing, friendship, safety, respect, and support, I was able to build a relationship with her and see her without the constraints of her multiple diagnoses and treacherous past.

To illustrate what I mean when I talk about "working around her agenda," I will use an example. At this time, Angie had just managed to get a job independently and was waking up at 5 a.m. every morning to commute for an hour to get to her job. I was working nights and it was around 4–7 a.m. when I would finally have a chance to have a nap. My agenda was to get some rest. I also had an idea in my head that if Angie asked me nicely, I *might* give her a ride to work every now and then. Of course, things did not go as I wanted and, instead, Angie would wake me up at 5 a.m., demanding a ride.

Sometimes the youth that we work with don't always "ask politely" or use socially acceptable strategies to get what they need. However, I want to suggest that we need to put away our egos and we need to *listen* to what they are asking for. It might not be sugarcoated, and they might not use the language that we prefer, but we still need to *hear* them. We need to look at the bigger picture and ask ourselves how we can provide support for youth regardless of whether or not we agree with their attitude. The times where I was able to get past my ridiculous expectations and really hear the message in Angie's communication were the times I could really give her the support she needed. Angie was not demanding me to give her a ride to work every morning because she was being oppositional; it was the way that she knew how to make things happen for her. I stopped trying to change who she was and started acknowledging that this was the way that she was able to survive and succeed. After I was able to put my preconceived ideas to rest, I was able to provide consistent and ongoing support for Angie. I know that this support, along with her fierce and resilient personality, contributed substantially to the strong, determined, and successful young lady that I see today, someone who is capable of blossoming within the context of support and acceptance.

Angie is now living independently, going to school, and planning to enroll in university. She has a job and is involved in healthy supportive relationships. She is a role model and leader for her friends. Her innate stubborn streak, ability to argue until the cows come home, her rigid sense of her rights, and her inability to tolerate the word *no* are all innate characteristics that now mostly work to Angie's advantage instead of her disadvantage. The Angie I see now has a marked sense of strength and autonomy. She continuously advocates for herself to receive the respect and care that she has a right to and deserves. She never lets anyone walk all over her and, most importantly, she believes that she can do whatever she wants.

The youth we work with are so much more than the behaviors they exhibit or the tough exterior that they try to hide behind. They are vulnerable children who need support and guidance. A personal relationship with youth allows us to have a safe place from which we can positively influence and support them. In the absence of a supportive relationship, behavior management only goes a short way. We need to get past our own egos and our own preconceived ideas of how things should be and seek to understand how things are for them.

Sometimes We Gain More than We Think We Will in Our Work with Children and Youth

I am a pediatrician, child and adolescent psychiatrist, and psychotherapist. After some years of training in a variety of specialty areas, I was a department head of social pediatrics before finally settling into private practice 10 years ago. Presently, I also work as a lecturer and clinical supervisor in the training of psychologists, psychiatrists, and psychotherapists.

In our daily work, we are three child and adolescent psychiatrists who run the practice corporately. Although we cover a wide range of services across a number of specialty areas, we have an emphasis on certain disorders such as early developmental disorders, autism, and ADHD. We work with therapists according to social psychiatry guidelines and provide clinical diagnostics as well as medical treatment, psycho-education, and psychotherapy. I have seen quite a number of patients throughout the years; still, there are some that I won't forget. Murat is one of them. He provided new insights for me at a point when I believed I already knew most of what I needed to know.

"Oh no, not again—I know this kind!"—that was one of my initial thoughts when I first met Murat. He came to me for a consultation on a Friday morning, probably happy to be able to miss school and probably without much motivation for change.

I live and work in the capital of my country, a metropolis with a markedly multicultural population where the Turkish population actually constitutes one of the biggest. The Turkish people in my city typically demonstrate strong and successful efforts with respect to integration and adaptation relative to the local environment and culture. Still, there is a substantial subgroup of the immigrant population who seem to reject or avoid any kind of adaptation or integration and who usually live in certain "problem districts" of the city.

As a mental health professional, I would usually only get to see people with this background when someone in authority would refer the family to me or someone else in my profession. However, at this time, I had already seen quite a number of Turkish adolescents from this kind of situation. Typically, they were male, aged from 13 to 17, presenting with problems of school achievement and severe conduct disorder, and being disrespectful toward people in their school (particularly nonmale adult authority figures). They were usually involved in violent or criminal acts, usually coming "for therapy" but more often in search of some kind of acceptable medical or psychological explanation for their behavior as a way to avoid imprisonment or judicial proceedings. Most of the time, they would come with a parent to ensure they kept their appointment. Unfortunately, many parents accompanied their child with the hope of avoiding deeper consequences for their child rather than to seek therapy. Sometimes the adolescents would come by themselves later in the process because the court had decided they had to undergo a therapy as part of their probation. Moreover, they did not want to feel more awkward or humiliated by being in the presence of others who would know they are receiving therapy. They typically did not want anyone to know about their situation.

Hence, I was not really enthusiastic in working with these youth who were unmotivated to change and who did not seem to have much respect for my time and efforts on their behalf. In fact, they failed to show up for appointments beyond the initial appointment and just wanted a written confirmation that they had been there. However, Murat was different. He was 17 years of age when he came to my office. He spoke German with a typical dialect of a Turkish-oriented subculture used in certain problem districts (in these areas, even minority groups of German youth adapt this slang to better survive on the street among the Turkish majority). Murat presented himself as someone with severe conduct problems and school underachievement. He was enrolled in the tenth grade of middle school, after 2 years of seventh grade, and was about to be expelled from school because of his violent, impulsive, and oppositional behavior (toward his peers and teachers). He considered his teachers to be responsible for most of his difficulties. According to Murat, his teachers were incompetent and unfair. Moreover, he thought he was considered the reason for others breaking rules in the school.

He had been charged with assault and robbery, but explained in detail that he had only been into this due to his association with the "wrong friends," and that his assaults and robbery were incidental and would not happen again. He also stated problems in his peer group, which, according to him, was a typical Turkish gang. Although he described it somehow different, it seemed that he didn't feel respected, accepted, or supported by the other members of the gang.

Although Murat was typical of adolescent Turkish gang culture, he was also different in some ways. For example, Murat came without any company. He was not sent by anyone; it was his very own idea to come to me for help. And—most important—he let me know that he would not quit. I don't know how I managed to give him a feeling of respect and acceptance in light of my reservations during this first session but he came back. It turned out that he would see me regularly for a period of almost 3 years.

Psychological tests demonstrated average intelligence and no specific learning problems. Case history, especially school history demonstrated increased motor activity, attention problems, and distractibility as well as impulsive and oppositional behavior from early on, finally leading to the diagnosis of ADHD. However, specific treatment in terms of stimulant medication was (apparently) only needed for a rather short period.

Murat described his father (48) as an unemployed worker, as a "person commanding respect," and as someone who often had beaten him. His mother (38) was described as a housewife, and considered a "person of trust," but it was revealed that he had practically never talked about his issues with her or sought advice from her. His older sister (20) would often argue with him. He stated that his younger brother (15) would disrespect him and even sometimes beat him. It was only with his younger sister (3) that he acknowledged an affectionate relationship. Although Murat's family was often referred to in our conversations during the first few months, I never got to know any of them throughout my 3 years with Murat. He continually described ongoing familial conflicts, which very often seemed to focus on him. He would say that he felt like the black sheep of the family. However, the significance and influence of his family seemed to fade during the treatment process until he only mentioned them by chance from time to time.

I visited his teacher to get some more background about Murat and was somewhat surprised that Murat supported my efforts in this regard. He was not expelled from school and with some support he was able to avoid severe conflicts with teachers. However, he had to leave school at the end of the school year because he did not achieve the necessary criteria for continuing. Through his own efforts and without familial support (his father wanted him to get into an unskilled job as quickly as possible to earn some money), he managed to get admitted at another secondary school and finally graduated with a surprisingly good result 1 year later. Although it was apparent to me that the school problems were the most important issue in the beginning, Murat's problems with peers filled most of our time. He constantly questioned his own role in his peer group, and although he talked about not being respected by his peers at first, it more and more turned out that he rather did not respect himself. He did not like that he could not control his impulsive behavior. Importantly, during our conversations, his self-reflection continuously increased. Although this was a good development, it also increased his suffering because he could more clearly see his part of the problem, yet was still not able to change it. It took him more than a year to find out that he would not be able to change his behavior in this environment and decided to distance himself from this group.

Murat did not have a girlfriend when we first met. It was after he was at his new school that this topic came up. In our conversations, he often demonstrated a very typical macho attitude toward girls and women (which also clearly represented the common attitude in his peer group). He had been interested in several girls and had been in some short sexual relationships. He eventually did have a girlfriend, but rather than having a romantic relationship with her, seemed to see her as a kind of status symbol (she was good-looking and from a Turkish family). One day, without any obvious incident or reason, he came to me and asked me my opinion about whether or not he treated her properly and how she might feel if she knew about other sexual relationships he was having. I asked him what he believed and it turned out that he already had a clear idea. After that, I never again heard about other affairs and I got the impression that their relationship became more romantic and meaningful to him.

As noted earlier, I hated Murat's "slang." It marked him as an uneducated and inarticulate problem youth. It took me only a short time to recognize that Murat was a much more self-reflective young man than I had expected. It appeared that he understood what I thought about his communication and presentation. In this regard, I cannot tell exactly when or how fast his slang faded, but it had completely disappeared during his more successful year at the new school and by the time he had quit hanging around the problem peer group. By that time and after this change, his personal appearance and presentation was quite comparable to many other German youth, which really impressed me.

Having a good average degree at the end of school, Murat started to apply for positions of apprenticeship training. As one would expect, it turned out that his Turkish provenance was unhelpful in the beginning, which he clearly realized. Although he sometimes would get angry, he never seemed to get discouraged. He tried again and again, while earning some money in an unskilled job. During this time, Murat wanted to intensify the frequency of the appointments, because he would question himself and needed more time for our attention to this self-reflection. Finally, he had an offer from a successful insurance salesman to pass an uncompensated internship for a half year, with the chance to be employed for an apprenticeship after that time. Although he clearly knew that there was no guarantee that he would be employed after the internship, he quit the unskilled job and accepted the offer. At that time, he mentioned his family again. Although he already had earned his own money, he still lived with his parents and had to provide some of his money to his family. His father could not understand or accept that Murat decided in favor of an uncompensated occupation. Still, Murat kept at it and became an apprentice after 6 months. Interestingly, being Turkish turned out to be an advantage for him because this was seen by his boss as a way to get new clientele. Now, Murat often came to the appointments wearing a suit and tie, and presented himself as a proud "businessman." Nevertheless, during this time, he still had to cope with one of his biggest challenges. In this regard, he had to work with a young German female employee who treated him as the unpaid Turkish intern. With his cultural background, there were several points that made this situation additionally difficult for Murat. Not only was she a woman but she was also in a position that allowed her to tell him what he had to do. This alone was a big problem for Murat. However, to his credit, Murat realized his difficulty with her was more his problem and fault than hers. Nevertheless, this relationship was difficult for Murat because he considered her as a person who gave him mindless orders or instructions. He had to overcome his tendency to act harshly and impulsively toward her, and he needed more support during that time. But, in the end, he successfully managed to get along with her and form a good working relationship with her. It was quite an effort for Murat to finally learn to respect her and even discover some good positive qualities in her. From that time on, the frequency of our appointments was reduced more and more until Murat finally decided that he would not need my support any more.

Throughout all this time, I saw myself as a rather cautious and reserved partner in our conversations (therapy). I rarely gave him concrete advice. More often, I provided questions for him to think about such as what he believed others might think about his thoughts, decisions, and actions. Although having psychodynamic training, I utilized a nondirective, person-centered approach. My view was that Murat needed me as a reflector or mirror while considering his own thoughts about how he should feel, think, and act with respect to things, and about where he wanted to go and what he wanted to be. Despite this approach, Murat strongly felt that I was the one who had given him all the advice he needed and that I was responsible for his development. Reflecting with him that he was the one who had done all the therapy work and clearly had had some goal he wanted to achieve was a final big step for his self-confidence and independence.

And how did this experience impact me? Well, during this long process with a young man whom I initially did not like in the beginning, he more and more evoked some kind of paternal sympathy within me. I developed a great respect for his constant struggle for change and development and I acquired much admiration for his changes in attitude and behavior. This therapy process really was something I never would have expected in this case, and it strongly influenced within me more positive feelings and understandings with respect to Turkish youth.

The next time I heard from Murat was almost 2 years later, when he called the office. He communicated his best regards and that he was doing fine, professionally as well as personally. He also wanted to announce another Turkish adolescent, to whom he had strongly recommended our office. That one, however, did not show up.

Epilogue

From our perspective, some themes within and across many of the stories in this section include the following.

- Some parents of children with a diagnosis of conduct disorder want them to attend therapy (that is most often court-ordered) in the hopes that their child will avoid further legal consequences.
- It is difficult for some professionals to be enthusiastic about working with these youth as their behaviors often support the hypothesis that they have little or no interest in changing.
- Conduct problems of children and youth are the result of a complex interaction of genetic, temperamental, family, social–contextual, peer, social–cognitive, and school-related influences that are very challenging to understand and treat.
- Person-centered strategies within the context of family, peers, and school are important in the prevention and intervention of children and youth with conduct problems.
- Many youth with conduct problems are reluctant to seek and undergo therapeutic treatment to examine and adjust their behaviors.
- Children and youth with ODD seem to lack the ability to reason when confronted with demands from others.
- Children and youth with ODD have a mismatch between what they perceive they need to do to get their needs met and what parents, teachers, and others want them to do to meet their needs.

Specific thoughts expressed in the stories that we thought were insightful and we have heard parents and professionals express in similar ways over the years include the following.

- Children and youth with conduct problems seem to be unable to "connect to" the distress their behavior causes others.
- Children and youth with conduct problems seem to lack the ability to self-manage their aggressive behavior.
- There seems to be an innate lack of empathy for others that results in a disregard for how their actions are hurtful to others.
- There seems to be an inability for children and youth with oppositional disorder to understand that their behavior is disruptive to our routines and that their behavior frustrates everyone.

Part II

Emotional Disorders

3 Childhood Anxiety Disorder

Prologue

Anxiety disorders are among the most common psychological disorders affecting children and youth. The primary features of childhood anxiety are excessive tension, worry, fearfulness, apprehension, and concern about oneself and/or people and situations far beyond what would be considered normal for someone of similar age and development, and are associated with severe impairment in functioning relative to such things as peer relationships, school activities, and a variety of routines and tasks. Children and youth can be diagnosed with a number of anxiety disorders, including GAD (generalized anxiety disorder that involves intense and disproportionate apprehension and worry), panic attack (generalized anxiety disorder that involves individually distinct periods of discomfort and/or fear), agoraphobia (concern about being in situations or places where escape might be difficult), OCD (obsessive-compulsive disorder that is associated with recurrent, repetitive, unwanted, and unreasonable thoughts (obsessions) and/or rituals (compulsions)), specific phobia (persistent, excessive, and marked fear of specific objects or situations), social phobia (fear and subsequent avoidance of social or performance situations in which people are exposed to unfamiliar people or to possible evaluation by others), post-traumatic stress disorder (intense fear, horror, or helplessness from an experienced traumatic event that involved death or serious injury to self or others), and acute stress disorder (symptoms similar to those of post-traumatic disorder that occur immediately after a traumatic event).

In this section are nine stories: three personal stories, five parental stories, and one professional story pertaining to anxiety disorder.

The first personal story, entitled "Managing the Fear," is written by a young woman who recalls her experiences with OCD. She recounts her developmental journey that allows us to understand how she came "to grips" with her obsessions and compulsions that challenged her throughout her childhood and youth. The second personal story, entitled "A Little Bit of History Repeating," is written by another young woman who reflects on her childhood and youth with respect to her life with OCD. As she presents and explains her story, it becomes obvious that her journey with her disorder is much more than about repetitive obsessions and compulsions. The third personal story, entitled "Taming the Worry Beast," is written by a teenage girl who shares her struggles with her intense apprehensions and worries (GAD), and how she has handled it. The first parental story, entitled "A Minefield," is written by a mother who tells her story about how she "came to terms" with her son who suffered from an anxiety disorder that necessitated her to be not only loving and vigilant but also open, flexible, and adaptable. The second parental story, entitled "Learning to Trust Oneself," is written by a mother who shares her story of being a mother of a young child with social phobia. Her story reveals how she was as much a

Exceptional Life Journeys. DOI: 10.1016/B978-0-12-385216-8.00003-5

factor with respect to her child's disorder as she was a loving and affirming mom to her daughter. The third parental story, entitled "One Minute, One Hour, One Day," is written by a mother who lets us know how she learned to cope and adapt with respect to her son's social phobia and steward his development. The fourth parental story, entitled "Am I Going to Have a Good Day?," is written by a mother who reveals her struggles with understanding and dealing with her daughter's separation anxiety disorder and panic disorder. In her story, we realize and appreciate the frustration and exhaustion that parents undergo in their journey with their anxious children. The fifth parental story, entitled "Connecting Thoughts, Feelings, and Behaviors," is written by a mother and psychologist who share the parenting practices she undertakes with her daughter with anxiety. The final story in this section is a professional story, entitled "Strength Within Ourselves Often Comes from the Strength of Others," written by a clinician who shares her experiences of working with a young boy with GAD. Her story is a testament to the importance of collaboration between professionals and families in the treatment of childhood anxiety disorders.

Personal Stories

Managing the Fear

I do not remember the first time I realized I had OCD, but my earliest memories of having it is when I was 14 years old, the year my beloved dog died.

At the age of 11, my sister, mother, and I moved from London, England, to a small town in upstate New York. We brought our tiny and scruffy Yorkshire terrier with us on the plane from England. She was a real English dog. Her favorite meal was scrambled eggs with tea, the only foods she would eat when we brought her home for the first time. She slept in the kitchen every night, but in the morning, after my mother opened the kitchen door, I would hear her feet going pitter-patter as she ran up our wooden stairs and scrambled into my bed.

The horrible day she died, I came home from school and opened the front door to our house. Unaware that Penny had spotted a dog across the street and had been barking keenly at it for the last half hour, I swung open the front door: She bounded out. I only had time to glance at her as she circled her way around my feet, scurried down the porch, and onto the sidewalk. I ran after her and, as she hesitated for a moment by the side of the road, I called out her name. "Penny!" She heard her name, panicked, and ran out into the road. After the car that hit her sped away, I scooped her off the pavement where she lay convulsing. As I lifted her into my arms, she twisted her neck and raised it twice to look at me. Her mouth was drawn back, her teeth clenched, and then she just lay still in my arms, with her eyes wide open. It was not until she was on the veterinarian's table and he had finished examining her that I accepted she was dead. That evening, I placed a photograph of her next to my bed. Before I went to sleep, I looked at it and I said the following prayer:

> *I pray that all the kind animals in the world stay away from roads and that they are all having long and happy and healthy lives. Amen.*

Repeating this prayer is second nature now as I must have said it thousands of times. Each night I would say this prayer, and every time I rode in a car or walked down the street and saw an animal by the side of the road, I would have to repeat this prayer in my head. I say would "have to" because reciting the prayer became a chore, something I absolutely had to do, and had to do perfectly. Until I got it right, I would have to repeat it, and repeat it again.

Now saying a one-sentence prayer should not prove a challenge, but saying it and having it "feel right" to me are two separate things. And in order for the prayer to work, it had to feel right. While reciting it, I concentrated with great effort at thinking only "good thoughts": puppies running around with their mother, Penny when she was alive and well. If any negative thoughts—an image of Penny lying on the pavement, a deer by the side of the road—came into my mind, I would have to repeat the prayer. If I did not repeat the prayer correctly, something bad would surely happen. I soon dreaded saying the prayer, for as soon as I would begin, I would have to start over, and over again. I found myself closing my eyes when I was a passenger in a car so that I could avoid looking at any animals that might be at the side of the road. That way I wouldn't get started in the never-ending chore of that prayer. I soon also began to dread getting into bed. Sleep meant the ritual of saying that prayer, and the ritual began to include touching Penny's photograph at the same time as I recited that prayer. And that added step proved to worsen the situation. The "touch" also had to feel right to me, and this added to the difficulty. But why was it so difficult for me to get through that prayer? Surely thinking good thoughts shouldn't be hard? But it was for me. No matter how hard I tried, no matter how much I concentrated, bad thoughts would bombard my head and flash before my eyes. The only thing that would relieve my discomfort was a correctly said prayer. After saying the prayer with only my very best thoughts, relief followed and I could get to sleep or go on with my day. The problem was that it would sometimes take 40 or 50 recitals before I would get it right. I would lie in bed, sometimes for hours, saying that prayer and touching the photograph over and over again.

As my teenage years progressed, so too did my ritualistic behaviors. They evolved to include repeating that prayer not only while I touched the photograph of my dog but also while I performed ordinary tasks. Turning off the bedroom lights, closing doors, getting into bed: All these mundane ordinary actions became obstacles that needed to be overcome.

After turning off my bedroom lights, an unsettling feeling invades me. Suddenly, I have the urge to flick on and off the lights once again. I know that if I do not, this feeling of unease will not go away. On the contrary, with every second that passes, it will only grow stronger. The feeling is like a mental itch I have to scratch. Just turn on and off the lights, it tells me; then I will feel better. So I turn the lights back on and off. But that feeling has not subsided. I still feel anxious, like some impending doom might come unless I turn on and off those lights one more time. An image of my mother comes into my head like a foreboding warning. Yes, harm will come to her unless I turn on and off those lights just one more time. I try to resist by telling myself that it is a completely irrational thing to believe, but this suffocating anxiety will not leave me. I turn the lights on and off while I mentally picture my mother and I on a picnic in a beautiful park. The negative thoughts have successfully been blocked out by this image as I turn the lights off. I now feel at ease. The anxiety has gone.

My obsessions and compulsions carried some general themes. My compulsions had to do with touching things and repeating actions. My obsessions had to do with harm to my family, usually my mother. My OCD became part of my hourly existence. The connections between objects and danger became more and more elaborate. One day, the thought that a clock represents the time we have left in our lives came to mind. From that moment on, touching a clock became dangerous to me. It had to feel right when I touched it or I would have to touch it again. And it hardly ever did feel right the first time around. I would sometimes spend 20 min each night setting my alarm clock for the next morning. Eventually, while at university, I no longer set an alarm clock but instead relied on my roommate to wake me up in the morning. It was easier that way.

I have read that Parkinson's is a disease in which you cannot control muscle movements, whereas OCD is a disorder in which you cannot control your thoughts. Having these thoughts was obviously distressing and so was the anxiety that they created. And it is no ordinary anxiety that everyone would have while thinking of gruesome images. Rather, it is the overpowering type of anxiety that comes with believing something bad will happen that you have the power to prevent. I thought of myself as a rational person, yet turning on and off the lights was a completely irrational behavior. I knew how ridiculous I would appear if someone was only watching, yet I was compelled to do it. Resisting for me was impossible because the anxiety and self-doubt it would create (what if something bad did happen?) would be too much to endure. And I do not think of myself as a morbid person. Why then was I having these terrible thoughts? Was I losing my mind? Was I a bad person? I knew I would have to talk to someone, but whom?

My sister and I have always been close and so it was to her that I turned for help. She recognized that what I suffered from was OCD. The fact that I had my sister to talk to was a tremendous relief. She confirmed what I had come to believe, that I had OCD. Once I knew with certainty what I had, I knew where to look for information. I read all the books I could find on it. Every time OCD was a topic of discussion on Oprah, I would turn on the TV and sit glued in front of it. As I became more informed about my disorder, I became less fearful of it. I was not crazy, after all. I was in tune with reality, except during those moments when I could not resist my compulsion because I believed it might prevent something bad from happening. But even in those moments, I recognized my behavior was irrational. I knew that what I was doing was irrational, yet I could not stop. Knowing the name of what I had did not erase the shame that came with it. I understood what I had, but who else would? While immersed in my ritualistic behavior, I would become fully aware at how ridiculous I was to look at to someone watching. And that embarrassment and misunderstanding are precisely the reasons I vowed to never tell anyone else.

While growing up, my mother and father never saw any of my OCDs. Most of my rituals took place not only behind closed doors—in the bathroom and bedroom—but also throughout the house when no one was home. I never fully understood the fact that I could control my OCD better in front of other people. I think because

I knew that no matter what, I would never let anyone in on the secret, that mental itch I would get so often while alone simply did not kick in as much around others. However, I don't want to give the impression that it never occurred. I would have the urge to touch an object more than once, close the fridge again and again, and I often did give in to the urge; I simply disguised it. Opening the fridge four times could mean I kept forgetting foods that were left in it. Touching the heel of my shoe a few times could mean I was trying to remove something stuck to it.

It was not until my third year in college that anyone besides my sister was aware of what I had. After 2 years of living with my boyfriend, it became harder and harder for me to hide my strange behaviors. One day, while I placed the salt and pepper shakers on the kitchen table, I could not let go of them. I had the urge to touch them again, and as I kept tapping them with my fingers, more and more images of my mother in harm's way invaded my mind. I could not shake the "itch." My boyfriend suddenly spun around and said, "What the heck do you think you're doing?" My heart jumped into my throat and all I could utter in my defense was, "Nothing." I was simply mortified. My boyfriend let that instance go but he started noticing more peculiarities in my behavior: I tended to tap the floor with my feet a few times before getting into bed. I seemed to have trouble flicking off the bedroom light and it would flick back on and then off again. I would ask him to set the alarm clock each night even if I was lying closer to it. Rather than continue making my boyfriend believe he had a neurotic girlfriend, I started to fill him in on what was really occurring. I lightened the tune, calling my behaviors my "funny habits." I made jokes about OCD and he began to joke about it lightly as well. He acted like my idiosyncrasies were cute. He was therefore stunned 3 years later when I called him and told him my OCD was so serious that I was withdrawing from medical school.

Looking back over the years, I would say that my OCD has waxed and waned in its course. Some months were worse than others; some years were worse than others. If I was tired or stressed or drank too much coffee, my OCD would be stronger that day. If my mother was to fly in an airplane the following day, if my sister and I were to go on a road trip, my OCD would be worse. I never knew from one day to the next how severe it would be, but one thing I knew was that it was constant in my day-to-day life; it just varied in degrees. My decline into an existence dictated by my OCD started the year I started medical school. A typical day went along as follows:

My roommate awakens me. With the first glimpse of a new day, my heart sinks with dread, my body feels heavier and sinks in the bed, and I grow more tired. I know that getting out of bed will be a chore in itself, and then the obstacles of showering and getting dressed must also be overcome. As I get out of bed, I toss my bed covers aside. I will not make my bed. That will simply add to the rituals of the day. I walk into my bathroom. Immediately, that mental itch has started. It didn't feel right I tell myself. Go back out and in; otherwise something bad will happen to mother. Do it. It's better to just do it than to deal with the anxiety that I will eventually give in to anyhow.

As I reenter the bathroom, I start picturing a field of flowers. But no, flowers are what people place on graves. I go out of the bathroom and try again. This time I imagine a person in my class, someone who I know just as an acquaintance, someone who represents a neutral image. What bad thing could be associated with him? It works and I am able to enter the bathroom. As I shut the bathroom door and lock it, I continue to picture this student (just his face). He's not doing anything but standing. I try to maintain this image as I brush my teeth. One ... two ... I start counting my brush strokes. I must avoid the number six I tell myself. Fifteen ... sixteen ... there that's good. A multiple of four is always lucky. I rinse my mouth four times, I place the toothbrush in the cup four times, and with four motions, I pull my nightgown over my head. As I do so, I suddenly yell out "no" firmly, to negate any dangerous thoughts that may have entered my mind.

As I continue to think of this student, I quickly step over the bathroom rim to enter the shower. I see a grave with my mother's name on it. I know what I must do. I step out of the shower and back in again. I can no longer hold that image of that student in my mind so I picture yet another student in my class. I step back into the bathtub; again, the image of the grave flashes in my mind. I try to ignore it. I take four deep breaths but it's no use. The anxiety has risen and I am now sweating, not only from getting in and out of the shower so many times, but also because of my rising anxiety. I tell myself to ignore the feeling. Nothing bad will happen, I tell myself. But then the doubt creeps in. What if something bad does happen to my mother today? I picture her lying on the road after being hit by a car. What if something does happen to her? I will never know if I could have avoided it by the simple act of getting in and out of the shower. Better to just do it again.

I step out of the shower and while thinking a neutral thought of a classmate, I am able to block out any terrible thoughts from entering my mind: thoughts that would lend themselves to an itch that would not go away until it was scratched. I turn the bathroom taps on. I must do this with eight turns per tap. The water temperature is not quite right but I leave it as so. I pick up the shampoo bottle and I squeeze a drop out. Before placing the bottle down I must complete the ritual of touching the underneath of the bottle four times while saying out loud "no bad thoughts" as I place it down. I do this again and again until it "feels right." I soap myself with an even number of strokes (always avoiding the number six) and I place the soap in its dish. I must repeat this action until it feels right.

A soft washcloth remains hung on the shower nozzle. It has stayed there untouched since two months before when I bought it. Once, while using it, I hung it on the nozzle and an image of a hanging entered my mind. I have not bothered to use the washcloth since. I step out of the shower. It did not feel right to me. I must get back in. After twelve tries, I successfully exit the shower. By this time, I am cold and shivering. A simple shower has now turned into a forty-minute affair. Soon I will be late for school if I don't hurry. But hurrying just raises my anxiety and makes it worse. I'll just miss my first class, I tell myself. I take four deep breaths and enter my bedroom. By this time, I am exhausted by the repetition of my actions. Even more so, however, I am mentally exhausted by my constant effort to concentrate on certain thoughts in order to block out bad ones. I am anxious and tired, and it is only 8:30 in the morning.

While at school, I subtly carry out my compulsions. I take four small sips of coffee at a time. I count the letters in the words I write in my notebook as I take notes in class, repeating those words in my head that hold the number four, or a multiple thereof. My most difficult situation while at school is when I must go to the bathroom. Behind closed doors, I have less control over my OCD. I cannot tell myself I cannot give in to a compulsion because someone might see. No one in fact can see, and so I tell myself I indeed have the capacity to act out a ritual to prevent harm to my family. It is much harder to resist. I close the door to the stall again and again until it feels right. I lock the door more than once, all the time hoping no one can see me through the cracks in their stall. I flush the toilet eight times.

The ride home from school is not a pleasure. My anxiety rises because I know my OCD will take control once I get home. My car radio has been broken for months but I have made no effort to repair it. I'm actually glad it is broken because I can recall days when I would drive with one hand while continuously turning the radio knobs with my other hand. Better it's broken. By the time I make it to my room, I am exhausted again. I sit down in my chair. I'd rather sit in this chair and daydream than risk walking around my room "doing things." If I just sit here, I can relax. I don't want to pick up a pen, open a book, or open the fridge to find some food. I am too hot but I'd rather remain hot than turn down the thermostat. I just sit here daydreaming, thinking. … I eventually glance at the clock. Two hours have gone by. I finally open a book to begin studying, but reading has lately become increasingly difficult for me to do. As I read, I become more and more uncertain that I have understood what I just read. I begin getting caught up in repeating paragraphs, counting letters, counting words, turning and re-turning pages. The simple act of reading has now become one of my dreaded and most difficult of tasks.

I realized I had reached my threshold. I dreaded waking up in the morning; I dreaded going to sleep at night. Some mornings I would be late to class or I wouldn't go at all. This was because it was getting so hard for me to get through my morning routine. I dreaded taking notes in class, and I could no longer read properly. I had to read aloud so that I would know for sure that I got through a sentence. If I read in my head, it could take me 5 min just to get through two sentences. If there was one thing that helped me during this time, it was that we were studying psychiatry at school. We spent two glorious days covering OCD in class. The professors lectured about people that seemed even worse than me. One lecturer discussed how a man came to her office because his OCD had gotten so bad that he was making his 3-year-old son walk through doors. Another doctor told of a patient she had who was pregnant. She came to see the doctor because of terrible thoughts she was constantly having in which she would harm her child once it was born. This woman had never suffered any symptoms of OCD until she was pregnant, and now she was terrorized by thoughts of harming her baby. These cases gave me a great deal of relief. I was not the only one out there experiencing these obsessions and compulsions. Armed with this insight, I proceeded to investigate the treatments available, and I made an appointment with a psychiatrist.

The receptionist asked me to be seated. The only other people in the waiting room were an old woman and a younger man. They were obviously together, but I couldn't tell if the younger man was the woman's son. He looked ordinary enough, but there was definitely something odd about the woman. She sat slumped over in the chair with her hands fidgeting in her lap. Her hair was a complete dry matted mess, and her face was twitching. She blinked more than she should and she kept pursing her lips. Her legs were as fidgety as her hands; she kept crossing and uncrossing them. *What am I doing here?* I thought. *Do I belong here? Am I as nutty as that old woman?* The woman at the desk called me in to the doctor's office. The doctor was seated behind a broad table. There was a chair in front of his desk, which I sat down on. "Hello, how are you?" he asked. I had to stop myself from chuckling. What an absurd question to ask! I ignored it and simply said hello. "What has brought you to see me?" He asked. I interpreted that as "What has made you so nuts that you are now seeking the help of a psychiatrist?" I had never told anyone about my OCD other than my sister and boyfriend. How could I tell a complete stranger? Because he's a psychiatrist, I told myself. He's a doctor. He's probably seen plenty of more bizarre things in his career, and if he can't handle it, it's his problem, not mine. "I ... I have ..." The lump in my throat was so big that I couldn't get the words out. I took a deep breath, "OCD." With that word, a flood of tears followed. "Obsessive-compulsive disorder?" he asked. "Yes." I reached for the tissue box strategically placed on his desk within my reach and I pulled a handful of tissues out. "I've never been actually diagnosed by a doctor as having it, but I know that that's what I have," I said, my voice quivering. "For how long have you suffered from it?" He asked. *Suffer*, I thought, now that's an appropriate word. "I don't know," I replied. "My earliest memories are when my dog died. I was 14. He got hit by a car, and after, I couldn't stop saying a prayer in my head. It had to be perfect and it never was." He didn't hide his puzzlement well; it was written all over his face. *What's the use*, I thought to myself. How could I ever explain what was going on in that head of mine? He pulled out a sheet from his desk. "Could you please fill this out? It won't take long." He handed me the piece of paper. "Yale Brown Obsessive Compulsive Scale Symptom Checklist" it said at the top. The instructions read, "Check only those symptoms which are bothering you right now." The first category was "obsessions"; I placed a check mark next to the following:

"I fear I'll be responsible for something terrible happening.
I am bothered by intrusive mental images.
I have lucky and unlucky numbers.
I have superstitious fears."

The next category was "compulsions," and I checked the following:

"I have excessive or ritualized showering, bathing, or toilet routines.
I reread or rewrite things.
I need to repeat routine activities.
I have mental rituals.
I need to touch, tap, or rub things."

It is hard for me to explain how difficult it was for me to fill out this form. I did not trust this doctor. *How much did he know about OCD?* I asked myself. *Would he secretly laugh at my absurdities?* With each check mark, I felt more and more humiliated, more and more exposed. I passed the form back to him. He looked at me squarely. "Well, it seems to me like you're suffering from OCD." Tell me something

I don't know. At the end of the session, he asked if I would like medication. "No," I said. I was afraid about the side effects. "Isn't there something else I can do?" My head screamed with the words "behavior therapy." I had read in books about OCD that patients who did not take medicine could be helped all the same with behavior therapy. Surely this doctor would be capable of that. Or maybe he could refer me to someone else? "Prozac, I'll write you out a prescription. Take one tablet every night at bedtime." He handed me the piece of paper. I was speechless. He rose from his chair as his cue that the meeting was now over. That I must go. As I got up, he said awkwardly, "I hope everything works out for you." I couldn't believe it. Famous last words, I thought.

After I got into my car, I closed the door and I put on my seatbelt. It was raining hard and cold outside. I put the key in the ignition but my eyes welled with tears before I could turn the key. I gripped the steering wheel with both hands and I pressed my forehead against it. That lump in my throat was growing bigger. It started to rise up my throat, and it was making me gag. My stomach started to heave. *I'm going to be sick*, I thought, but then the flood came; I sobbed instead. I felt more alone now than I ever had. What was the point of me filling out that form if he wasn't even going to discuss it? It was a cruel joke. I pulled out the prescription sheet from my pants pocket. I stared at it and then I ripped it up into tiny pieces and stuffed them in the ashtray of my car.

I came home from school, threw my book bag down on my bed, and I slumped in my desk chair. Two days had passed since that meeting with the psychiatrist. My OCD was strong today, and I felt drained. *I can't go on like this*, I thought to myself. My life is a charade. I had done everything I could, hadn't I? I had seen a doctor about my OCD, and yet here I was, as bad as I ever was. The line between showering and not being able to shower, between getting out of bed in the morning and not getting out of bed, was drawing thin. *What would happen in the future?* I thought. In 1 year, I would start my rotations in the hospital, and after graduation, I would begin my residency training. All this meant more stress and less sleep. How would I cope? And how could I ever hope to have a normal family life one day with children? Children meant more anxiety, another person in my life to worry about and protect. I was afraid of my future. Any day now, I would lose what little control I had over my OCD and would fall into an abyss.

The next morning I called the office of student affairs and withdrew from school for a year. After leaving medical school and moving back home, I made an appointment with another psychiatrist. As I waited in the waiting room to be called, my heart quickened its pace and I began to shake slightly. *Why am I so nervous?* I asked myself. But I knew the answer. All my hopes of getting better rested on this doctor. *What if treatments didn't work? Try to relax*, I told myself. If he is not helpful, then I will simply have to try other doctors until I find one that can help. But I wanted so much for this doctor to be the one. My OCD was so exhausting that I just couldn't wait for the day that it might be just slightly better. If I had just a little more control over it, maybe then I could better get through my day. The secretary called my name. As I walked into Dr. P's office, he stood up from behind his desk to greet me. He extended his arm, which I shook.

"Hello, I'm Dr. P. Please have a seat." I sat down in a chair on the other side of his desk. "As I understand, you're here because you have some anxiety?"

"Yes, I have OCD. I've had it for many years," I replied. My eyes started to well and my voice quivered. Sensing I was getting very upset, he quickly changed the subject.

"I hear you go to school in Canada?" he asked.

"Yes. I'm a second-year medical student at Memorial University." He then explored my family and my childhood.

"You are having some anxiety?" he then asked.

"Yes. I have OCD," I replied."

"Okay. Could you talk a little about that? When did …?"

"Dr. P," I interrupted, "do you see a lot of patients with OCD?" I was testing his knowledge.

"Oh sure, I have had plenty of experience with patients with OCD. It's quite a common psychiatric illness, especially now that more people are coming forward with it," he said.

I was more than satisfied with that answer. Knowing he had experience with the disorder put me at ease. I felt comfortable talking to him: He wouldn't think I was nuts. I told him I didn't know for sure when it was that I first got OCD, but that my earliest memories of having it were when my dog died. I talked for quite a while, uninterrupted.

"How often do you have your symptoms?" Dr. P asked. "Do you have them five times a day, more than five times, every hour? How would you quantify it?"

"I don't know," I replied. "It's pretty much a constant throughout my day so it's hard for me to give a number. I would say there is not an hour in the day that I do not have symptoms. It's everything I do, Dr. P. Everything that I do." Just talking about it made me choke up. I began to feel very sorry for myself.

"Do you have unpleasant thoughts?" Dr. P asked.

"Yes", I replied.

"Do you believe you have to do certain things to undo them?" he asked.

Suddenly I was relieved: Dr. P was not yet asking for specifics as to my symptoms. He was exploring the issue in general terms. I felt comfortable talking to him. I went into more detail.

"Let's say I'm putting on my lipstick. All of a sudden I think something bad may happen to my mom, so I have to take the cap off it and put it back on again."

"You think if you don't put the cap back on, something bad may happen to her?" he asked.

"Yes, even though I know it's ridiculous to think that."

"But yet you can't stop?" He was completing my sentences. He knew what I was talking about. He understood. "And what is it you feel when you try to stop?" he inquired.

"Anxious. I feel incredibly anxious."

"The only way you can relieve your anxiety is by putting the cap back on the lipstick?"

"Yes. But usually it still doesn't feel right to me."

"How many times would you say you need to keep putting the cap back on: ten, twenty, fifty, one hundred times?"

"I wouldn't say a hundred, but maybe ten or twenty, but the number could get higher. For instance, when I flew to London I went to the bathroom in the airport and I pulled my lipstick out of my handbag to use it. After I used it, I put the cap back on. Because I hate to fly, I was anxious and it took me maybe five whole minutes just to put the cap back on. I was just standing there taking the cap off and putting it back on again."

"And what reassurance do you need to get? Does the action of putting the cap on have to feel right to you?"

"Yes, but it also has to look right to me."

"Okay, so touching it and looking at it both have to 'feel right.' What happens if it doesn't feel right to you?"

"I have to take the cap off and put it back on again. And I need to have good thoughts while I do that."

"Or you will get anxious?"

"Right, if it doesn't feel exactly right to me when I put the cap back on the lipstick, I think something bad will happen to my mom, for example."

"What do you wish would happen? What would help you?"

"To not ever wear lipstick." We both smiled.

"Do you ever resist the urge?" Dr. P asked.

"Well sometimes I try, but it's just no use. I get anxious and I start thinking all these terrible things will happen."

"Is there ever a time when you are able to resist? When you can just walk away?," he asked."

"No, not really. The anxiety it creates is just too great. It feels like I have this mental itch that just won't go away unless I repeat what I was doing until it feels right to me. If I try to resist, I get all these doubts in my mind. What if it does prevent something bad from happening? I think. The only thing that relieves my anxiety is performing the ritual, even though I know that it is a completely irrational thing to do."

"What I would like you to do before our next session is to write down all the rituals you perform in a day."

"I don't know if I can do that Dr. P I mean, I would have to basically write down everything I do from when I get up in the morning until I go to sleep at night," I replied.

"Try to do your best. Also, I would like you to put all those items in order, starting with the one that bothers you the most and ending with those that bother you the least. And I would like you to pick just one ritual and try to resist it as much as you can. Don't pick a difficult one. Instead, pick one that bothers you the least and write down how it makes you feel when you try to resist. Write down what thoughts or images go through your mind, etc."

My chest started to tighten. "But I can't resist," I said. I started to cry. "If all these terrible images didn't enter my head, I could, but how can I resist when I can't even control my own thoughts?" *Didn't he understand that?*

"If you can't, that's okay. I don't expect it to be easy or even possible, but just give it a shot. Listen, the OCD you have is severe. It is. And it's had a hold on you for a very long time. But it is now doing circles around you: You are basically a slave to it and you are obviously fearful of it. I can see that. But you must understand that the more you give in to your urges, the more you perform your rituals, the more you reinforce them. Sure, if you perform a ritual, you are gratified immediately, your anxiety level goes down, but the satisfaction is short lived. Your anxiety returns as soon as you perform the next action. The way to beat it is to break the cycle."

"But I can't. I just don't have the strength" I replied.

"This is why we're going to take it slow and one step at a time. This is just part of your treatment. I want you to try some medicine as well. Your OCD has a strong hold on you but medicine will give you the support you need, the strength you need, to resist. I want you to start taking some medicine straight away. Once the medicine starts to have an effect, you will be better able to control your symptoms."

"I don't know," I said, "I just don't see how medicine can help. My OCD is so strong; I just don't see how taking a pill can have any effect."

"These are powerful medications," he replied. "They do work. But you have to find the right therapeutic dose for you, and you must start at a very small dose and then build it up slowly. The most unfortunate thing you could do is to start a drug regimen and then stop taking the drug because you couldn't tolerate the side effects. That would be a real shame."

"What if the drug doesn't work? What then?" I asked.

"There are about a dozen different drugs you can try. If one doesn't work or you really can't tolerate the side effects, then we will try a different drug."

He wrote me a prescription. I thanked him for his time and left. On the subway ride home, I was excited. Dr. P had impressed me with his knowledge about OCD, and I felt at ease with him. I was relieved to have found a doctor I liked and I now had renewed hope. Perhaps there would come a day in the not-so-distant future when my OCD would loosen its hold over me. How I longed for that day.

"How are you holding up? How's your OCD?" my father asked me. "It's no different," I replied. It was true. I had been taking the new medication for 3 weeks now, and my OCD was still as bad as it ever was. "I have a meeting with Dr. P tomorrow afternoon, and then I can go a bit higher on my meds," I told my dad.

I looked forward to increasing the dose. As for my own effort at resisting, I wasn't faring too well. I had written down all the rituals that I did during the day, and it basically amounted to a laundry list of every action that a person does in the course of a normal day. "Getting into and out of the shower, using the shampoo bottles, walking through doors, brushing my teeth, turning on and off bathroom taps, flushing the toilet, getting dressed, opening the fridge, washing the dishes, putting the caps back on containers, pouring something to drink, closing drawers, and so on."

The list went on and on. I put getting into and out of the shower and walking through doors at the top of the list. These gave me the most trouble: I had to repeat them the most number of times. Closing drawers and putting on my lipstick lay somewhere in the middle of the list. It was hard to rank the items because how difficult they were for me to perform really depended on my mood and thoughts at the time. As Dr. P had requested, I chose an item that I would tackle, putting on my lipstick. Every day, I would concentrate on taking the lipstick cap off, putting on the lipstick, and then putting the cap back on. The goal was to do each action just once. As simple as it seems, this was almost an impossible task. When I was successful at doing each action only once, it usually required me to say out loud "no bad thoughts" and immediately walk out of the room (which created its own challenge). Each time I gave in to my impulses, as Dr. P had explained, I had reinforced the anxiety that would follow. In my next meeting with Dr. P, I told him I was not doing well.

"What I would like you to do between now and our next meeting is to keep a daily diary of your progress. Every day, I want you to give a score to your OCD. A ten being the strongest symptoms you experienced; for instance, how you felt when you left school in January. A score of zero would mean that you have no symptoms at all. Beside the score, each day I want you to describe some episodes you had during the day. Write down whether you were able to resist or whether you were not. This diary will help us monitor your medication and any effect it has on you. But remember, just tackle one item at a time. Take it slow." He then told me to increase my dose of medication, only after all my side effects were gone and at least 10 days had passed. Three weeks later, I would see Dr. P again.

As the days progressed, I became more and more convinced that my medicine might be having an effect. When I closed the taps to the bathroom sink, when I closed the fridge door, when I did the dishes, I noticed that mental itch was sometimes gone. I was conditioned to close the fridge four times out of habit but one day I stopped myself from doing it. My anxiety didn't rise; mental images of my family in harm's way did not enter my mind. I simply closed the fridge and walked away. But how could my medicine be working? I was only taking 15 mg. Surely it couldn't work at that dose?

The day I became convinced that my medicine was working was the day that I stepped just once into the shower. I was standing under the running water after I had turned the shower on with eight turns from each hand, when I realized that I had just stepped over the rim of the bathtub only once. It can't be, I thought. I couldn't remember the last time when I had stepped into the shower just once. My heart beat fast in my chest. It just can't be ... I told myself to not get too excited, that it was probably just a fluke and didn't mean a thing. I tested it. I picked up the shampoo bottle and squirted some out onto my hand. Do not tap the underneath of the bottle, I told myself, just place it back on the rim of the bathtub. I was able to without my anxiety rising. The urge to pick the bottle up again and touch its surface underneath was gone. It must just all be in my head, I told myself. The placebo effect, I thought. The real test, I thought, was whether or not I could get out of the shower with one try. I soaped myself, conditioned my hair, and the next thing I knew I was drying my hair with my towel while I stood on the bathroom floor. Oh my gosh! I thought, as I realized where I was standing. I had stepped out of the shower without even noticing. It couldn't be! But it was. That was the first time in 10 years that I had taken a "normal" shower. This was not all in my head. I knew that now. My OCD was much too powerful for it to just go away without my medicine working. What if it doesn't last? I told myself. But it did. It lasted throughout the morning; as I got dressed, it lasted all the while as I drove to my girlfriend's house. My heart was dancing in my chest.

I remember that morning so well. It was a turning point in my life. It was the light at the end of the tunnel that I had waited for so long. I couldn't wait to tell Dr. P and hear what he would say. The following week when I entered Dr. P's office, I was beaming. I told him about that morning and how every morning since had been just as good. "There's one thing, though" I said. "I feel that during the evening, my OCD comes back. It feels as if the medicine has worn off during the day. It's amazing the difference I feel."

"Well, you should aim to be at a higher dose but we will work on that," he replied.

"It's just amazing, Dr. P. I'm not anxious like I used to be. Every day since that morning, I have taken normal showers. If I get the urge to touch something, I tell myself to resist. And, unlike before, I don't get that rush of anxiety. Also, my mind isn't flooded with all these awful images and thoughts. That rational part of my mind that knows my rituals don't have any power seems to dominate now. I can talk myself into resisting. But in the evening, when my OCD grows stronger, I could spend 20 min turning off my bedroom light. All the anxiety comes back, all those terrible thoughts, and I am just as bad as I have always been."

"Are you still taking 15 mg?" Dr. Pierce asked.

"Yes," I replied. "Would you like to see my journal?" He nodded and I pulled it out of my purse and handed it to him.

"Some days are worse than others?" he inquired.

"Yes," I acknowledged.

"You must realize that you will have relapses. Some days will be worse than others; you can expect that. But we will work at increasing your dose to a point when your symptoms are consistent from one day to the next. Keep increasing your dose by 5 mg every week and keep writing in your journal." He stood up to shake my hand and say good-bye. I stayed seated.

"Dr. P, I feel this is too good to be true. What if my OCD all comes back?" I asked.

He sat back down. "I want you to realize that you will never go back to the way you were before. You will never go back to how you were when you left school. You are a different person now than when I first saw you." Dr. P told me, "You are one of the lucky ones. Some people wait far longer to get help. Others have to go through many different drugs to find the one that they can tolerate. You are on the right track. You will never go back to how you were before." Hearing those words was music to my ears.

During this time, I was adjusting to my new life without the crippling hold of OCD. For the first time in months, I was able to pick up a book and read through its pages without the sense of mounting anxiety that I was accustomed to. I was able to climb into bed at night without the unsettling feeling that I must touch the floor again with my feet. Performing simple tasks of daily living were becoming just that simple. As time passed, however, and the contrast between my new level of functioning and my old one became less stark in comparison, I came to realize that my OCD had not left me. While performing daily tasks, the urge to ritualize remained, but it was a much weaker force. I had the power to resist my urges, to cast them away as irrational nonsense, at least most of the time. Sometimes the itch would be quite strong and I would give in to my urges. But, unlike before, my compulsions were less intense and time-consuming. I would repeat an action maybe four or five times, but never the numbers I had before. My medication allowed my OCD to become what Dr. P had predicted: manageable. Within 6 months of leaving school, I felt my OCD was under control. I contacted my school and informed them that I would be returning in January.

It is now 10 years later, and I am a child and adolescent psychiatrist. Looking back at my journey with OCD, I realize that I had bought into the stigma of mental illness. I was ashamed to be in the waiting room with that disheveled old lady. I was ashamed to be thought of as "crazy," as I thought of her. Through my training, I now understand how pervasive mental illness is and how shame should not be associated with it. But having OCD reinforces to me how easily it is to feel ashamed, because with OCD, you retain the insight that your obsessions and compulsions are irrational.

I recall a 13-year-old boy I assessed during my fellowship in child and adolescent psychiatry. He was seeking help for his OCD. During the assessment, I asked him what he did to block out bad thoughts. He responded with shame and embarrassment. He shifted in his chair, he stalled, and he repeated over and over again how answering this question was so hard for him, so embarrassing. Finally, he said he thinks of a bear to block out bad thoughts. I remembered how I thought of movie actors or faces of students in my medical school class and I remembered how his shame felt. No longer do I feel his shame because I understand how "normal" it is to have OCD. But I empathized with him, and I commended him for his courage to talk about it, to ask for help at such an early age.

Helping others with mental illness satisfies my desire, established at an early age, to help others. Mental illness can be devastating to those afflicted, but recovery is possible.

A Little Bit of History Repeating

My name is Natalie. I'm very close to 30. I'm a wife to Chad and mother to Shaelagh, our spunky 4-year-old daughter. I'm in the home stretch of my master's degree in applied psychology, readying myself for my internship. I am a special education resource teacher at the same French immersion school I started my career at 8 years ago. For someone who has to repeat three or four times what an average person would only do once, I've somehow accomplished quite a lot. I have managed OCD since I was about 8 years old, and although I've since learned to control the irrational thoughts and compulsions that once ruled my day, OCD is still very much a part of my life. Many people talk of "being OCD" when they wash their hands too often or can't leave the house without checking to see if the door is locked, but sufferers know that there is much more to OCD than repeating actions.

As a child, I was very attached to my mother, who was much milder and calmer than my distant father, who would go on to receive intensive therapy for his own mental illness when I was a teenager. My mother offered the stability needed by a shy, emotionally fragile child like me; she was reassuring, fawned over me when I was upset, and met me at the bus stop long after most parents stopped doing so.

The first signs of OCD occurred around age 8. I didn't realize until recently that the appearance of these thoughts coincided with my mother's return to work, a position she'd placed on hold to stay at home with me. Now, my elder brother was preparing to leave for university, and the impending tighter financial situation warranted two incomes instead of one. Because it was completely acceptable to do so back then, and my brother typically raced to an after-school job without stopping at home first, my parents decided that I was mature enough to handle staying home alone for the 30 min between my own school bus drop-off time and the time they arrived home after work. My mother gave me a key to the house (which I kept around my neck on a shoe-lace at all times), had me practice unlocking the door for what seemed like a hundred times, and reminded me to call her to let her know I had made it home. The process went off without a hitch for about a week, until one day, I couldn't get through to my mother's office when I arrived home at my usual time. When my parents' car failed to pull into the driveway at the usual 4:15, I panicked. Not being able to connect with my mother some 20 min earlier was one thing, but both her and my father not arriv-ing home on time seemed like the end of the world. Despite being so young, I knew that heavy traffic or road construction were unlikely causes for their delay because such events occurred so rarely, but I really had no idea what was holding them up and found myself convinced that they'd been in a car accident. Was it the only alternative my 8-year-old mind could come up with? I'll never know, but I remember being told that I was in such a frenzy that when they came in the door at 4:35 p.m., repeatedly apologizing for the delay, I barely heard that they'd been stuck in a drive-thru line for our dinner, at the restaurant where my brother had a part-time job.

Despite a very reasonable explanation for their late arrival home that day, my usual, albeit higher-than-average anxiety about typical childhood concerns morphed into intrusive thoughts of my parents in violent car accidents. At first, these visions occurred in isolation, and were relatively easy to push out of my mind, even though I didn't share them with anyone. After a few weeks, I was compelled to carry out bizarre compulsions to counteract the violent thoughts. These compulsions took the form of odd bedtime routines, taking up to 2 h to carry out, complete with intense attention to the number of steps I took to the bathroom to how many times I rubbed my hands together while I washed my hands. Any deviation required the whole pro-cess to be restarted, or else I was certain the visions would push their way back into my mind, or worse, that they would become reality. My parents were baffled by my behaviors, and sometimes tried to reason with me and have me see the unlikeli-hood of such a situation occurring. In the end, they grew frustrated at how little I responded, moving on to punishment. They tried spanking. They took away toys and time with friends. They tried locking me in my room to keep me from carrying out my routine. Anyone with OCD knows that being kept from carrying out a compul-sion makes the whole cycle worsen, so I replaced my overt rituals with ones that could be carried out in the privacy of my bedroom: walking the perimeter of my bed five times before quietly opening and closing each of the four drawers of my dresser. If I lost count of how many times I'd walked around the bed or made what I thought was too much noise opening a drawer (I knew any sounds coming from my room after bedtime would tip off my parents), I'd have to start over again.

Little was known about mental illness in my small town in the late 1980s, let alone something like OCD. My family doctor was nearing retirement and cared very little about psychiatric concerns in children and refused to refer me to the psychi-atrist who visited town monthly, on the belief that I was in control of what I was doing and had to decide on my own to stop. Persistent, my mother approached the outpatient mental health unit of the hospital, signing me up to see the only counselor who "agreed" to work with children. Although I don't know what he discussed with my mother when I wasn't present, he didn't ask about my visions or rituals during our sessions. I remember him turning on what I interpreted to be relaxing music and asking me to close my eyes and think of happy things. Needless to say, little progress was made from these sessions, and my parents grew frustrated and stopped seeking support completely. My OCD took many forms in the years that followed, but strangely, the intrusive thoughts and subsequent compulsions were always far worse at night, greatly interfering with my sleep, resulting in many missed school days. In fact, if it hadn't been so debilitating at night, I believe my OCD would have interfered very little with my daily functioning, because troubling thoughts always seemed less intense and easier to "correct" with a few openings and closings of a book or a quick walk to the school washroom to wash my hands multiple times. Although I'd always known that the terrible visions were just that—imaginary—as a child, the prospect of your parents being in an accident is a very real and concerning possibility.

My OCD diagnosis came much later, at the age of 19, when an across-province move with my now husband caused my symptoms to escalate to a point where I couldn't leave the house. Ironically, the doctor stationed in the tiny community recognized my symptoms right away, asked why I'd gone without help for so long, and set out a treatment plan. Although I'm by no means symptom-free a decade later, I have learned numerous strategies for coping with intrusive thoughts, where my husband and child have replaced my parents as key figures, and my compulsions, interestingly, are still typically carried out in the bathroom! I don't know if I understood the lack of connection between my routine and the thoughts I fought to suppress, or that washing my hands or only taking a certain number of steps had no bearing on whether or not my parents would be in a car accident, but in retrospect, I wonder if maintaining control over my actions helped me feel more "in control" of a potentially traumatizing incident. To this day, I often wonder if on top of the inconsistent and inadequate professional support I received, that my immaturity and existing anxiety made it particularly difficult to overcome the compulsions that consumed me.

About 5 years ago, I was transferred out of my fourth-grade classroom teaching position into a full-time special education resource teaching position. Feeling unprepared in terms of education and experience, I asked the principal to attend most of the standard beginning-of-the-year parent meetings. One such meeting was for a family who had been part of our school community since I'd been teaching, and although I'd never taught either of the two children, I also hadn't heard much about them requiring any exceptional services. You can understand my surprise when the third-grade student's mother called to request a meeting to share her and her husband's concerns, in a tone that resembled the tired, helpless voice I'd heard my own parents use when describing my odd behaviors to doctors. Not wanting to get into detail over the phone, the mother mentioned that she needed advice and any support we could offer. I agreed, not knowing if I even had the resources or knowledge needed to provide any kind of support. Regardless, I had no idea what to expect.

The meeting was emotional from the beginning. The father shared his daughter Sarah's thoughts of the house burning down or someone breaking in and harming the family. Her accounts were fairly graphic according to Sarah's mother, who cried as she explained, having no idea why Sarah felt so unsafe in such a caring, protective environment. Sarah's family had never experienced a fire or a break-in, nor had any other of their close friends. Furthermore, such events were infrequent in general in our small community, so hearing about them in the media was unlikely. Sarah's parents went on to share how she lined up her shoes at the door in an effort to keep these incidents from occurring, and when the smoke alarm happened to go off after her mother forgot dinner in the oven, Sarah saw this as an early sign that her compulsions weren't going to be effective in protecting her family much longer. She moved on to a complicated bedtime routine of opening and closing every window on the house's main floor, always in the same order, and always repeating the cycle twice. When the summer temperatures warranted leaving some windows open during the day, Sarah's anxiety levels rose as she struggled to decide how to complete a process that revolved around first opening, then closing each window. Worse still, if Sarah's parents insisted on leaving some of the windows open overnight, Sarah would grow so frantic that she wouldn't sleep, pacing the house for hours until her parents conceded and allowed her to perform her ritual. Looking back, Sarah has missed an alarming number of days of school, but these absences were always explained as illnesses. When asked, Sarah knew deep down that it wasn't normal to be so distracted by thoughts of her family being harmed, or spending at least an hour of her day opening and closing windows, but didn't know how to stop being so worried. The family's doctor offered little direction, and Sarah's parents didn't know where to turn beyond his office.

At this point, my own emotions began to surface, but instead of feeling helpless like I had so many times before my own diagnosis, I felt I had something to offer to Sarah's family: my understanding, a reassurance that they weren't alone in what they were going through. I briefly shared my own experience with childhood OCD and how I finally learned to manage my symptoms through the support of a fantastic local therapist, who welcomed children with various disorders. I provided the family with his contact information, and within 2 weeks, I received a "Thank You" card that I find myself referring back to quite frequently: Sarah had seen the therapist I recommended, and through his collaboration with the family doctor, they had developed a treatment plan that was already reducing Sarah's symptoms dramatically. Sarah began attending school regularly, mainly due to her improved sleep schedule.

The meeting with Sarah and her family took place long before I began my master's degree in psychology, but knowing that my experience could be valuable in promoting change in others cemented my decision to further my education and embark on a career change. I've often heard that people with OCD shouldn't discuss their obsessions and compulsions with other sufferers, but I think providing therapy based on personal experience and engaging in appropriate self-disclosure is beneficial in cases where people feel alone in their suffering, much like I'd seen with Sarah's family. In a world where mental health concerns are so prevalent yet continue to be perceived so negatively by the general public, anyone who can do their part to promote awareness and growth should be encouraged to do just that.

Taming the Worry Beast

Hi, I am a preteen girl who loves to be active, and live a great life, but worrying has always been my problem. This is my text on what I have worried about, how I handled it, and how I got rid of worrying. It all started a while ago. ...

About three or four summers ago (at my cottage), I was very sick. I had an ear infection in both of my ears, and I was pretty dehydrated. Family friends had come that night for dinner and to visit. After our dinner, I didn't feel so well, so I went to lie down. Now I knew why I didn't feel good ... because my sickness was acting up, and I had to vomit. That was a scary moment for me, and vomiting is my #1 worry.

After that incident, I did not want to go anywhere. I wanted to stay with my mom or dad everywhere I went. I was so scared of vomiting; I didn't even want to play with my friends or go to their homes. School was even harder. I didn't want to go to class, so I chose to stay in the principal's office for the entire day. I felt like I was safer in the principal's office, but it actually made me feel worse. It made me think that I could stay in the principal's office and worry, instead of going into the class and worrying ... but at least I would get my work done. I felt terrified and embarrassed. I felt really dizzy; I had hard time breathing, and a nervous tummy all the time, when this happened. I was really frightened. A lot of this happened ... until I transferred schools.

As I was still in fourth grade, there was still some "bad days" for me at my new school. I didn't have the guts to go into class without missing my mom and dad, and worrying that I might vomit. You see, the vomiting worry made me worry about other little things, like thinking that if I eat too much then I might vomit, or how long I have to eat before I have to do exercise, or even where I ate. I always wanted to eat with my family, because I always worried that I might vomit at the people's house I was eating at. I would much rather vomit at my house than at someone else's house. Those worries made me not want to eat at all, so I lost some weight. That was not good, because I have already lost some of my weight by vomiting. That was a really terrible worrying stage; so finally, my mom and dad decided to bring me to a place called "The Child and Family Services." This building is a place where families, children, and/or adults go to for extra help. What you do is meet with a special anxiety worker, and get to know each other. Then when you feel ready, you tell your worker what your "disability" is, and then you work through it together.

Thankfully, I had a fantastic worker. Her name was Judy. She taught me some techniques to do when I am worrying or when I am stuck in tough situations. The technique that helped me the most was deep breathing. Some more good techniques were listening to a relaxing CD; it told you to wiggle your toes, arms, and legs (and so on) to get all of the worries out of your body. Another technique that I used wasn't necessarily a technique; it was more of a tool. It was a book called *What to Do When You Worry Too Much*. This book helped me make smart decisions about worrying, and helped me ignore the worry beast. Another technique that was super helpful was what I call ... the worry dolls. I used the worry dolls before I even went to see Judy and Natalie (another fantastic worker). The worry dolls are small dolls that you whisper into them your worries, and then they take them away. One last technique that was helpful was very similar to the worry dolls. It was called the worry stone. What you do is hold a stone, any stone that you want, firmly, and then you think (or say) your worries. Your worries travel from your body all the way into the stone. Then, all of your worries are in the stone, not you. Those were some of the best techniques I have ever used. Judy and I also played some board games together about worrying. There were different questions we had to read, and then answer. There were also some activity sheets that I had to work on. They were really helpful, because they helped me through my worries, step by step. Some of the sheets were stories, and you had to answer some questions from the story like, how do you think this person felt when this happened? That really helped me, because even though it was a story, it felt like I wasn't the only one with worries … (I know I am not)!

Unfortunately, Judy had gotten another job at "The Child and Family Services," and I wasn't ready to go on my own yet, so we decided to see a different anxiety worker. Her name was Natalie. She was just as nice as Judy. She taught me some different exercises, and some of the ones that Judy and I did together, too.

In fifth grade, I still had some bad days at school. Some days I wanted my parents so bad, that I would threaten the principal that I would run home without asking. That's how bad some of the days were. Some other times, I would feel dizzy, and worry that I was going to vomit, so I didn't want to go to class. When I didn't want to go to class, some days it was also because of my "not alive" feeling. Eventually, those bad days had stopped. I am so proud! This year, I had no bad days so far! Thanks to the help of Judy, Natalie, my principal, teachers, and my parents, I have had no bad days. These very special people have strongly encouraged me!

During the springtime of fifth grade, Natalie had invited me to be in a group called "The Friends Group." The Friends Group is a group with about five or so people with different (or the same) worries who come and join together to learn many things, like you're not the only one that worries, how to handle your worries, some techniques that you can use when you need them, and much more. Together, we worked on some sheets with problems like Natalie and I did. Sometimes we even went outside to do yoga to relax. I think that the Friends Group was a great program for me, because I really didn't have to hide my worries. Also, as I was there, I noticed that all the people there looked, well, "normal." I always worried that I stood out, and that people knew that I worried. After this, I felt even more confident in myself!

Soon after the Friends Group was over, and I had had some more meetings with Natalie, and I decided that I could finally go on my own now. Thanks to everyone's help! To me, that was a moment of freedom, a moment where it felt like nothing else could get in my way. I was so proud, and so were my family, and Natalie.

Even though I was ready to go on my own, I still have little worries I deal with. One of the big "little" worries I have is feeling what I call "not alive." It is a feeling that is probably caused by worrying. This feeling makes me feel like I am out of my body, and my eyes get all weird, kind of like you just got something really big, and you're in shock, but not really in a good way. You feel really tired, and you feel like you know what is going to happen in the near future. You feel like you don't want to do anything but go to bed. No one can really take away that feeling, except for me. All I have to do is focus on what I am doing that moment, and carry on with it. I am almost professional now! You have to really put your mind on it, and then eventually, the feeling goes away.

Another thing that really takes my worries away is ringette. Ringette is a really fun sport. It is almost like hockey, but with a stick and a ring, and it's a whole lot more fun. Ringette is a skating sport (like I said, almost like hockey). It really takes away my worries, because it takes my mind off of things, and it just makes me think about what is happening. Ringette is really exciting too, so it makes me feel happy. After the games is really good too, because I am pooped out and I can get a good night sleep.

Being a person with worries is pretty much the same as any other person's life. Sometimes you feel different, but you are … because everyone is different in his or her own unique way. Although some people who worry might not want to do a certain thing for a "silly" reason, and the other people might laugh because they think it's funny, but it isn't to the worrier. I know how it feels. It's because the "non-worrier" doesn't have the same feelings as the worrier. Also, sometimes when I see something frightening in the newspaper or I hear something on the radio (or by someone else), I get kind of worried. For example, say someone at my school had gotten cancer, I would be scared that I might get it, or say I felt something weird on my skin that was in my body, I would think it would be a tumor or something. Another example is that, say that someone in their eighties had a seizure: I would be scared that I might get one. For one last example, if there were a bad train or plane accident, I would not want to go on one of those again. But sometimes I fight my worries, and do what I need to do. Also, worrying that I might die is pretty scary for me. You know, people might think that these are really dumb things to worry about, but they are not, for me. They are just the usual worrying.

Although I still do worry, I know how to handle it and behave properly. I am so proud of myself for doing the best I can and succeeding. I will continue to do my best and stay on the right track. I know that whenever I am feeling worried, I always have my family to help me, and my tools to use. I want to give out a special thanks to my parents (family), Judy, Natalie, and my principal from the last 2 years. I want to thank them for helping me through all of my tough times. I will always remember them … forever.

Parental Stories

A Minefield

It wasn't until Jim was old enough to stand and talk that I noticed something different about him. He was in his third year of life and the most beautiful blonde-haired toddler you could have hoped to lay your eyes on. Jim was a happy boy from the moment he was born. He was like a light in the room—always smiling and giggling with the most loving disposition. Strangers were drawn to him and friends adored him. One day, when Jim was around 3, we were standing in line at the grocery till and a woman came up to me and commented on what a beautiful boy I had. Jim looked up to her and he was furious. He hissed at her, "Stop looking at me!" His behavior was so uncharacteristic of him, and I was stunned that such venom came from this happy little boy. Similar behavior began to occur more often. When strangers looked at Jim and commented to me, he would rudely tell them to leave him alone. On one particular day, we were greeted at church by the minister who shook my older son's hand hello and reached down to tussle Jim's hair. The venom once again came hissing from his lips, telling the man not to touch him. The anger in his eyes was fierce, and there was no reasoning with him.

When Jim reached kindergarten, his outbursts started to worsen, and I noticed his strange behaviors branching out into new directions. He was continually holding up the school bus driver in the morning by running in and out our front door to ensure that he had closed his bedroom door. He would get to the bus door, stop, and throw his hands to his head in frustration and then run into the house to his room to touch the doorknob and examine the seam of the door to ensure it was closed. Some days he would only have to do this once or twice. Other days he would end up in a collapsed pile of tears and exhaustion and I would tell the bus driver to go on without him and end up driving him to school once I was able to calm him down. And on all of the days, he would call out to me, "Make sure my door is closed!" The other ritual that he started to do was wash his hands obsessively. I would catch him in my bathroom with soap up to his shoulders and he would be scrubbing as if he was being eaten alive by ants. By mid–school year, his hands were cracked and bleeding, but I was unable to comfort him with creams because the greasy feel of them would only create another frenzy of scrubbing.

I had a good friend in my late teens who ended up marrying a man with OCD so it didn't take me long to recognize that my son was definitely a victim of this monster. But like a lot of mothers, I thought I would be able to cope with it and love him through it. I really believed that the disease would be manageable. It wasn't until 2007 that my husband and I realized his disease was progressing to a point that we were no longer capable of dealing with it on our own, and we sought professional help. He was 8 years old. Jim saw a private psychologist for almost a year. She was helpful in winning his trust and giving him some strategies toward certain behaviors, but things weren't getting better like we had hoped. In the fall of 2008, my family doctor referred Jim to the Mood and Anxiety Disorders Clinic for a proper diagnosis. After a lot of paperwork and interviews, a panel of highly qualified professionals officially diagnosed our son with OCD and ADHD. The former was not a surprise to us, but the later was. It was a new diagnosis that we weren't expecting; however, it did help to explain some of the inappropriate social behaviors that Jim was exhibiting. It explained why mornings were hard to get Jim out the door on time for school. The attention span he has in the mornings is close to none.

After consulting with our "team" and weighing our options, we decided to put Jim on the drug Adderall to help with the ADHD during school hours and began an aggressive plan of cognitive behavioral therapy sessions. We had hopes that the drugs would help him to focus better in school, therefore lessening the stress that he was feeling from trying to hide his OCD from the other kids. In Jim's case, he had the reverse reaction, and his anxieties increased. His anxiety peaked daily around the time that school ended, when the medication started weaning from his system. His rituals in simply trying to get from school to home—a two-block radius—was tormenting for him. For a few months, an obsession kept telling him that if he didn't lie down at that exact moment, then his spine would break. So in the middle of winter, he would drop onto his back while the school kids would stand and stare at him and ask me if he was okay. I had horrified mothers tell me that I shouldn't allow my child to lie in the middle of the road when there was such heavy traffic. The judgment was immense and critical. If it was humiliating to me, I can't begin to imagine how it made my little boy feel. This spell lasted for a couple of weeks and then transferred to a different ritual of something else. Always one obsession/ritual would "puff into thin air," only to be replaced by a new and unrelated one.

The broken spine ritual was replaced with a decontaminating one, wherein in the midst of winter when it was so cold that your skin would freeze within minutes of exposure, I would find Jim in my front yard after school, stripping out of his "contaminated" clothes. The dogs would have to be put out of the house before he entered and beelined it for the bathroom. If for some reason he entered the tub the wrong way or the water wasn't at a certain level, or the phone rang at that exact moment, or something else threw off his ritual, he would run out the front door, bare naked and soaking wet, to start the ritual all over again. I never knew what would set him off. None of it made any sense to me, and our home became a minefield of temper tantrums and tears that were being ruled by his disease.

My husband would come home from work to find me so exasperated and exhausted from the battles of the day. I myself dreaded when the clock struck 3:30, and I knew he was coming home from school. I never knew what to expect when he walked in. Sometimes his tantrums from trying to decontaminate would last for 2 h, in which I couldn't do anything right and was always in trouble with him. If I tried to help him, I made things worse so I would stand in my kitchen, listening to the front door slamming over and over again while he struggled to perform the ritual "right" so that he could get on with his evening. Eventually the ritual would finally be finished and Jim would surface totally exhausted. When he tried to relax by playing on the computer, a new set of rituals would start with that.

During the days when he was at school, I would find a temporary reprieve from the stress he created. My house was always quiet. No music, no TV. I would sit and delight in the silence of him being gone. The minefield was quiet for 6 h, and I could breathe easy until that dreaded 3:30 would strike on the clock once again. This was a very far cry from what I had imagined motherhood would be like for me. We were so consumed in the chaos of Jim's disease that we didn't realize that our older son, Derick, was virtually becoming invisible in our home. Because he was the "normal child," he took less effort, and because he was so self-sufficient, he started to slip past our radar. Eventually, we noticed a lot of sibling rivalry between the two boys. Derick started to become abusive toward Jim—hitting him behind our backs or hurting him in some way. It started to become a division in our family. Nothing that was apparent in the calm moments of our daily lives, but when the boys got into it, my husband would rush to the aid of his younger son while I tried to defend the frustrations of our older one.

Derick was embarrassed many times with Jim's behavior in front of his friends, and we noticed that he was coming home less and less because of it. Once Derick hit junior high school, we saw little of him. He tried to stay away for as long as possible. And quite frankly, because of the friction between the two boys, I didn't mind that my firstborn wasn't coming home. There was enough tension living with Jim and his OCD. Derick only added to the tension.

With consultation from our team at the Mood and Anxiety Disorders Clinic, we all agreed that what we were doing wasn't working effectively. We agreed that it was best to take Jim off the medication, as it seemed to be increasing his anxiety. We also agreed that Jim was resisting the therapy and that perhaps it was time to take a break. His resistance was something I completely understood. It took an extreme amount of effort on all of our parts to get him through an obsession without performing the ritual. It would throw him into a tantrum of tears and exhaust me in trying to help him through it. So, selfishly, when we decided that Jim wasn't quite ready to stand up to the disease, I was somewhat relieved.

My husband and I continued with therapy to learn strategies with which to help Jim. In the interim, Derick began acting out in school, and got suspended and eventually expelled for fighting and doing drugs. We are currently in counseling with him to sort through his anger and frustrations. Jim's therapy is still "on hold." We plan to return to it when he seems a little more mature to deal with his issues. Our goal is to eliminate any unhealthy anxieties that currently exist, such as the stress that our older son is causing, to provide a more stable and stress-free environment for Jim to work in. For now, we simply try to support Jim in any way that we can and have developed little code words to use when we need to stay out of his way and let him perform his rituals. The minefield still exists. I keep reminding myself that it's progress, not perfection. We have a long way to go!

Learning to Let Go and Trust Oneself

I was the second of five children born to our parents. My mom was 16 when my older sister was born, my dad 23. Both of our parents were from large and impoverished families. They learned little about love from their parents, and I would suspect they learned (as I did) about love from their children. I didn't always know what I know now. My life makes more sense to me now as I am able to reflect on the accumulation of experiences and knowledge that form who I am.

Today, as a registered psychologist, I recognize that anxiety, among other things, can develop as a learned behavior. When I reflect on my early childhood, I appreciate that I had learned many things that my parents did not intend to teach me. I spent most of my time with my siblings, my mother and my mother's oldest sister and her children. Our fathers worked as carpenters in "the city" while we resided in "the country." Weekends, as I remember them, were often spent at our home where many of our relatives gathered and partied. I learned early that I could not trust the adults in my life to take care of my needs; thus, my world, much like my mother's, was cloaked with fear. As in most families where alcohol is prevalent, I learned to passionately protect my family and not to allow anyone to see who I truly was. Parts of my life reflect severe anxiety and a very low self-concept. My children's father was the boy I dated from the time I was 15 years old. He was my first boy-friend and the son of an alcoholic. Peter's parents were both professional people: his mom a nurse and his dad a policeman. Peter was the eldest of five children.

We married after being together for 7 years. He was 25 and I was 23. Melissa was born 2 years later, and Luke 16 months after that. Our marriage was based on traditional roles, with little conscious awareness of values. I took care of the children inside the home while Peter worked to support us financially. Alcoholism and the effects of alcohol worked its magic in our lives: Communication was poor, and we unknowingly neglected each other's emotional needs. I tried to fill the void by working with and for people who made great promises but delivered disappointment. Peter, like many functional alcoholics, spent most of his time doing favors for and drinking with friends, expecting me to be there to meet his basic needs while he lived an otherwise independent life.

When Luke was about 3 years old, I experienced a hallowed moment. I was standing in our living room, Luke hanging on my nightdress and kicking me, saying "I'm on daddy's side." It was in that moment that I knew I needed to find the courage to change my life and the course of my children's lives. It took me 2 years and an affair before I found the courage to leave. Nine months later, Peter died in a single-car accident while drinking and driving.

Ironically, Peter's mom had left her marriage the year Peter and I married. She worked tirelessly to understand alcoholism and its effects in her life. She was my biggest fan and that one person who truly believed in me. We spent many hours talking about what she learned and, although we didn't really know it then, together we changed my life and the lives of my children and their children. Barb died almost 5 years ago of esophageal cancer, with a legacy that is priceless. Within a year following Peter's death, I made the decision to go to university. Twelve years and many life-altering experiences later, I was officially a registered psychologist. Shortly before registration, I met Bill again. He was an old friend from junior high school, and my now husband. My journey was shared with my children and, together, we have grown into beautiful, healthy, authentic adults.

As a young child, Melissa was taught through my behavior to feel unsafe, a lesson I did not intend to teach her but did. Some of my insecurities were expressed in anger. Sadly, my children were my safest outlet, and although I knew it was wrong, it took me a few years to consciously change my behavior. I think that because Melissa was the oldest, she experienced more negativity. Melissa was a beautiful and sensitive child. She had a pretty great first 5 years. Her father was a good dad. She loved her brother and mothered him from the moment he was born. It wasn't until she started school that her anxiety started to manifest as problematic. Prior to school, there were times when she'd hide behind my legs to avoid direct contact with people she did not know well but that behavior did not register on my strange behavior scale; I just thought she was shy.

Soon after Melissa started school, she started to refuse to go back. Every morning for the first week, she cried when it was time to get on the school bus. Not knowing what to do and feeling pressured to do what I thought was the socially responsible thing to do, I picked her up and put her on the bus. Fortunately, after the first week,

Melissa's significantly older second cousin who was on the same bus began greeting her as she got on the bus and sat with her for the ride to and from school.

Within weeks of starting school, Melissa began to verbally express her dislike of her physical education class and refused to go. I thought that after the stress we had both experienced with getting her to school that allowing her to miss gym was an easy solution. I wrote a note to the school giving her permission not to go. Within a week, I was in the principal's office with the school principal and an angry physical education teacher. I was torn between these educators telling me I was wrong and forcing her to be there and participate, and Melissa's explicit communication of her needs. I revoked my permission and, for a month or more, Melissa continued to show her distress.

Throughout her years of school, there were what seemed to be random and unpredictable times when Melissa would refuse to participate. There were also times when her participation made no sense to me. She absolutely refused to answer the telephone (regardless of who she inconvenienced) and she would not order a meal on her own at any restaurant, but she performed violin on local TV and did a solo part in her school's musical. I just could not see any rhyme or reason.

Life got more difficult for Melissa starting in junior high school. Her anxiety contributed to the problems she had with organization and math. I saw her as intelligent and capable. She saw herself falling further and further behind in her work. She knew that not completing her assignments would reflect poorly in her mark but was far too intimidated to talk to her teachers about the work. Melissa would procrastinate (it seemed) to the point where she couldn't recover. I understand procrastination differently now and acknowledge that Melissa didn't always have clear knowledge of what was expected of her and was afraid to ask. True procrastination would involve knowing and having all that was needed but "putting off" doing.

Junior high and high school were very difficult, as Melissa needed to work with dealing with many personalities and sets of expectations. Teachers who knew Melissa worked with her to support her needs while others took her behaviors personally and penalized her. The further Melissa got behind, the more pressure she would put on herself. She seemed to create a lot of catch-22 situations, whereby she felt damned if she did and damned if she didn't. Much of Melissa's self-talk was circular.

The social scene in junior and high school was difficult, as Melissa was what I would now consider very morally mature. She was greatly affected by friends betraying friends. She did not understand gossip or why girls would hurt their girlfriends to be with a boy, and so she avoided as much mainstream interactions as she could (choosing not to go to dances, have a boyfriend).

Melissa is and always has been very artistic and creative as well as musically talented. She participated in band and played violin and double bass. Melissa loved her school choir and received great support from all of her art and music teachers. Melissa also found another passion. She attended a fifth-grade program called Mysterious Encounters Earth, an earth education program. Melissa became a junior leader, and later a paid employee with that and other nature-based programs. She met her husband while employed with the hosts of a Canadian environmental conference.

I did talk to my family physician about my concerns for Melissa. He recommended I take her to see a friend of his who was a psychiatrist. The psychiatrist worked with adults but was going to see Melissa as a favor for my doctor. Again, I didn't realize the message that would give to Melissa. I can see how she would have felt a bit hopeful but at the same time it affirmed that her problem was hers and that it must be really bad. The message I got from the psychiatrist was that I didn't show Melissa enough tender care and that if I hugged her more often, there would not be a problem. She diagnosed Melissa with dysthymia and prescribed antidepression medication. At some level, I hated that she was right. As much as I tried to comfort Melissa, I had become part of the problem and she rejected my attempts to comfort her as a result.

Melissa was probably in eleventh grade by the time my education and her ability to articulate her feelings intersected. I had studied and understood enough about anxiety to label hers as social phobia. She had completed an art project that helped me gain insight into her personal struggle. She entitled her drawing "Social Phobia ... Inside and Out." The drawing was a self-portrait of her looking into a mirror. One image was a depiction of a shy beautiful girl. The image reflected in the mirror was of a gray-shadowed outline of Melissa surrounded by swirls of red and black. Within the image, she wrote the following words:

> "I can't. I am so stupid. What will they think? What if something happens I can't control and it looks like my fault? What if they yell at me in front of everyone? I think I did it wrong, I better just say I didn't do it at all. If I'm not crazy, why do I have to take pills? I'm so behind but if I ask for help, they'll think I'm so stupid and they won't ever want to talk to me ever again. I know my fears are irrational but they're still there. Why can't I just go and have fun instead of thinking about what could go wrong? None of my fears are based in sound judgment. They make no sense and aren't based on past experiences. I don't get why I just can't do things other people don't think twice about. The possible scrutiny drives me nuts. What if I shake and they notice. What if I say something stupid? I'm afraid."

I learned a lot of things and started to do things quite differently. I learned not to let my fears get in the way of my kids' life. I had become so focused on Melissa's problems that I let Melissa become the problem. There is just so much more to her than anxiety. I learned to affirm her fear rather than minimize it. She learned to trust that I understood and was a safe person and even though she was in twelfth grade, I also started to advocate for her at school. She needed to know I wasn't going to abandon her again.

Melissa's fears of answering the phone, of taking the bus, or of ordering food started to make sense when I started to think about it in terms of a fear of being embarrassed. When I started to think from her perspective instead of my own, things made so much more sense. I realize how difficult it is for parents of children with anxiety to know when to protect and when to push. Anxious behavior very often feels aggressive and defiant. I understand that more now in terms of escape and avoidance. Sometimes disappointing me was the lesser of what Melissa saw as two evils. It was better than having to do that which caused her even more distress. I see that as a back-handed compliment these days and recognize that it was only because she knew at some level that I had never left her.

I didn't know how difficult it was for her do things that seemed meaningless to me. It was only when I could recognize her internal struggle that I understood the trade-off she was making and that how I felt was irrelevant. I learned that anxiety is a very self-focused disorder that is often an affliction suffered by people who are seemingly selfless. People with social anxiety constantly worry about what others are thinking about them. It seems like it's about other people, but it isn't.

Anxiety in general is a consequence of negative focus whereby one worries about the minor probabilities of what feels like impending gloom and doom and never about possibility of having a great day or winning a million dollars. The fear of being out of control is a common trait in people with anxiety. I think that that's the learned part, whereby our early experience teaches us that we cannot trust our self to keep us safe and we develop a relationship with self that is insecure. When we figure out that we are enough and that most situations that arise on a daily basis are ones we can manage, the others we can deal with if and when the time comes. Learning to trust one's self is key.

I've learned that life is a process and that my children are not me. They need to make their own mistakes, and they need to know that when they do, I will love them and not judge them. I need to affirm their experiences if I want them to share, and when we are sharing, we can have conversation that is real and meaningful through which we can learn to trust ourselves and each other.

Today, Melissa is a confident woman who has learned to not take herself so seriously. She knows herself well and likes herself. She is an awesome mother, wife, and active environmentalist. She is sensitive to the needs of herself and others and intentionally works to accept and modify her anxious thoughts. Melissa understands how she can continue to grow while concurrently working to prevent her child from carrying her legacy.

One Minute, One Hour, One Day

Our middle son, Chris, had always had bouts of anxiety that led to issues at school. In tenth grade, the anxiety became debilitating.

As a toddler, he often didn't want to go to Sunday school or preschool, screaming for me not to leave him. He loved soccer but wouldn't go on the field with the team. Instead, he'd have a tantrum on the sidelines, crying because he couldn't make himself play when everybody was watching. I was the only mom that, 2 weeks into kindergarten, was still walking her child into the classroom. In second grade, his teacher had to put him, struggling and screaming, into a bear hug to keep him from bolting out the door while I walked away in tears. Once I was gone, he usually settled in fine. Everyone, including family, friends, and child psychiatrists, told us he needed more discipline, was more stubborn than his brothers, or was simply a holy terror.

In fifth grade, his behavior at school deteriorated further. He had a teacher who liked to yell, and even though she wasn't directing her wrath at him, it made his heart rate accelerate and often triggered a flight response where he'd get up and leave the class. If someone tried to stop him, his fight response would take over. He could get aggressive, knocking over a chair or calling the teacher a name. Chris was also the victim of bullies. His quick fight/flight response made for spectacular results when teased, pinched, tripped, or harassed. His anxiety-fueled reaction would often lead to the other student getting hurt worse than Chris. The principal labeled him a problem child. The school brought in a psychologist to evaluate his behavior. The principal was sure it was anger management. After meeting with Chris several times, the psychologist's recommendations were to have a calmer environment in the classroom, perhaps play soft music prior to class, and to have Chris tested for food allergies. Maybe no one diagnosed the anxiety because we found he did have several food allergies that gave him stomachaches.

We got through the elementary years by using the premise of baskets as outlined in Dr. Ross Greene's *The Explosive Child*. Issues that were not worth fighting over for two or more hours would go in basket C, for a while, like saying no to a light snack before supper. Issues that needed to be addressed and that required some education on how to deal but might still cause a meltdown went in basket B. If he could handle a reasonable conversation, it was dealt with right away. Safety issues went into basket A. They were worth the fight. By using this system, we picked our battles. If there was something Chris couldn't do, we would let it go and try a different approach. We did our best to educate his teachers on the theory and how to handle his meltdowns without making them worse and starting a battle of wills. With a change of diet, including taking him off of all food dyes, and a change of schools from elementary to junior high, we noticed marked improvement in his behavior. He began curling with a group of friends and joined a lacrosse league. In the summer, he went mountain biking and, in the winter, snowboarding. He got a job washing dishes at a local restaurant.

In ninth grade, the final year of junior high, he was still curling and playing lacrosse, and his marks were good. In November of that year, he was assaulted by a stranger. Chris and some buddies were walking across a parking lot and an impatient driver simply drove through them. After sending one of Chris's friends flying, the adult driver got out and assaulted the kid. Chris jumped between them and the driver turned on him next. The driver was charged with several offenses, including assault. Chris wasn't badly hurt physically, but he changed after that.

By February, Chris was withdrawing from school, his activities, and his regular friends. It was painful and stressful to watch him be excited about a lacrosse game, but physically slow down as the time to leave for the game approached. Some days he couldn't even put on his shoes, let alone go out the door, and yet 20 min prior he'd been enthusiastic and full of bravado. He lost contact with his teammates. Soon, he had found a new social group, the kids who also had problems at home or with the law or within themselves. He got into more fights and was suspended from school. He started smoking.

We sought help, yet again, from a family therapist. Why was our son skipping school, hanging with a bad crowd, and giving up everything he enjoyed doing? She told us it was a phase. She affirmed our parenting style, gave us a list of books to read, and sent us on our way.

By the end of the school year, Chris had barely passed his classes, and was antagonistic and defensive all the time. He was in trouble with the law. However, once school ended for the summer, his behavior and attitude improved. He laughed, hung out with the family, met his curfew, and did his chores. We thought the worst was behind us, but underneath was still a current of anxiety. He had a meltdown when we enrolled him in a summer football training camp, and walked the 6 km home, still in his cleats; he wouldn't answer the phone; and he never looked anyone in the eyes.

The first week of tenth grade didn't go well. He complained that all of the kids he had been in fights with were staring him down and laughing at him, that everyone was whispering about him behind his back. He felt that the teachers didn't like him and the courses were moving way too fast. He wanted to quit school. After a conference with the administration, Chris changed teachers in several of his classes to see if teaching style would make a difference. It didn't. He started skipping school and quickly falling behind. Phone calls from the school became a daily occurrence. I was so frustrated—what did they want me to do? He was 6 feet 2 inches and over a 150 pounds. I couldn't make him stay in school; I could barely get him to school. I was worried about what he was doing. Was he getting into trouble again? Was he back with that dangerous crowd?

At home, if we asked Chris to participate in his usual activities like school, lacrosse, or something with the family, he'd get agitated. He'd argue with me for hours, going round and round the same subject. Chris would go on about how he didn't need an education. He could live in his car and just work enough to survive. Why did he have to go to school? Or about how everybody was judging him and he just wanted to be left alone. Why couldn't he stay in his room forever? If I tried to stop the argument by being silent, he'd accuse me of not listening to him, of not reacting to him. If I tried to make sense of his arguments, he'd yell that I didn't understand. Often these tirades ended with him punching a wall and walking out, sometimes in only his stocking feet. Soon our house looked like Swiss cheese; hardly a wall was without a fist-sized hole.

We had another conference at the school, this time with the counselor. She heard our story and talked to Chris. He wasn't doing drugs or drinking, he told her. He simply had this huge pit in his stomach all the time. Coming to school made it worse. The counselor gave him an anxiety level test and he scored 38 out of 50. She looked at me, and said, "You don't have a problem child, you have a child with a problem and here's what you need to do." She gave me phone numbers for mental health associations and support groups. We immediately arranged for psycho-educational testing to find out more about him and what might be causing his anxiety. Did he have ADD or was high school too overwhelming? Did he have a learning disability that made classes difficult? The results showed he was gifted, with slow processing. The psychologist that tested him also noted a high level of anxiety. We were put into the cue to see a mood and anxiety psychiatrist.

In the meantime, Chris stopped going to school. One day he simply refused to get out of bed. I pulled off the covers, sprayed him with water, bribed him, everything I could think of, but he would not move except to pile his desk, chair, and bed in front of the door. He screamed at me that he'd rather die than go to school. What he did was crawl back into bed and cry.

I phoned a mental health professional, asking for help. She told us that we should tell Chris that if he didn't go to school we would kick him out of the house and that we should do so for a minimum of 3 days. I couldn't believe it. Was she really telling me to put my "in-crisis" 15-year-old son on the street? We'd been working for the past year to keep him from giving up on everything and doing just that. Now I was supposed to facilitate it myself? How was that going to help him get better or help us cope with his meltdowns? We didn't follow her advice, and we're thankful we didn't.

Chris's school was extremely supportive. They set us up with a system specialist who met with Chris to find a place in the school system that would be more comfortable. There were a few mental health options, but Chris wouldn't have anything to do with going to school with strangers or being singled out as the mental health case in a larger school. But there was a school he thought he could handle. He'd heard from the few friends he still had that the teachers were nurturing, understanding, and supportive, that the students were nonjudgmental, and the class sizes small enough to not be intimidating. However, as a gifted student, he didn't qualify. It only offered lower-level academics and was mostly vocational in focus.

Chris was uncomfortable at home. He felt that we were always judging him. When we'd ask if he'd brushed his teeth, he'd think we were saying his teeth were yellow. If we suggested he wear a warmer coat, he'd think we were telling him he made bad decisions. Because of this, he actually spent a great deal of time away from home with a core group of four friends. They were an eccentric bunch and never judged him or made fun of him for his fears. One young man's eccentricities included walking outside in slippers or a housecoat. Another would make weird noises or do imitations at any time or in any place. In this group, Chris could hide and still be able to participate in some activities. With them at his side, he could go to the mall or a movie. If he wasn't sequestered in his room by himself, he was out with these friends. The flip side to this was that he also had difficulty standing up to their demands for his time. His fear of making them mad outweighed the routine of a family dinner at home or visiting grandparents.

Traditionally, he and his childhood best friend went skiing every New Year's Eve day. When it came time for this excursion during his tenth-grade year, Chris was torn between going with his friend as planned or staying with the three buddies that weren't going. He felt he would be letting one group or the other down and that they would be so disappointed in him that they would hate him. Did he express this clearly? No. The night before the event, he argued with me for 4 h instead, breaking down several times because he couldn't decide what to do. He was so afraid of getting in trouble with one group or the other, even though I knew they would support him no matter what. In the end, we made the decision for him, told him to put his gear in the car, and we took him to his friend's house to go skiing the next morning. Once he was there, he was happy and no one suspected the turmoil we'd been through. We, however, were ready to fall apart. What was happening to our son?

When we finally met with the psychiatrist at the Mood and Anxiety clinic, she diagnosed Chris with social anxiety disorder. He had an intense and debilitating fear of social situations and new people. He couldn't handle being made fun of or being judged. She advised cognitive behavior therapy. Unfortunately, Chris would have nothing to do with speaking to strangers. She told us that we had to get Chris back to school. If he wouldn't do formal therapy, then going to school would be his therapy. She also reluctantly suggested medication. The psychiatrist told Chris that if he didn't do therapy or at least take the medication, he would soon self-medicate and likely be a drug addict by the time he was in his twenties. That was a scary thought for all of us. Armed with her recommendation, we applied at the vocational school and filled a prescription for the medication. Although he wouldn't talk to a stranger, he would take the medication. He truly wanted to get better and have his life back.

During the application interview for the school, our socially anxious teen, who hadn't looked anybody in the eyes for over 8 months, who couldn't make a phone call to save his life, and who never talked to a sales clerk, told his desire to the assistant principal. He told her that although he was gifted and should be aiming for higher-level courses, because of his mental health, he wasn't really that high functioning and needed to gain his confidence back and succeed at something. The school accepted his application.

As much as he wanted to go to school, he still had a hard time making himself actually attend. Each morning, he'd pace back and forth at the bus stop for 20 min, trying to talk himself into getting on the next bus, and yet he'd let bus after bus pass. He knew he had to go to school, and he wanted to go to school, but he simply couldn't make his feet take that first step to board the bus. I'd end up driving him and still he'd pace in front of the school doors until he could force himself to enter the building. I'd park around the corner and secretly watch him, trying magically to add my will power to his. Sometimes, I just sat there and cried.

His first semester, he took two courses, social studies and art. He missed so many classes that his social studies teacher called. I told her that this semester wasn't about passing, but about going to school. He attended more of his art classes, but most days he couldn't manage to stay longer. He passed art. The next semester he signed up for four classes, quickly dropped one—physical education—because he still wasn't ready to participate so openly. He passed two of the courses. Chris never took a lunch to school. He'd rather have gone hungry than eat in front of people. We are so thankful to his teachers, the administration, and the other students of that school. They never judged him, though he shook from his toes to his nose, or disciplined him for being late or missing school. He was simply welcomed back to class when he made it. He needed a nurturing place to heal, and they gave him that.

Although Chris would not participate in therapy, the anxiety clinic offered to work with my husband and me so we could gain coping skills, and hopefully they could help Chris through us. As well, I joined a parent support group offered by our local mental health association. We learned to talk about our own anxiety around Chris so he understood that everyone suffers from anxiety once in a while. We'd tell stories of people we knew who'd been anxious about something and what they did, and how it all turned out so he'd know it was survivable. The therapists gave us strategies on how to get him to participate in activities and kept our expectations in check. Maybe it was too much to expect him to play lacrosse, but we could still ask him to go to the games. Maybe he couldn't make a phone call, but if we dialed maybe he could talk on one. With their help, we developed a figurative box of tools. We put together an anxiety cheat sheet with the situations that caused him anxiety and how to deal with them that he carried around. We helped him develop a mantra he could repeat in times of anxiety to make himself move forward. After a few months, we were able to get Chris to attend a few sessions himself. He wanted to quit smoking and, as it was directly related to his anxiety, we convinced him he had to deal with that first. It was through this one-on-one time that the psychologists noticed that Chris also had ADD. We added new medication, and he was able to cope with even more stressors.

Humor became a major tool to get through the day. My morning routine was to search the newspaper for a funny or ironic story, comic, or picture. If we could get Chris laughing, or even just smiling, in the morning and then keep it going on the drive to school, he had an easier time moving forward. Something that bothered him a lot was big, loud trucks. He thought they were all mad at him because of their deep sound and intimidating grilles. He was sure every aggressive driver was demonstrating how mad they were at him. I'd make up stories about the drivers and their motives, putting a funny spin on it. *That guy cut us off because he wet his pants and needs to get home to change. That driver gave us a funny look because I didn't brush my hair this morning.* At first Chris would just nod, tight-jawed, but eventually we started having full conversations in the car and not always about the other drivers.

As parents, we had to adjust our expectations. We had to realize that there are many ways to get an education, and maybe Chris would take a different and longer road than his brothers, but he would get there eventually. Our priority changed from getting him to go to school to getting him well. It eased the pressure on him and the anxiety we too were feeling about his future. We also developed a mantra: "One minute at a time, one hour at a time, one day at a time, and when it all blows up, start over." The first year our mantra didn't get past the 1 min very often, but now we hardly say it all.

Mental health disorders in teenagers make it difficult to differentiate between regular teenage conduct and mental health issues. Many healthy students skip school, hang out with the wrong crowd, or fail classes. It's hard for professionals and parents to see past the behavior and into the cause of the behavior. Once we had a diagnosis and educated ourselves, parenting was easier. We were able to push a little on the regular teenager concerns like curfew, and work through the problems that caused the anxiety, like attending school.

Along with going to school as therapy and taking medication, we also used incentives to motivate Chris. He, like most teenagers, wanted to learn to drive. However, we weren't giving a teenager with fight response issues a 2000-pound fist. That was a major prize we held out to him to work toward getting better. If he'd take his medication every day, go to school at least 80% of the time, pass all of his courses, and work with us to learn to control his meltdowns, we'd support him getting his license. He's now 17, and has met and exceeded all of our expectations. He hasn't had a meltdown in months; he's going to school every day and is on the honor roll. He also interviewed for a job and is playing lacrosse again. Chris is now the proud owner of a learner's permit.

Recently, he had to do an assignment for language arts class based on the *Pursuit of Happyness* by Chris Gardner. He wrote about what makes him happy now and in the future. He outlined his goals, including family, career, and hobbies. In the paper, he talked about how he used to care what people thought about him and how unhappy he had been, and how he's learned to live with it and not care about being judged so much. It was wonderful reading.

We no longer worry about his future. Thanks to the correct diagnosis, useful therapy and medication, and a nurturing school environment, he now has a future and is capable of making it everything he wants it to be.

Am I Going to Have a Good Day?

I am the mother of a beautiful 11-year-old girl who excels in school and at sports and has many friends. She has a loving and supportive family, with mom and dad and an older sister at home, but for Tara, every day has its challenges. Decision making is difficult—what if she makes the wrong decision and something "bad" happens? What if she eats the wrong thing and gets sick? Tara worries about everything, and this is her story told from my perspective.

When Tara was 3, I watched a group of children playing on a big snow hill, jumping off into a fresh heap of powdery snow and giggling uncontrollably. Tara stood at the top for a while, analyzing every possible outcome. What if there was a stick in the snow? What if there was a rock there? What if she fell the wrong way? What if she got snow in her boots? After 15 min at the top of the hill, she turned around and climbed down.

When Tara learned about Terry Fox in school, she had a sore leg for months and thought she had cancer. If she watches the news (which we now try to avoid), it opens a whole plethora of possible scenarios that she can visualize for herself. Windy days might bring a tornado, traveling in the car could result in an accident, and so on. This is the way her mind works. And truly, many of us think these things, but we are able to weigh how likely certain outcomes are and move on. This is the story of how we have been trying to help Tara manage her worries, and to not let them control her life experience.

In the early years, we knew Tara was a "thinker." She also exhibited signs of separation anxiety, which we attributed to the fact that I worked at home so was with her much of the time. We put her in nursery school a day a week just to encourage some time away from me. Leaving her there each day was a challenge; she sometimes cried and other times found 101 ways to delay my departure. And some days were just fine.

Then elementary school started. For the most part, Tara loved school, but adapting to change did not come easy. The first few days each year were dicey, but we learned to minimize the trauma by visiting the classroom and meeting the teacher in the days before school started. Once she knew where her locker and desk were, that eliminated some of the "unknown." We also tried to ensure that she was with at least one of her closest friends to maintain some sort of continuity from the previous year.

Although Tara showed some signs of anxiety in the "early years," her father and I were both hopeful that she would "outgrow" it. We also knew she was more "clingy" and probably more careful than many of the other children we knew, so we adapted. For the most part, she seemed to be a very happy and well-adjusted child and, at this time, she was able to stay overnight at the homes of a couple of close friends or relatives. She was attending school every day, although on her way to school in the morning, she would ask about my schedule for the day: where I would be, when I would be home. She took the bus to school with her older sister, and generally once she was on her way, things were OK. The tough part was always the actual point of separation—but she did it.

We went through a time in first grade where going to school became more difficult. I tried different solutions to get her there. I tried driving her to the school. (That was NOT an optimal solution; I would leave her in tears, but at least she was there). I tried giving her a locket with my photo in it so that she knew I was always "with" her. She later told me she used to go into the bathroom, look at the photo, and cry. So again—not the best solution! Although there were challenges, the key to these younger years was that we always seemed to manage.

I want to fast-forward to the summer between third and fourth grade. We were at our cottage having a get-together with family friends. We didn't know it at the time, but Tara had a serious ear infection that caused her to vomit rather suddenly after dinner. To control the ear infection, Tara required antibiotics. The medications prescribed unfortunately resulted in stomachaches all summer (and in the end, she needed medication to counteract the effects of the antibiotics). The vomiting episode, in my opinion, was the event that initiated Tara's downward spiral. It was from this point onward that she let her fear of vomiting and her separation anxiety dictate her behavior and her day-to-day routines.

Because Tara was now terrified of vomiting, every time she ate and then experienced a sore stomach, she would stop eating to avoid the possibility of throwing up. Over the course of the summer, she lost a significant amount of weight, and I honestly was at a loss to help her. She was fanatical about eating only "healthy" food, and no "junk," as she said. She would constantly ask for reassurance about the foods she could eat. She would not exercise within a certain amount of time after eating so that her food could properly digest—or at least she would ask for reassurance that her food had digested. Eventually, we would just tell her what she wanted to hear so she would get on with it and go play with the rest of her friends. But she then became dependent on this constant reassurance. Although she still goes through phases of seeking reassurance (excessively), we have learned to turn the question around and ask her what she thinks. We want her to take control.

By September of that year, I was starting to fall apart. I was exhausted with worry about Tara's worries and about both her physical and mental health. I talked to our family doctor to try to set up an appointment for her because, by now, she was also realizing that she was worrying too much but she couldn't control it. The doctor tried to explain in very simple words to Tara about why the body vomits. "There is something bad inside that it wants to get rid of." Well, that just started her thinking about what kinds of bad things there could be, and the overanalyzing continued.

She became overly concerned about germs, due dates on foods, and so on. I finally requested the names of child psychologists so that we could perhaps see someone, but in the city where we live, resources are limited. I did however manage to get on a waiting list to see a clinician at Child and Family Services, though I was told it could take up to 3 months.

About this time, we began to notice the onset of several obsessive-compulsive behaviors. It was almost like she recognized that so much in her life was out of control, that there were certain things she could control—like the order in which she would get ready for school in the morning, or having to always use the same sink in the bathroom. She went through a phase where she had to turn her bedroom light on and off a certain number of times. The great thing (at least I think it was great) is that she knew she was doing it and was consciously trying to break the patterns. There are still some days and weeks where this becomes noticeable—largely when there is some sort of stress, such as a break from routine or she is very tired. These are the times when managing her emotions are most difficult.

As a parent, watching your child struggle with mental illness makes you feel very helpless. I was constantly second-guessing myself. Although I felt I was reading the right materials, and using some good strategies, there is always a feeling of "Am I doing the right thing?" One of the most difficult things was fielding comments from friends and even family about "Tara playing games for attention" or that we should just employ some "tough love." People need to understand that, although it seems simple enough to tell the child to "smarten up, everything is fine" or to relax, in that child's mind, nothing is OK. What they are feeling is very real to them. I very often felt judged by those who tried to solve my situation. I suppose that, by the time others knew there was something going on, we had already been dealing with it for several years. I had done all of the reading—there wasn't much people could tell me that I hadn't already tried. And there is the guilt. I would be lying if I said there weren't times when I had to walk away because I couldn't listen to Tara screaming and freaking out about going to school. It broke my heart, and I did not want her to see me crying. One incident I recall is when my husband came home from work to bring Tara to school. He was more able than I was to separate the raw emotion and be more "matter of fact." And, the thing is, eventually she would go with him.

Fourth grade was not a good year for Tara. Every morning she would leave for school and ask me, "Am I going to have a good day today Mom?" What could I say? I always reassured her: "Yes Tara, today will be a great day." But, some mornings, she would fall apart on the doorstep. As a mother, how can you be firm? I wanted to take her in my arms and suck all of the worries out of her body and just hold her until she felt better. But, at the same time, I did not want to give in. I knew that as soon as I let her stay home, I would lose the battle. So, I did what it took to get her to that bus stop. Then, for a variety of reasons, she began to have difficulty going into her classroom at school. This was something new. In the past, once she had gotten to school, she was able to get through the day. I believe there were several reasons for this. The teacher changed halfway through the year. The new, young, male teacher did not have good control in the classroom, and things were chaotic. Her anxiety then caused her to have a panic attack at school one day. She felt short of breath, dizzy, and "not alive." Of course, getting these strange sensations also made her feel like she might throw up. That then made her panic more. Eventually, she worried that these panic attacks would continue to happen, and she was afraid to go into class. This was compounded by comments from her principal like, "If you don't stop crying, I will call an ambulance" or "If you don't get into the classroom now, you will fail fourth grade." I am not clear how either of these threats could be seen as helpful. I was both very angry and completely discouraged at this point.

The worries continued to take over. Tara stopped eating her lunch in the classroom; she was terrified of germs and throwing up. She worried each and every day that "today might be the day that I throw up again." She stopped going outside for recess with her friends. She stopped going to sleepovers. She withdrew from so much, but continued to play her favorite sport, ringette. In many ways, that was her only outlet. When she played, she did not have to think of anything else. At one time, we had tried to use that for leverage: "If you don't go into class today, no ringette tonight," but we learned very quickly that she *needed* to do something she loved. Playing ringette got her "happy juices" flowing, and she could totally lose herself in the game. The best part was that she always felt good about herself after she played. We learned to never take away social activities or sporting activities that would boost her self-esteem.

When Tara could no longer cope and get into the classroom, I called Child and Family Services directly and told them we were in a crisis situation. My husband and I could no longer handle the situation on our own. I have to say that even my husband was somewhat skeptical at the beginning, thinking that maybe seeing a clinician would be like opening a can of worms, so to speak. Tara picks up on such subtleties that we were a little worried that the clinician might bring something up that could initiate a new worry, but in retrospect, it was the best decision we've made for Tara. We were very lucky. We got a meeting with a clinician within about a week. I was reassured to know that I had already read the books they recommended to us, so I was doing something right after all! I was just so happy to have an "expert" to talk to and validate what we were doing to try to help Tara. When I tried to give the contact information to the school so that we could all work as a team to get Tara back into the classroom, and so that we could develop a toolbox that could be used consistently by all parties, I was disappointed in the result. After 2 weeks, the principal had still not made contact with the clinician. Unfortunately, she was "busy" with other items. In my mind, I was furious: What was more important than trying to help this child return to the classroom where she belonged? All I was asking her to do was make a phone call.

Eventually, Tara's "case" was passed on to the special education teacher at her school, who I have to say was very compassionate. She tried to work on some schemes to get Tara back to class, and although there were glimmers of progress, I think that with all that had happened, Tara was just not comfortable, and this was now compounded by the fact that the other students knew something was going on. That then created a new worry … because that is how Tara's mind works. In retrospect, I also think that the special education teacher was too compassionate. By that I mean, she offered too many "outs" for Tara (e.g., out-of-class free cards for good behavior). The goal was to keep her IN the classroom!

In the end, we made the decision to look for a new school environment that had more to offer in terms of extracurricular activities, somewhere that there would be things to look forward to outside of the classroom, where Tara could stay busy and think less. We did not take the decision lightly, and because of Tara's difficulty with adapting to change, we knew that it would not likely be easy. But we were extremely blessed to find a school that fit the bill and a principal who was willing to face the challenge head-on, working directly with me and with Tara's clinician. In fact, I am certain that had we not made the change, we would not have progressed to where we are today.

We switched schools in the month of May, mainly to allow Tara to make some social connections before the summer, and because there was a huge slate of activities lined up for May and June. Was there a chance that school might be fun again? (As an aside, I should note that four other girls transferred from the previous school to this new school immediately after Tara. As well, her cousin was already at this school and in the same class. There is no doubt that these things contributed to a smoother transition.) The rest of the school year was very good. Tara missed minimal class time and got involved in track and field, and the book club—among other things. We also developed a new morning routine, where her father drove her to school. It seemed to be easier for her to say good-bye to me at home rather than in the schoolyard. There were also some days where she walked to school with her cousin. We had to get creative because there was no bussing to this school.

Throughout this time, we continued to meet with Tara's clinician, who was helping her to develop a toolbox of ways to control her anxiety. I have to say that Tara was very receptive to these meetings and always seemed to feel good when we left. Our goal was to hopefully control the anxiety with cognitive behavior therapy and not use medication unless it became essential. I have not ruled out that one day it may be necessary; however, I am encouraged by her progress today.

The summer was good because Tara was at home, although she was still not back to her usual self. After the vomiting incident, she did not venture far from our house. She did not eat meals at other people's homes. She would almost always start getting her "not alive" feeling in the evenings as soon as she would start to get tired, and this feeling always caused her to "shut down" and want to separate herself from everyone else. I don't want to paint the picture that she was sad all the time—she certainly wasn't. She had lots of fun with friends, as long as it was close to home … just in case. She also had one close friend at whose house she was very comfortable—just not overnight.

Along came fifth grade. The year started well, but as we got into November and December, Tara started to regress and have difficulty going into the classroom again. When asked, her fears seemed so illogical—but in her mind, so valid. She did not like to go into class in the morning because someone threw up one time during "O Canada." So … obviously, that could happen again—and—it might be her this time. Again, I am so grateful that her principal was on board with our approach to helping Tara. Essentially, our goal was to make it less attractive in the principal's office (she would go there to escape) and more attractive to stay in class. I tried to avoid the school at all costs, because if I appeared, she would want to leave with me.

As they say, things definitely got worse before they got better. I had mornings where I had to literally pull her out of the back of our van because she refused to go into the school. Ultimately, my goal was to get her into the school, and once inside, it was the principal and teachers' jobs to keep her there. On her absolute worst days (there were a handful), she was actually a flight risk. I can't say if she really would have run out, but when Tara was having a panic attack, she was not herself. My shy, quiet daughter became a screaming and illogical terror. It was like someone else took over her body—and her only objective was to get the heck out of that school. There were a couple of occasions where she had to be physically held. During those times, there was no way to talk to her—nothing seemed logical to her. We quickly learned that it would take about 2 h after one of these episodes to be able to sit with her and talk about what happened. It was truly like witnessing an out-of-body experience—that was not my child.

The scariest part of Tara's darkest times for me was the comments she would make. For example, "I don't want to live my life like this." I knew that it was critical that we continue with the cognitive behavior therapy so that she could not only develop the skills she needed to manage her worries, fears, and anxiety but also to learn how to use them *before* they would take over and she would lose control. Some days she is prone to falling into what I call a pattern of negative thinking ... the glass half-full syndrome. We always try to focus on the great things in her life, and swing her attention that way. I am concerned about depression in the future, so the more tools she can develop now, the better.

January of fifth grade was another turning point. During that time, Tara had both her worst episode as well as the beginning of her climb to where we are today. It was like she turned a corner. She told me that she thought, "What is the difference between feeling worried in the office or worried in the classroom?" and she decided she'd just go to class. It was as if, all of a sudden, a lightbulb went on, and she decided, "I can beat this thing." For the rest of the school year, she had no major relapses. She continued to worry—but the key difference was she was able to manage the worries better through breathing, talking back to the worry, and staying active. At one point, we had to change clinicians as well (due to a job shift). Because it was during a time when Tara was doing well, we were told we could discontinue the service at that point; however, I was skeptical. I knew that Tara was not ready. She had still not conquered the fear of vomiting and still couldn't be away overnight. Although we were making good progress, there was still work to do. At first, Tara was apprehensive about the change, but after meeting her new clinician, she was very happy. One of the new tools that worked extremely well for Tara was a "worry jar." She had different sized pom-poms she would use to fill the jar. Every time she would beat a worry on her own, she would put pom-poms in the jar. Depending on the size of the worry, she could add more or less pom-poms. We would pick a reward to work for, and when the jar was full, she would collect the reward—usually something like a special lunch on her own with mom and dad or some kind of fun activity. She loved this tool, and continues to use it during the "rough times."

Tara is now in sixth grade. Even though this school year has been remarkable so far, I still hold my breath every morning. Will she get out of the car? She has a new principal, who knows only the very basics of what has happened because Tara did not want to make a big deal about it. I think this gave her a sense of power. Her teachers are somewhat aware of her journey because they have been with her now for two and a half years. The second half of the year has just begun, and there have been a few little bumps in the road. We had one morning where we had to drive around the block a few times, and then ended up coming home for about an hour. At the time, I was weighing the pros and cons of making her go into the school, risking the very real possibility of a full-blown panic attack and losing her for the day, against missing 1 h of school to do some breathing, relaxing, and focusing. I think I made the right decision, because she was able to go in later that morning without a fight. We don't want to encourage certain behaviors, but sometimes, there is a lesser evil.

Having a child with anxiety can be frustrating and exhausting. Unfortunately, there is no cure for it. At best, we strive to help our daughter develop the tools she will need to move through life without being crippled by it. Surrounding yourself with people who are understanding, compassionate, and positive is critical. Those who are quick to judge have not lived a day in your child's shoes, or in your shoes. They have not been there when your child is weeping in your arms because they "hate their life." There are times when I have said things I wish I could take back because I was tired or impatient and I know it created more anxiety. It isn't easy to always think twice about how you say something so as not to cause any undue stress. In our case, we also have to consider how all of this affects our other daughter. There are times when our concern for Tara's mental health and well-being has been all-consuming. We have to be attentive to her sister's needs as well and take into account her feelings and perception of what life is like with a sibling who has special needs. We are very open about dealing with Tara's challenges, and we work hard to provide her sister with the one-on-one time she deserves as well.

I am very thankful for the help we have received along the way, especially from the best principal in the elementary school system (I might be a little biased, but without her, I know we would not be here), and from Tara's clinicians, who were able to convince her that she is "normal" and that many kids out there are just like her. And close friends showed us nothing but love and concern in the darkest moments. If I can stress anything, it is that it is absolutely necessary to have an open dialogue between the home, the school, and the clinician. During the trying times, Tara's principal would e-mail me daily with updates—good and bad. Our ultimate goal, as we travel this winding road, is to teach Tara how to navigate through life while managing her worries. We want her to be able to navigate the impending teen-age years with as little difficulty as possible, so arming her with the correct tools now is essential. We know the road ahead will be a winding one—there will be good days and bad—but we are ready. Recognizing her challenges without paying undue attention to them will continue to be a key to success. The more successes she has in "taking control" of her thoughts, the more confident she feels and the more willing she is to take on new challenges.

We are thankful for this opportunity to share some of our personal experience. There are so many children out there experiencing different levels of anxiety for a wide range of reasons. No matter how serious we think it is, it is very real to them. No one should be made to feel like he or she is different, or that having a "normal" life is hopeless. Guess what? Everybody's normal *is* different. There is hope that we can all have a good day.

Connecting Thoughts, Feelings, and Behaviors

As a psychologist working at a Mood and Anxiety Disorders Clinic, it is somewhat ironic that I have been blessed with a child who experiences excessive worries. I believe that the experience has been enriching and ultimately has allowed me to be more helpful to the anxious children and families that come to the clinic. Of course, my work experience has also influenced my parenting approach. The story I plan to share, hopefully, will provide some examples of typical struggles parents of an anxious child may face and some important strategies for parenting this type of child.

As a preschooler, my daughter was sensitive, shy with strangers, and had a slow-to-warm-up style, even with extended family. She was very outgoing with familiar people. At age 4, she started showing separation anxiety. She suddenly, seemed to want to avoid going to preschool class, the day home, dance, or swimming lessons and birthday parties. Intermittently, she would voice worries about "what-if" scenarios (e.g., what if a tiger and bear escape from the zoo and come to our house?). Starting kindergarten was difficult for her, as with most children at that stage of life. It took months however, rather than weeks, before she adjusted to this new experience. When school staff announced that there was a coyote seen in the vicinity of the school, she worried at length, despite reassurance that coyotes are fearful of humans. Later in the school-age years, she showed reluctance to take ski lessons, not because she was fearful of skiing, but because she worried that she would not be able to find her parents among the big crowd after class. She worried then also of robbers breaking in at night. For a while, she worried about tornados. She showed extreme reactions (screaming and running) when she saw a bee near her, after being stung twice by bees in the summer. She eventually developed a real needle phobia after almost fainting during a routine vaccination, so much so that she began worrying in third grade about the fifth-grade vaccinations. Weekly, she would announce that she did not want to take the vaccinations in fifth grade, hoping desperately that we would allow her this exemption. She was reluctant to go on sleepover trips with guides in different cities, because she feared becoming separated from the group and being lost. Over the years, she has also shown a tendency to be a perfectionist (anxiety's cousin). This has led to homework assignments taking a long time and stressed feelings when she perceives that there is insufficient time to do the job super well. These are only some of the examples of my daughter's past anxiety.

My daughter is now in eighth grade and continues to have a tendency to worry (e.g., if parents arrive home late, for dance exams and school tests, ninth-grade vaccinations). Anxious, unrealistic thoughts will trigger anxiety intermittently on what seems to be a less frequent schedule than when she was younger. However, how she copes with the worries and anxiety has definitely changed over time. I attribute the improvements in coping to the strategies she has been encouraged to use over the years, rather than to maturation, because many children with anxiety disorders seen at the clinic do not automatically cope better with anxiety as they get older.

Often parents wonder how it is that their child is so anxious. I know from my training that it is due to the combination of *genetic predisposition* (someone on one or both sides of the family with anxious tendencies) and *life events* that influence whether a child may display anxious tendencies. The genetic predisposition certainly applies to my daughter and was evidenced by her shy, slow-to-warm-up style. She had a number of unfortunate events (some of which I will share) that also contributed to the manifestation of her anxious, worrier style. She was accidently locked in a vehicle when she played with the key remote as a toddler. She was left behind in a park by day camp leaders, was stung by bees, and almost fainted when getting a needle. Her separation anxiety, worries about being left behind, and fear of needles and bees make sense given the experiences that occurred. If there were no genetic predisposition though, the life events would have produced only short-term anxiety rather than a longer-term style.

The strategies I used in parenting my anxious daughter, and that I recommend to children and families from the clinic, are based on the information I am about to share. The information, hopefully, will provide a context that promotes parents to try the strategies more than once or twice. Because the strategies often do not produce immediate and lasting change initially, parents may want to give them up. I know that, even with my training, I found myself second-guessing the approach and rereading books in the hopes of figuring out what I was doing wrong. Over time, it became clear that the strategies do in fact work, if used consistently over time, but perseverance and patience are important.

All humans have an automatic response mechanism or *anxiety alarm* for dangerous situations. Children (and adults) who have anxious tendencies or experience greater-than-average worries and anxiety tend to have an anxiety alarm that is malfunctioning. Similar to a smoke detector that goes off when a match is lit (suggesting that there is some fire danger, when there is not), anxious children have anxious thoughts that set off their anxiety alarm when there is a low probability of the negative event happening.

When the anxiety alarm is triggered by thoughts—"the bee will sting me; the coyote will bite me; robbers will break into our house; I will not be able to find my parents after the ski class; I will faint in front of my classmates when we get vaccinations; my work needs to be perfect; if my parents are late, they must have been in an accident; I am going to fail that test" and so on—the child feels anxious and displays anxious behavior. Usually children will try to avoid the situation that causes the anxiety (flight) or become angry and lash out (fight). This behavior leaves parents with a dilemma of how to respond. Most parents will try more than one approach (letting them avoid or making them do the scary thing). Often parents report that it seems as though both those approaches are unhelpful at resolving the problem.

The strategies that I used, and that worked over time, are based on the underlying principle that there is a strong connection between one's *thoughts, feelings*, and *behaviors*. Therefore, my daughter's anxiety alarm would be triggered by the anxious thoughts listed previously, and she felt and acted in an anxious way. When this occurred, she needed guidance that helped her realize her unrealistic, anxious thoughts, and she needed help to replace those thoughts with *positive, realistic thinking*. She also needed to replace anxious behavior (avoidance) with *facing her fears in small steps*, while using some calming techniques. Replacing anxious thoughts with realistic ones and facing fears in small steps will inevitably create reduced fear because the thought–feeling–behavior connection is being converted in anxious situations.

So how did I apply this information to my daughter's worries and anxiety? When she voiced what-if scenarios about tigers and bears escaping from the zoo, I explained that the animals are locked up, and even if they escaped, they were unlikely to get all the way to our house, which was far from the zoo (realistic thinking). I explained that robbers are not likely to rob a house at night because that increases the chances of being caught and reviewed the facts that our neighbors, friends, and family had never had that happen to them. I provided evidence that showed that her anxious thoughts were low-probability events. When she got older, I explained the anxiety alarm concept and the malfunctioning smoke detector. We could then say, "Oh, that's just your anxiety alarm going off when it shouldn't. Let's review the evidence and come up with a more realistic thought."

I did not allow her to avoid the birthday parties; ski, swimming, or dance lessons; or going to school, because I knew that anxiety would increase significantly over time if this became her coping style. Frequently, I would use this script to help my daughter face her fears in small steps: "If I let you avoid_____, I would be helping anxiety grow. I don't want to help anxiety, I want to help you." When dealing with her fear about getting lost after ski class or going away with guides on a field trip, I ensured that she attend by explaining the low likelihood of her fears, but because her own life experience provided evidence that this could happen, we needed to develop a reassuring detailed plan in the event that her fear came true. We had a specific meeting spot for pickup after ski class and a place to go if we were not at this place. If lost with guides, she would use a phone to call us, and our phone would always be on. To calm her overreaction to bees, we discussed the fact that screaming and arm flapping increases one's chances of being stung by a bee.

For many of the situations, she would be encouraged to use self-talk that included both realistic thinking and coping statements (e.g., "Stay calm, you can do this, bees are not likely to sting if you don't run and scream," or "Robbers don't usually come at night," or "Just because you fainted once during a needle does not mean it will happen every time."). As she grew older, she was also encouraged to use relaxation techniques like deep breathing and going to a happy place, using as many of the five senses as possible during times that she was feeling anxious. She was given opportunity to practice these when she was not anxious (e.g., before bedtime), and then was reminded to use calming techniques when agitated and anxious.

These days, my daughter knows not to avoid anxiety-producing situations and to use self-talk and relaxation tools when feeling anxious. There are some times (not often anymore) when she will need coaching in the moment. Recently, for midterm exams, I needed to prompt her to say these realistic thoughts while driving her to school: "I have studied for the test. I usually get good marks. I have never failed a test (facts/evidence). So I will likely do well. I am not going to let anxiety get in the way of me getting the mark I deserve." As she is saying this, she is encouraged to take deep breaths and go to a happy place/memory (favorite ride at Disneyland). These strategies have become almost habit for her now. I expect that, as new stresses and life events occur, she is not always going to be able to use these skills independently and she will likely need more guidance of ways to boss away unrealistic, anxious thoughts, but I do believe that if she does not abandon the tools, she will be able to cope well with future anxiety.

Another important point to discuss, related to parenting an anxious child, is the internal struggle parents feel when they see their child in distress while feeling anxious. Some parents may become frustrated by what seems to be an overreaction to a mildly fearful situation. Others may feel overwhelmed when they have to push their child to face feared situations. Parents' internal struggle is influenced by whether they struggle with anxiety themselves and by their parenting style. Some parents are more "kick butt" when parenting, whereas others tend to be "rescuers." Knowing where one falls on the kick butt–rescue continuum is important, because it will influence how well a parent can coach an anxious child in distress. Ideally, parents want to avoid being at either extreme of the rescue–kick butt continuum because both ends can lead to conveying unintentional messages to the child. The rescuer who lets the child avoid frequently or intermittently is conveying (unintentionally) that they do not believe the child is capable of handling the fearful situation. The parent who disregards the child's distress/anxiety and expects no avoidance, tears, and so on, is unintentionally conveying the message that the child needs to always be tough or strong, and is not allowed to feel fear. Both these messages can lead to poor coping in anxious situations. A middle-of-the-road parenting style involves expecting the child to face fears in small steps, showing compassion for the child's distress, and promoting gradual independence in facing anxiety-producing situations. With this parenting style, the underlying message is: "You can do this. I will help at first." It is rare that any parent can use this approach consistently. To complicate things further, it is unlikely that both parents (whether or not living in one household) can consistently use the same approach when dealing with an anxious child. This makes coping with anxiety even more challenging for a child. My husband and I have each had to make modifications over the years, and we continue to work on this goal. Our daughter, fortunately, seems to have made good progress, despite our needing to learn to how to apply modified parenting styles. As she continues to grow, our hope is that she feels empowered by her struggles with anxiety, because she is learning that she is able to handle stressful situations. She is learning that worrying and avoiding do not help decrease anxiety but they take away her strength.

Professional Story

Strength Within Ourselves Often Comes from the Strength of Others

Mental illness can be such an intimidating phrase. For some, it stigmatizes the individual and can threaten one's personal identity. It can act as a shadow, looming over life and threatening the innocence of it all. It is especially challenging when mental illness affects children. As adults, we simply want our children to grow up healthy and lead a happy and fulfilling life. We want to protect our children from such challenges as mental illness, and we can feel helpless when we cannot do so. However, as difficult as mental illness can be, with support from family, friends, and the community, a child can become someone outside of the disorder. Each child is unique and offers something special to the world. Understanding children for who they are can offer strength in difficult moments and help us support each other toward happiness.

These thoughts are within me as a family member of someone living with schizophrenia and another with anxiety. They also find their way to me as a clinician in a children's mental health agency. Mental illness is all around me, just as it is around all of us. It affects everyone in a different way, but it is there nonetheless. I have found that the greatest strength of all is within the child and his relationships within his or her support circle. This narrative is thus a story of one child's undeniable strength, and about the impact of his parents' love and determination. From my perspective, this story illustrates how support from family, friends, and the community is the key to happiness, even when the shadow of mental illness seems overwhelming and the future appears dark.

Let me introduce you to Nathaniel. Nathaniel is an 11-year-old boy who loves to draw and build model planes. He is athletic and plays on a soccer team during the summer and on a hockey team in the winter. He is a popular young boy who gets along well with his peers. He has two best friends with whom he spends a lot of his spare time. He grew up with these two boys, and their families have been friends for years. These families go to church together every Sunday and take vacations together during the summer. At school, Nathaniel is well liked by his teachers. They describe him as an intelligent, ambitious, and focused student. The principal refers to Nathaniel as a model student, as he is always pleasant and polite.

From this introduction, Nathaniel and his family appear to have the perfect life. What possible problems can this family have? Well, although Nathaniel has unmistaken positive qualities, he has serious challenges regarding anxiety. His troubles with anxiety though are not straightforward. In fact, it was not until a few months ago that Nathaniel was diagnosed with an anxiety disorder. For most of his life, this young boy struggled with anxiety, but displayed it through anger and aggression in the private setting of his home. This "secret struggle" weighed on the family for years as they constantly tried to cover up the reality to protect the positive aspects of their life. They feared that their family's struggles would threaten their image and lead to negative judgment from those around them.

We can certainly relate to this stigmatizing fear. Although there is a lot more awareness around mental illness, misconceptions remain. In this family, the parents blamed themselves for Nathaniel's aggressive behavior. They felt like failed parents who could not control their child. As for Nathaniel, he saw himself as a bad person and was drowning with guilt. These thoughts and feelings eventually became so overpowering that the positive qualities previously described lost their meaning and jeopardized their familial bond.

To fully appreciate this story, let us go back 2 years. Two years is how long I have known this family. Their struggles certainly date back more than this time, but I can only speak to the time I have been involved as their clinician. The family presented themselves to the agency where I am employed, seeking counseling for their 9-year-old son. They hoped that counseling could help their boy gain insight into his behavior and learn how to express and manage his anger in a more appropriate fashion. The parents were quick to list Nathaniel's qualities. They boasted Nathaniel's success in school, at church, and in his sporting activities. As such, it was mentioned that Nathaniel had recently been awarded the student of the month award for volunteering as a reading buddy to a first-grade student. They shared that their son absolutely loved school and enjoyed learning. The parents cautiously went on to confide that this remarkable child was unmanageable at home. They spoke of daily angry outbursts accompanied by the increasing use of verbal and physical aggression. The father stated that the last incident took place only the day before. Apparently, Nathaniel was at the kitchen table doing his homework when he became frustrated with his mother for not being able to clearly explain a math problem. He shouted and accused his mother of speaking unclearly and being dumb. He used curse words and knocked a chair over. When his father tried to intervene, Nathaniel refused to go to his room and pushed his father away when he tried to gently lead him down the hall. The argument went on for well over 1 h until Nathaniel's anger finally faded and he stormed outside to play in the backyard. The mother then shared that their son returned inside a half hour later, appeared exhausted and sad, and apologized for being aggressive. All was well for the remainder of the evening.

According to the parents, this type of breakdown was an almost daily occurrence. They cried revealing their frustration. It was apparent that the family was struggling to keep "above water" and that the parents were "walking on eggshells" to have some peace at home. The parents also shared that they both struggled with mental illness and were afraid that Nathaniel may also have some similar disorder. They had many suspicions about what may be troubling their son. They wondered if genetics were at play, if they spoiled him too much, or even if they didn't show enough love. The self-blame was crushing and was impeding on their ability to see the light at the end of the tunnel. They had little hope in counseling, but were at a loss about what else to do.

At this point, I need to emphasize that, despite this family's troubles, their inner strength and commitment to their family's well-being were clearly evident. They may not have been able to acknowledge it at this time in their life, but it was no doubt there. Underneath it all, the parents tried their best to act as team and focused all of their energy on helping Nathaniel. Their hope was for their son to work through these feelings of anger so that he could have a happy childhood and attain his full potential.

When I first met Nathaniel, I was not surprised to be greeted by his smile. He was pleasant and talkative. He spoke of his talent in soccer and hockey and of his interest in reading and drawing. He participated very well in the engagement activities and was cooperative through the session. Yet, during the initial assessment, Nathaniel was unable to voice any concerns or set any "therapeutic" goals. He stated that he was happy and had no problems. This in itself did not raise any red flags, as many children this age show limited insight into such issues and thus disagree with their parents regarding what is considered a problematic behavior or situation. What struck me the most was that this intelligent and well-spoken 9-year-old boy had a very limited "feelings vocabulary." In fact, Nathaniel could only use the words *happy* and *upset* and became extremely agitated if other words such as *angry, anxious*, or *sad* were suggested to him. This pleasant and cooperative boy spontaneously became physically agitated and extremely anxious at the thought of feeling anything but happy or upset. The mere label of these other feelings brought on such intense emotion that tears were followed by aggression and then by remorse for his behavior.

It was apparent that Nathaniel's outbursts of anger were deeply rooted in his difficulty to identify and to communicate his feelings. Imagine not being able to tell people how you feel and, worse of all, to not understand your own emotions. It has to be frightening to feel such intense emotion and to experience the accompanying physiological sensations, and to not understand what they are, what causes them, or how to manage them. It is a wonder that he was able to function so well at school and in other settings outside of the family home.

Given Nathaniel's poor insight into his feelings and his low tolerance of frustration, the individual therapy began by focusing on gaining his trust and slowly building his awareness regarding emotions. The other equally important component of the therapy was to be delivered through the parents. Both the mother and father were open to viewing the problem of angry outbursts through a different lens. They tried to step away from seeing the anger as defiance or as a sign of bad parenting. By doing so, they were able to be more attentive to their child's true feelings. As such, the parents began identifying the triggers of the anger outbursts. They soon realized that the underlying feeling seemed to be anxiety. They noticed a wide array of worries including a need to have everything perfect and in order, a preoccupation with cleanliness, and fears regarding getting ill, not doing well in school, and not being liked by others.

Meanwhile, Nathaniel showed slow but sure progress in individual therapy. We built a trusting relationship and he developed an interest in attending sessions. Although he still did not acknowledge any significant problems, he participated well in the therapeutic activities. He was learning to express various feelings indirectly through play, which signified an increase in his "feelings vocabulary." However, he could not yet apply this new knowledge to his own emotions. Consequently, there was little change at home, and the angry outbursts continued to intensify. At this point, the parents and the clinical team (clinician and parent support worker) decided to seek consultation with a child psychiatrist in the hopes of gaining more insight into the situation. Sadly, the consultation triggered Nathaniel's anger and compromised his participation. Consequently, the psychiatrist could not properly assess his mental health and suggested that we continue with the current treatment plan until Nathaniel could be more open about his thoughts and feelings.

During the next 9 months, the parents continued to participate in the parenting support program, which taught them how to manage the aggression and to address the anxiety. This task proved to be challenging for the parents as they felt that their efforts were not making a difference. Nevertheless, they persevered and continued to apply their new skills. They also sought marital counseling to preserve their marriage and their own mental health. They understood that they needed to take care of themselves and their relationship to have the strength to help their son. The parents' continuous hard work and devotion, coupled with Nathaniel's participation in individual therapy, finally offered hope to this family. During the last few months, more than 1 year after therapy began, the family has been noticing positive changes. Nathaniel recognizes that he feels strong anxiety and that these feelings lead to anger outbursts. He wants to learn how to better manage the anxiety and to control his behavior. He is beginning to take ownership of his thoughts, feelings, and behavior rather than blaming his parents for upsetting him. Most importantly, he appreciates the progress he has made as well as the effort he needs to continue to put forth to meet his goal.

Nathaniel's outbursts at home are less frequent and intense. He finds it very difficult to control his anger and to manage his anxiety. A few weeks ago, Nathaniel agreed to meet with the child psychiatrist. This second experience turned out to be a positive one in many regards. I saw the tenderness between Nathaniel and his parents without the overhanging conflict and hurt. In the end, Nathaniel was diagnosed with an anxiety disorder. The diagnosis not only confirmed our suspicion, but offered solid answers and direction to the family. Nathaniel agreed to continue therapy to address the anxiety and to consider medication if needed down the road.

Nathaniel and his family acknowledge that the work is not done, but they feel relieved that they can now see the light at the end of the tunnel. They understand that there will be other obstacles, but they are confident that they will be able to overcome them. The parents are proud of Nathaniel's progress and of their parental commitment. They shared that the community and social support they received throughout the last year has helped preserve their strength, which allowed them to work through their problems and grow as a family.

As a clinician, this family's story stands out because they remind me of the importance of collaboration. The parents, who struggled with their own mental illness, sought professional support to care for their mental health, which gave them the strength to help Nathaniel reach for his full potential. In time, Nathaniel also accepted the support offered by others and set his own personal goal for wellness. On the community support level, we are lucky to have access to a variety of professionals. Throughout the last 2 years, this family was supported by a clinician, a family support worker, a psychologist, and a psychiatrist. As well, the family maintained close relationships with their friends and family, which offered a break from their troubles and the opportunity to focus on more positive aspects of their family. Through this collaboration, Nathaniel continues to excel at school and in sports and, most importantly, is a child outside of the anxiety disorder.

Epilogue

From our perspective, some themes within and across many of the stories in this section include the following.

- There are many factors that contribute to childhood anxiety, including genetics, temperament, psychosocial, and parental.
- The primary problems of children and youth regardless of the type of anxiety disorder is an unrealistic worry and a feeling of being out of control.

- Children and youth with anxiety disorders benefit from supportive professionals with whom they make positive and trustworthy connections and with whom they do not feel judged.
- Childhood anxiety not only affects the child but also significantly affects their parents and their interrelationship.

Specific thoughts parents expressed in the stories that we thought were insightful and we have heard expressed in similar ways over the years include the following.

- There is much more to my child than her anxiety; we have learned to affirm her fears rather than minimize them.
- Things have made much more sense about my child once I understood my child's perspective because it changed my way of thinking about her and my way of parenting her.
- My child's anxiety sometimes looked more like aggression and defiance.
- As parents, we had to adjust our expectations for our son.
- Watching your child struggle with a mental illness makes you feel helpless as a parent.
- One of our major questions throughout our journey with our daughter and her suffering from anxiety was, "Will she always have excessive worry"?

Specific thoughts professionals expressed in the stories that we thought were insightful and we have heard expressed in similar ways over the years include the following.

- Families often feel like they are walking on "eggshells" and often try to avoid triggering an emotional event to have some peace at home.
- Often, realistic self-talk along with coping statements and relaxation strategies can help children and youth deal with the anxiety-provoking situations that confront them.
- With professional help and the support of family, friends, and the community, many children with anxiety disorders can become much more who they are outside of the disorder.

4 Childhood Mood Disorder

Prologue

The primary feature of childhood depression is an expression or constellation of unhappiness, irritable mood, worthlessness, lost of interest or pleasure, and hopelessness, which disrupt typical developmental processes and contribute to impaired functioning in one's association with others across situations and contexts (e.g., school, work, and social relationships). Moreover, it can co-occur with anxiety disorders, disruptive disorders, and ADHD. Two major types of depression in children and youth are major depressive episode (can last within a 2-week period and can include depressed or irritable mood, loss of interest or pleasure, diminished ability to concentrate and/or make decisions, feelings of restlessness, inability to sleep, decrease or increase of weight, recurrent thoughts of death and or suicidal ideation), and dysthymic disorder (can last between a 1- to 2-year duration and can include being sad or depressed, having an irritable mood, having low energy, self-esteem, and interest, undereating or overeating, experiencing difficulty making decisions, and/or having poor concentration, self-criticism, sleep problems).

In this section are four stories: one personal story, one parental story, and two professional stories pertaining to depression.

The personal story, entitled "The Prison of Oneself," is written by a middle-aged man who reflects on his childhood and youth with respect to depression and shares the effects this disorder had on him during his early life, as well as later in his adult life. The parental story, entitled "Staying 'in Touch,'" is written by a mother who recalls her life as a parent of her daughter who suffered from depression during her childhood and youth. In her story, she shares her challenges, reflections, and resolutions. The first professional story, entitled "Mad, Bad, or Sad: Learning Is a Lifelong Learning," is written by a psychologist who shares her experiences while working with an 8-year-old boy and a 14-year-old girl who both had dysthymic disorder. She shares her focus of involvement, which was to help create a supportive and caring school environment for them while also helping them to cope and succeed in the classroom. The last professional story, entitled "Meaning Is All You Need; Relationship Is All You Have," is cowritten by a psychologist/psychotherapist and by a clinical psychologist who share their therapeutic (integrative and meaning-centered) journey with a young girl with major depressive episode.

Exceptional Life Journeys. DOI: 10.1016/B978-0-12-385216-8.00004-7

Personal Story

The Prison of Oneself

As a person who has experienced disabling depression during significant periods of my life, I would like to share some of my childhood experiences and how these contributed to my future growth and development. The period of my childhood was in the 1950s and 1960s. I grew up in the postwar baby boom at a time when there was a large population of children in our veteran's residential housing district. I was fortunate to have a lot of continuity in my life in that we stayed together as a family in the same house for most of my childhood and I attended the local schools with a hoard of other children. Although I was surrounded by people, I was all at once a sad, frightened, and lonely child. I had incredibly low self-esteem and saw myself as an unworthy person who failed to meet the expectations of others.

During the 1950s and 1960s (the period of my childhood), there was relatively little awareness of mental health issues in children. The focus at that time was on parents providing the physical necessities of life and a supportive environment in which children could receive education and prepare themselves for adult life. My parents were struggling financially and emotionally. They had endless arguments and little time or energy to invest in developing supportive relationships with children. I have two sisters and a brother. I was the oldest child with a sister 2 years younger and another who is 14 years younger. My brother is 9 years younger than I am.

The first thought of my childhood that comes to mind is that of a scared little boy. I seemed to always be afraid and anxious. One time I recall riding my bike and scared to go home because my aunt and uncle were coming to town and I was not sure if they would like me. Being liked and being accepted by others was a key driver in my childhood life. A key feature of my childhood life was anxiety. It seemed that I was always anxious about something—going to swim class, trying out for soccer, going to school, and meeting other people. I recall that, from an early age, I had sleeping problems. I said to mom, "I'm worried that I can't play soccer like the other boys." Mom replied, "That's silly, you don't need to worry about that so just go to sleep now." As a young teenager, I was disciplined a lot at home for "acting-out" behavior—primarily yelling, running away, and hitting my younger sister. Mom reported these concerns to our general practitioner who prescribed Librium to control the outbreaks and stabilize my mood.

During the period of my childhood, little attention was given to parent–child relationships. Certainly I did not have a close emotional bond with my parents. They always seemed to have problems of their own, and from the earliest age, I recall them constantly arguing with one another. I can't recall any cuddles from mom, nor did I see much of dad. He preferred to work the evening shift at the freight sheds where he worked as a warehouse man. It is not my intention to blame parents or anyone else for my mental health issues. Rather, in the absence of any close bond with adults, I experienced a strong sense of "aloneness." I walked alone, I played alone, and I experienced fears and sadness on my own. I would cry quietly to myself quite a lot of the time when I was on my own. I would cry to myself and ask, "Why am I such a failure? Why am I different than other children? Why do I feel so sad?" I recall that my cousin and I attended a class to learn to swim. The goal was to be able to swim a width of the pool. My cousin swam the distance and got a certificate. I never made it, and so there was no certificate for me. I was so ashamed that I ran from home over to the hockey sheds at the local community center. I cried and cried about my failed swimming attempt and how upset my mother would be. I was afraid to go home.

In those days, a lot of attention was given to "what will other people think?" As a child, I thought it was important not to let the parents down. After all, mom was so concerned about what her sister would think and if her mother (my grandmother) would approve. My mom and dad did the best that they could as parents but that did not include providing emotional support for children. This is unfortunate but not surprising as they were products of "emotionally deficient" childhoods. It seems to me now that sadness and crying were a way of nurturing myself. I cried a lot and I spent a lot of time on my own. This pattern continued through my childhood but changed somewhat as a teenager. The change that took place came through the discovery of alcohol. I found the effects of drinking beer in the back alley to be emotionally liberating. I recall the glow of the snow and the gentle background sounds of the city. When we were drinking, all the worry and feelings about not being liked and accepted went away. I was comfortable as one of the boys and thoroughly enjoyed the exhilarating experiences I had when we were drinking.

One of the best parts of my childhood was as a somewhat delinquent adolescent. This was a period when I was discovering myself and making some progress in building a sense of self-worth. I was fortunate to be brought up in the Salvation Army where I learned to play a brass band instrument. In fact, by the age of 14, I was expelled from the Salvation Army band for smoking on a band trip. I then joined other community show bands and concert bands. This gave me a great opportunity to make friendships and to further my interests in smoking and drinking booze. I had a lot of cars and a lot of independence in my adolescent phase. My parents were quite happy to leave me to take care of myself. By age 15, I was working for the carnival in the summer seasons and even managed an illegal working trip to the United States. Acting true to form, I was picked up in the United States for illegal underage drinking and (informally) deported back to the Canadian border. As a young adult, I even spent a night in jail after being picked up in an inebriated state for being unlawfully on a construction site.

Despite the anxiety and depression that I chronically experienced throughout my life, I have been a successful person in so many ways. I attribute this to my own inner strength to do my best but also to the many significant others who guided and influenced in a positive way. There were certain teachers, bandleaders, adults, and childhood friends that connected with me and affirmed my goodness as a person. The literature on resilience has a great deal to say on the importance of these relationships with significant others. I think that these relationships outside of the family assured my success and my ability to overcome the odds.

Although I had a detached relationship with my parents, I was fortunate to have supportive and positive relationships with other adults. I especially recall the parents of a friend who invited me to join them and their only son on day trips and weekend excursions into the mountains. I built confidence through achieving at outdoor events such as skiing, boating, and hiking. I had some supportive teachers also who I think saw my emotional fragility and supported me behind the scenes to participate and experience success. One of the biggest contributions my parents made to my child-hood growth was through getting me involved in brass bands. I learned to play an instrument at age 10, and the band provided me with a great context for developing friendships and building my self-esteem as a moderately capably backseat musician.

Looking back, the key feature of my life was that I was living in an emotionally detached world. I would spend a lot of time alone and literally cry out for attention and for acceptance by others. I never had much in the way of hugs and no opportunity to relate at an emotional level to my mom and dad. I would never even think of tell-ing them how I felt or how afraid I was. I had a lot of trouble sleeping, which I think contributed also to my depression and difficulties coping. Mom saw me as a problem child and sought medical help from doctors who prescribed a range of medication to calm me and help make my behavior at home more manageable. My sister and I fought a lot and had little in common. I was "of" the family but not "with the family."

My main coping strategy in childhood was to conceal my feelings from other peo-ple. I would self-regulate to assure that I was doing all that was required despite my overwhelming sense of anxiety and sadness. Being alone became important to me. When alone, I could nurture myself with tears and let go of emotions through crying silently. I was careful never to cry in front of others but rather I would walk or bike away to someplace where I could cry by myself and even put my arms around myself to ask, "Why am I like this? Why can't I be like others instead of being afraid and crying all alone?" It felt to me as though I had a split personality in which there was an internal me (a scared little boy) and an external me (the delinquent adventurer).

I desperately wanted to have a girlfriend in my adolescence but never managed to achieve this until after high school. I considered myself to be unworthy and was very anxious about the idea that I would be rejected if someone found out how disturbed and inadequate I was. This situation changed enormously, however, when I met my first true love. I met her in a pharmacy where I was working after school. I recall the excite-ment and the enormous emotional connection that I developed with this girl. Never had I been so happy and so close to another person. Being able to kiss, to hug, to play, and to share feelings with her made me unbelievably happy. At age 18, I developed the first deep emotional friendship of my life and had a chance to get close to another person in ways that I had never experienced. Not surprisingly, I became overinvested in this rela-tionship, with the result of her breaking off the relationship after a year or so.

In a desperate attempt to hold on to the relationship, I attempted suicide by taking an overdose of painkilling medication. Fortunately, I took this at her apartment and so I was found later and taken to the hospital for emergency treatment. The doctor told me that this large overdose would have been fatal without timely intervention. I was put into the psychiatric ward at the local hospital for a period of 2 weeks. The recovery period was extremely rough but I was determined to make it and to demon-strate to this woman that I could succeed in life. My depression deepened after this time, and I barely was able to function. My therapy was to write to myself. I spent a huge amount of time writing down and analyzing my feelings. I made a commitment to myself to succeed no matter what. I held on to this relationship and my admiration for this woman for the rest of my life, despite two marriages and a family.

This first relationship was incredibly significant and represented a turning point in my life. She encouraged and nurtured me at a time when I needed to find a direction for myself. This led me to junior college and university studies in social work and psychology. I was a success academically and got a start on a career. This led me to work in the fields of rehabilitation and education. I began a long and successful academic and research career. I went on to develop long-term relationships and successfully functioned as a partner and a parent. However, depression remained a big part of my life.

Drinking alcohol has been a significant challenge for me throughout most of my life. The anesthetizing effects of alcohol brought me temporary relief and so drinking became a big part of my life. I suspect that this is true for many people like myself who suffer from chronic depression. Ultimately, alcohol and drugs generate more problems and less ability to cope. Through participation in Alcoholics Anonymous, I came to understand that I was not alone and that fellow alcoholics were also dealing with mental health issues.

It was not until the age of 40, when I experienced a midlife crisis, that I was able to leave depression behind and essentially live a near-normal life. Having entered a new relationship, I became overwhelmed with guilt and anxiety about leaving my previous wife and young children. I couldn't get through the day and just had a tremendous feeling of dread, feelings of hopelessness and wanting to kill myself. At this time, I saw a psychiatrist who prescribed Ativan in the short term and recommended a long-term treatment with Prozac (fluoxetine). I have been taking this medication for 20 years now, and it proved to be life changing. I no longer suffer depression on a daily basis and, although I have a lot to contend with, I feel in control for the most part and able to deal with whatever happens in life.

Reflecting back, I think that the lack of attachment and emotional connection with my mom and dad was a huge contributing factor to my childhood depression. Fortunately, I met some significant others along the way who were able to guide and comfort me. It was these relationships with others that built my resilience and enabled me to cope with life. The most significant danger that I faced was the attempted suicide, which could have been lethal in my case. I still think that depression can be the silent killer and that there are many children like me who are emotionally lost in an internal world of sadness, anxiety, and self-doubt. Today, there is increased awareness of mental health issues in children. I believe that there is a need, however, for us to stress more on social and emotional development in education.

We need to give more attention to teaching students about feelings and social interactions that can enhance their learning and emotional well-being. We need to provide age-specific lessons about optimism and empathy. A lot has now been written on the subject of emotional intelligence, yet there is a need to put this into educational practice. In the process of actively engaging children in how to listen to their own emotions, we will create self-understanding and opportunities to express feelings and to acknowledge emotional needs. For me, I created a private world where the only safe way to express my emotions was through aloneness and crying. It was a sad and isolated world in which I experienced a lot of emotional pain and overwhelming anxiety that made me sick to the stomach. I often contemplated ending my painful and inadequate life, but I am glad that I had the resilience to make it through and to experience personal satisfaction and career success later in life.

My life story underlines the importance of being public about depression and suicide. Teachers and parents need to know how to support students with these and other mental health issues. Children at risk need to be helped in the process of making sense of it all, through understanding feelings and the role of emotions for improving well-being. The troubling gap is the lack of courses and learning units on social–emotional development for both new teachers and established teachers. The best programs are likely to be long term and integrated into the school environment. These need to be continuous and ongoing just like other essential learning areas.

Social–emotional learning programs are optional in most school districts, often depending on teachers to champion their development and implementation. However, as research builds upon mental health issues in childhood, greater importance is now being given to the early identification and intervention for depression and anxiety. Several professional advocates now consider that emotional and social development belongs in the curriculum. It is possible and desirable to integrate these learning experiences into curriculum areas. For example, students in nature studies may be instructed to pay close attention to the sounds, sights, and smells in their environment. They may discuss their feelings and what they saw in science class, or write a poem about their feelings for an English class. Students can work collaboratively to create solutions for their fears, brainstorming how to confront stage fright, or overcoming anxiety about social acceptance by others.

There is no prison greater than the prison of oneself. As my story illustrates, emotional development should not simply be regarded as a vicarious aspect of human development but one that is central to teaching for effective learning and personal adjustment. The current focus on student diversity and the creation of caring and accepting learning communities is promising. Recent research studies have shown that depression-prevention courses and social–emotional development courses can lead to improved personal adjustment and learning. Finally, there is huge value in combining antidepressant medication with educational interventions. However, more remains to be learned about the combined use of medication and social/ecological approaches that can promote the well-being of students at risk of depression and suicide.

Parental Story

Staying "in Touch"

My husband and I are both psychologists. We have three children and four grandchildren with another grandchild on the way. I worked part-time when the children were small. After the children grew up and left home, I started my own private practice. My husband works for the government. We live a pretty balanced life, get regular exercise, believe in the importance of nutrition, and have nurtured our marital relationship with regular communication and weekly date nights. When the kids were little, we even had weekly family meetings to discuss what was working and what wasn't. My husband and I feel very blessed in many ways, including having a strong faith, wonderful friends, and regular contact with our family at a weekly family supper.

My husband and I had been married for 5 years by the time our third child arrived. She was the first girl after 2 boys. From the earliest days of our pregnancy, we knew that something was amiss. The obstetrician couldn't hear our daughter's heartbeat at 3 months of gestation and he wondered if she was a "missed abortion." That meant that the baby had died, but had not yet aborted. He recommended that we take two ultrasounds a month apart to see if there was any growth in the fetus. There was growth, so it was decided that the baby was still alive. We should have known then that the adventure had begun.

Our daughter is now 32 years old, has just completed her master's of fine arts and is looking forward to a career in her area of passion. Life has not been easy, but she continues to persevere, and she is an inspiration to many others in the process.

After the first speed bump during my pregnancy, we had two and a half fairly peaceful years until she was diagnosed with cardiomyopathy—which means a weak heart muscle. We now realize that that is why the doctor had not been able to hear her heartbeat in utero. Her heart also happened to have a hole between the upper two chambers. This was called an "atrial septal defect." The good news was that it was reparable with a few stitches. The only hitch was that because her heart was weak, the cardiologists weren't sure whether or not she would survive the operation.

The first time open-heart surgery was booked, it was cancelled at the last minute because, according to his secretary who called to inform us, the doctor "didn't dare to do it." Eventually, we decided to go to a larger center to have the surgery. Believe me, this was the most difficult moment of my life—taking a living, breathing child in for surgery from which we were uncertain she would actually return.

After our daughter had survived the heart surgery, we thought we were home free until she had a stroke at age 6. This event left her as a "hemiplegic," which means that she has limited use of one side of her body. She walks with a limp and is basically a one-handed operation. She first experienced depression at this time. She was, understandably, very frustrated with what she could no longer do. It was a "reactive depression" because she was reacting to the changes in her life situation. We enrolled her in play therapy on a weekly basis for several months, where she had the opportunity to express some of her anger at the functions she had lost. She also formed a very positive relationship with the play therapist. The therapist also noticed that I didn't talk about the future in my interactions with our daughter or in discussions with others. That was a useful observation and I attempted to modify my language. It was hard, though, because we really didn't know how long she would actually live.

When our daughter was 9 years old, we moved from a large city to a small town in another province. We were there for 3 years and, during this time, she was rejected, teased, and bullied a lot by her peers. This experience was extremely stressful, and she felt that it "changed her personality" from a friendly, outgoing person to someone who decided that she had better learn to be happy by herself.

When she was 12 years old, we moved to another big city. I had been working full-time for several years, but wanted to settle in the place we would stay, so that I could begin to establish a private practice. I obtained full-time employment first, so I moved with our daughter and our second son. We rented an apartment close to the children's school. My husband stayed in the smaller community with our oldest son, so that he could finish twelfth grade. The house needed to sell and my husband needed to find a job. We had both become temporary single parents and were looking forward to the time when we would all be together again in our new city.

It was around this time that our daughter, now an adolescent at age 13, began to talk of suicide. The methods she talked about were so unrealistic, that I didn't take it very seriously. She would talk about just letting her bike crash while going off a curb. She now tells us that she also thought about jumping off the roof of the apartment building. However, my colleagues at work encouraged me to take her in for a psychiatric evaluation to Children's Hospital. The doctor didn't think that our daughter was in any danger of committing suicide and didn't really have any recommendations.

However, I also consulted with another professional, a psychologist who I had known since my graduate study days. He had been trained in psychoanalytic methods and had been very helpful at several times in our lives when we had run into challenges earlier with our daughter and later with my mother. This time he suggested that we take our daughter's threats seriously. He recommended that I begin to have a weekly "private time" with her, in which we would go for something to eat and just have fun together. There was to be no discussion of any kind about concerns we had about her behavior or schoolwork. It had to be something that was fun for both of us. This seemed to help her a lot. We began a practice of weekly private times that often consisted of having supper together followed by a movie. Not only did this intervention seem to result in our daughter feeling better, but also I believe that it contributed very positively to keeping the channels of communication open during her adolescence. Occasionally, when a few weeks would go by that we didn't do this, there seemed to be a lot more tension in the home. We kept up this tradition for years, as she went onto university and when she was a young adult in her early working days while still living at home.

At age 16, she needed surgery for "scoliosis," which was curvature of the spine. Again, this was a risky operation because of her weak heart and previous stroke, but I called all over North America to find a surgeon in whom we had confidence. Eventually, we went to another province where she had a successful surgery. Her spinal curve was reduced from 80° to 40° and she was in less pain; however, she was left with a significant hump in her back.

At age 18, our daughter was hit with depression again. It seemed to come out of the blue as far as I was concerned. But she says now that she struggled for quite a while not wanting to live but was too afraid to actually end her life. The only way I could understand it was to blame it on the fact that her nutrition had gone to the dogs and she was eating a lot of wheat. Wheat had been responsible for my depression and I was still sensitive to it and I suspected that she might be having a similar reaction to this food. She was now responsible for what she ate, and it included pizza pops for breakfast, hamburgers for lunch, and pizza for supper a few times a week.

I felt powerless to change her behavior and frustrated because I was confident that a change in her diet would help her to feel better. I now realize that, in reality, she was dealing with a lot more than diet—the physical challenges, the worry about her heart failing, the bullying, and the surgeries. Her greatest desire, like every teenager, was to "fit in" and, above all, not be different. That wasn't happening.

During this time, my husband was a great support to her. He was patient and available for listening in a way that I could not be. He remembers spending many evenings when she would cry and express her anger at how unfair life was and how difficult it was to be in her situation. I basically gave up for a few years, until she had another stroke and really hit bottom. It seemed that she was then ready to seek the advice of health practitioners. I took her to a naturopath who explained that she was sensitive to wheat and milk and other substances. Because she still wasn't ready to give up wheat, we took her for special allergy treatments, called Nambudripad's Allergy Elimination Techniques (NAET). Sometimes when she notices her "negative self-talk" coming back, she will arrange for another allergy treatment to wheat and she will feel better.

There was another part to this depression that I have just learned about in the last several years. Apparently, she now tells us that during adolescence she had a constant dialogue with herself regarding her sexual orientation. At that time, she wasn't ready to admit that she was gay, even to herself. She now tells me that she felt completely overwhelmed with all of the things she had had to deal with (e.g., heart, stroke, scoliosis, ADHD, and a disorder of written expression) and she just couldn't face another issue that meant that she was even more different. In actual fact, it took her more than 10 years or so to accept it within herself and "come out," which she finally did at age 26.

"My life journey has been one of dealing with childhood, adolescent, and adult depression, as well. I was depressed from age 8 until age 24, when I finally decided that it was a good time to kill myself. I had been wanting to do this since age 8. In fact, I can remember walking home from school and wanting to kill myself by running out in front of the traffic. But I somehow realized that my parents would be very upset if I did. I knew that they were good parents and doing their best, and I just couldn't hurt them in this way. As I grew older and stayed depressed, it got more depressing. As far as I knew, I had no real reason to be depressed. I had friends, a faith in God, was doing well in school, had a loving family, and couldn't understand why I cried every few days and felt suicidal. To the outside world, I looked happy and capable, but inside, I felt insecure and miserable. My symptoms were consistent with what might be called a "masked depression."

I finally decided to kill myself after I had been married a few years and had a 1-year-old child. I figured that my husband would take very good care of our child and that he wouldn't blame himself for my suicide, because I had been this way ever since we met. I had been looking for methods that I could carry out and finally decided upon taking aspirins. I thought that fifty would do it. But instead of becoming drowsy, as I expected, I became more alert and thought that I had failed. Now I know that if I had only waited a little longer, they would have done the trick. Instead, I woke up my husband and told him my attempt had not worked. I was taken to the hospital, had my stomach pumped, and was forced to see a psychiatrist.

During my visits with the psychiatrist, I really had nothing to complain about. However, I did tell him that I thought that there might be something biochemically wrong in my brain. At the time, orthomolecular psychiatry was becoming known. He agreed with me and suggested that I begin taking megadoses of vitamin B3—niacin. It seemed to help and I stayed on this until I found out that I was actually allergic to wheat and milk. Once I went off these substances, my mood drastically improved. My depression never returned, and now I cry once every 4–5 years instead of every 4 days."

Our daughter's story involves more than simply adolescent depression. There are many significant experiences along the way related to her challenges in her physical health, learning and attention issues, and the social rejection she experienced. Believe it or not, her sexual orientation seems to have been pretty smooth sailing so far, from our perspective.

However, in the realm of adolescent depression, there are two significant experiences that come to mind. The first has to do with my feeling helpless to do anything for her. She continued to eat what she wanted to eat despite the fact that I felt that it was causing her harm, and she was also not ready to listen to the health practitioners that were basically telling her the same thing. It was hard to watch her suffer and not be able to help her. She was an adult and entitled to make her own decisions. The second significant experience was the relief I felt when she finally hit bottom and asked for help. I then accompanied her to a variety of helping professionals, so that she could hear their opinions and decide what she was willing to do. Although it was hard to listen to them say what I had been saying all along, I was very grateful that she appeared ready to listen to some of them. Of course, I felt very relieved and hopeful as she started to feel better.

The most difficult challenge related to our daughter's depression as an early adolescent was in assessing the severity of the depression and then in finding appropriate help. We were fortunate in that we were able to get a timely consultation with a psychiatrist at Children's Hospital, but our daughter did not relate well to this individual and, in the end, we didn't agree with her evaluation that our daughter was not at risk. We were fortunate to continue accessing our resources, until we found someone who gave us a strategy that worked better for her.

For the depression she experienced in later adolescence, the most difficult challenge for me was watching our daughter make decisions that I believed were hurting her, but being powerless to influence her behavior. The biggest barrier to her getting help was herself.

The most successful strategy for the early adolescent depression (when she wasn't eating wheat) was having a weekly private time with me. It was so effective that we continued to do this for as long as she lived at home. The most successful strategy for the later adolescent depression was finding a way to treat her allergy to wheat. I can see now that she was in the process of accepting her sexual orientation as well. It is very likely that the hours she spent crying on my husband's shoulder were an important support and helped her get in touch with and express her anger toward God for what she felt was an unreasonable amount of challenges in her life.

Being told that she had to stay away from wheat and milk and other favorite substances was a strategy that didn't work for our daughter, even though this was probably good advice. We just kept going until we found interventions that she was willing to cooperate with (e.g., acupuncture and NAET allergy treatment). A low point was watching our daughter make decisions that were hurtful to herself and not being able to do anything about it, except wait until she came around. It certainly was a "high point" when we discovered how well the weekly private time strategy worked when she was 12 years old. It has actually been a bonus to our relationship that has now morphed into an annual girls' weekend. We now continue this every fall, no matter where she lives. During this time, we take 2 days to watch six movies, have a hot tub, swim, order room service, play cards, and just have a very intense special time together.

Some things I have thought about over the years about my daughter's depression are:

- When it comes to depression, there is probably a lot going on to cause the depression that the adolescent may not even be consciously aware of at the time.
- It was a good idea to trust my instincts in seeking the kind of help that is needed, until we got a suggestion that was effective.
- Adolescents are in a place where they are entitled to determine *how* they will live. The most I can do as a parent is share my feelings (e.g., "I feel worried when I see you eating things that are hurting your brain."), and use the language of choice (e.g., "When you choose to eat in a way that is good for you, the consequence is that you will feel well. When you choose not to eat in a way that is good for you, the consequence is that you will not feel well".).
- You can always pray that your adolescent will see the light.
- The most remarkable success for us with our daughter was in waiting until our daughter was ready to take responsibility for her health, and for her father and I to be able to support her in that journey. We did this in very different, but complementary ways. I have seen her grow into an incredibly responsible young woman who takes care of herself in many ways and continues to face her challenges with hope and courage.

What I have learned in our journey with our daughter is:

- Trust your instincts if you are worried about something, even if the professionals say that your child will grow out of it or that there is nothing to be concerned about.
- Persevere until you find the help that feels right to you.
- Find professionals that treat you with respect and with whom you can act as a team.
- Accept support from family and friends.
- Realize that everyone is on his or her own journey, and you can only do so much to alter the decisions your adolescent is making.
- Find ways to maintain your relationship during the times when your adolescent is making decisions with which you disagree.
- Make sure that you continue to nourish your relationship with your partner (e.g., weekly date nights), because it will be just the two of you when all the kids leave home, and if you don't know each other anymore, your relationship may be in trouble.
- Make sure that you are spending time and having fun with the other kids in the family. It is easy for them to be overlooked.
- Remember that all you really have is the present moment, so live as though every moment counts.
- Find ways to have fun and remember that this too shall pass.

Professional Stories

Mad, Bad, or Sad: Learning Is a Lifelong Learning

I left graduate school with a suitcase full of textbook knowledge and a toolbox full of prepackaged strategies, ready to begin my career as a psychologist. After spending so many years in an institution that focused on learning, researching, and studying, I naively felt that my learning was over and it was time to start doing. Little did I know that my learning had only just begun.

In terms of my personal background, I am a recently registered psychologist who works for a public school board. The experiences I have chosen to share took place during the course of the 1,600 provisional hours I completed to become a registered psychologist. I completed my provisional hours at the same school board for which I am currently working.

In an attempt to help expand knowledge and break down some of the misconceptions surrounding childhood depression, I have decided to share not one but two experiences. In my opinion, one of the most interesting or perhaps complicating elements of depression in children and youth is the actual presentation of this disorder. Historically, adult and youth depression have been viewed as one and the same, meaning that a diagnosis of depression in children was based upon adult criteria. As we all know, or have experienced in our lives, children are not the same as adults. In general, adults tend to possess the cognitive capacity to verbally express their thoughts, feelings, desires, and needs in a coherent and comprehensible fashion. Adults also have the ability to think abstractly and no longer require concrete objects to make rational judgments. Further, adults are often capable of hypothetical and deductive reasoning. Children, on the other hand, tend to be concrete thinkers who are still developing their abstract reasoning skills. Rather than verbally expressing their feelings, communication in children often comes in the form of behaviors, including both internal and external expressions.

Closely tied to the notion that childhood and adult depression are distinct is the concept that depression may also present differently in children than it does in adolescents. Specifically, although children and adolescents with depressive disorders may share numerous symptoms, their symptoms often manifest in very different manners. It is thus my hope that by sharing my experiences of working with both a child and an adolescent with depression that parents, teachers, and other professionals may come to understand the various manifestations and complexities of the disorder, so that they may best understand and support each individual.

The first experience I would like to share involves an 8-year-old boy in the second grade named Emmitt. I first saw Emmitt while observing another student in his class. His teacher pointed him out and informed me that he was next in line for an assessment, as his behaviors were becoming out of control in the classroom. I came back to Emmitt's classroom the following week, but this time to observe him. He was a small boy for his age, with big green eyes, and wore an angry expression on his face. Emmitt sat by himself in a desk at the very front of the classroom, right next to his teacher's desk. He was described by both his teachers and parents as a defiant child, with an irritable mood, who frequently threw tantrums, which included vocal outbursts and crying. Further, he was often angry and withdrew from most social situations. As I sat in Emmitt's class, I observed several noteworthy behaviors, including a refusal to comply with his teacher's requests to complete math questions, which led to an episode of laying on the floor crying, along with demonstrations of anger, including making growling noises and yelling. When the bell rang for recess, Emmitt's teacher approached me and said, "See, I told you he had oppositional defiant disorder." ODD is a behavior disorder characterized by a pattern of uncooperative, defiant, and hostile behavior toward authority figures that seriously interferes with the day-to-day functioning of the individual. Children with ODD are often viewed by others in a negative light. Although I had observed defiance and anger in Emmitt, my observation did not provide near enough information for me to conclude that he was, in fact, suffering from ODD. I told his teacher that I would need to spend more time with Emmitt before I could make any formal diagnosis and that I would be back tomorrow to observe him again. I left the school that day with a lot of questions about Emmitt and wondered what was really going on.

I came back to the school the next day, but this time decided to observe Emmitt over the lunch hour recess. The child I saw on this day was different than the one I had observed the previous day. While the other children talked with one another, laughed, played games, and climbed on the jungle gym, Emmitt sat by himself on a park bench. He kept his head down toward the ground and rarely looked up. Even when a soccer ball was kicked in his direction and landed at his feet, he just sat on the bench, never looking up or making any effort to kick the ball back to the other children. As I watched him sit so sadly on that bench, I realized that there was much more to this child than met the eye. In this very moment, I had a self-realization that learning was not a process that ceased upon my final walk down the doorsteps of graduate school, but rather was something that would continue as an integral part of my professional career.

Over the course of the next few weeks, I worked closely with Emmitt, his teacher, and his parents to see if together we could get to the bottom of what was going on with this little guy. This process was not a simple one by any means. I conducted clinical interviews, continued to observe in the classroom, and had Emmitt, his parents, and teacher complete diagnostic rating scales. I was met with, and faced, a tremendous amount of frustration from all parties involved. Frustration came from the school, for the fact that I would not simply diagnose this student with a behavior disorder in order that they could begin providing the appropriate programming for him. Frustration came from Emmitt's parents, who were worried about their child and were unsure of how best to support him in the home. Frustration came too from Emmitt, because the constant attention he was receiving was unwanted. There was also frustration on my part, as all my textbook knowledge and prepackaged strategies were not providing me the answers or solutions I was so desperately seeking.

Finally, after compiling all of the data and consulting with my supervisor, we concluded that what Emmitt was dealing with was not a behavior disorder, but was in fact dysthymic disorder. Dysthymic disorder is a chronic depressive disorder. The symptoms of children experiencing this disorder are often long-lasting and recurring, although of less severity than those with major depressive disorder. Although adults or even adolescents with dysthymic disorder often display symptoms of depressed mood, eating and/or sleep disturbances, fatigue, low self-esteem, and feelings of hopelessness, children with the disorder often display different symptomatology. Specifically, children with dysthymic disorder tend to be irritable or cranky, engage in aggressive behaviors or outbursts, avoid or withdraw from social situations, and often express vague physical complaints.

Providing a name or label for what Emmitt was experiencing initially posed a significant challenge for him within the school. Dysthymic disorder is not a common term used in the school setting, and Emmitt's teachers did not appear to understand what I was talking about. Similarly, when I attempted to explain to Emmitt's parent and other school personnel that Emmitt was experiencing a type of depressive disorder, there was a lot of confusion. This confusion appeared to stem from the fact that Emmitt's symptoms often presented as externalizing, or outward expressions of behavior, rather than internal ones. Emmitt's teacher appeared to be particularly perplexed as to how the boy could be depressed and yet also display symptoms similar to other children she had worked with in the past with behavior disorders such as ODD. More specifically, the reaction I repeatedly received from his teachers was that Emmitt was not *sad*—he was *bad*.

With some time and careful explanation of how depression often presents differently in children than it does in adults, eventually all parties involved seemed to be on board with the results. To help facilitate an understanding of depression in children with Emmitt's teacher and parents, I provided them with several relevant books. Once Emmitt's disorder was understood, the next step in helping him was to devise a treatment plan. The plan we devised was comprehensive, and involved the work of a multidisciplinary team. Specifically, Emmitt's family was referred to a pediatrician, and were also given the names of several parent support groups and community agencies from which they could obtain home support. A mental health therapist, specializing in cognitive-behavior techniques, was assigned to work individually with Emmitt, and I was given the role of building teacher understanding in the school on how to work with students displaying symptoms similar to Emmitt's.

Over the course of the next several weeks, I worked with school personnel to remove the stigma and shed some light on depressive disorders, help school personnel create a supportive and caring school environment, and also develop a protocol for reaching out and responding to students who were potentially experiencing depressive symptomatology. I also worked directly with Emmitt and his teacher, specifically focusing on implementing strategies to help him cope in the classroom.

Throughout my journey with Emmitt, there were many positive experiences and examples of success. Chief among those was the day I went out to observe on the playground, and saw Emmitt playing soccer with another group of students. The sad little boy I had seen sitting all alone on a park bench a few months ago was now smiling and laughing with his friends. At the same time, it is important to note that despite positive outcomes such as this and our ultimate success, we still faced numerous challenges. One of the biggest that I personally faced was the sheer length of time and effort required from myself and Emmitt's teacher before we began to see noticeable improvements. Not every strategy that we tried was successful, and even strategies that started off as promising sometimes ended up failing. There were difficult days when Emmitt was defiant, or days that he cried and had tantrums when asked to do an assignment. We overcame these challenges by staying focused and committed to working with Emmitt. For us, a strategy that often proved successful was remembering that we needed to celebrate any type of success, regardless of how small or large the accomplishment seemed. At the end of the day, Emmitt was making progress, and was slowly learning how to successfully cope in the classroom.

Looking back, I think the memory I cherish most from my experiences of working with Emmitt and his school are the changes I saw in teachers and the school as a community. Although I was initially met with resistance, this resistance soon transformed into acceptance and understanding. It was as if the whole school took on the project of learning as much as they could about depressive disorders, and worked to become an environment that supported all children, regardless of but not specific to the nature of their symptomatology.

The next experience I would like to share involves a 14-year-old, ninth-grade, teenage girl named Sophia. Unlike Emmitt, Sophia had been previously diagnosed as having dysthymic disorder. In addition to her issues with depression, she was also experiencing significant learning concerns in the classroom. Initially, I was requested to assess Sophia's learning. This request was later expanded to include working with her in an academic capacity, specifically to help her enhance her academic self-esteem.

One of the striking differences between my experiences with Emmitt and those with Sophia is that, in contrast, Sophia, her teachers, and her parents were acutely aware of the depressive symptoms she was experiencing, and how those symptoms were impacting her daily life. In addition, Sophia's teachers had already received support and guidance on working with her, and the overall tone of her school was supportive—even nurturing. Despite her personal awareness of her disorder and the support she was receiving from her parents and teachers, Sophia continued to experience considerable difficulties in school.

When I first met Sophia, she presented as hard and gruff on the exterior. She came into my office with a frown on her face and her arms crossed over her chest, in a defensive-type stance. She slouched down in her chair, without making eye contact, and informed me that this was the last place she wanted to be. In spite of this, by the time we were halfway through the clinical interview, Sophia began to warm up and let her guard down. She began talking about the challenges she faced and indicated just how difficult it was to make it through each day. She talked a lot about her overwhelming sense of fatigue, and also of her difficulty completing even simple activities, like showering and brushing her teeth. She indicated that most days she struggled to get out of bed, and that she rarely found interest in school or home activities or even in spending time with her friends. As she spoke, her eyes began to well up with tears, and I was able to see that the image Sophia was portraying was actually a mask she was using to hide her true feelings and to keep herself from appearing vulnerable and weak.

The assessment of Sophia's learning was one of the more difficult ones that I have encountered. A learning assessment similar to hers usually took me a few hours to complete; however, with Sophia, it took several days. The cause of this was that she appeared to have a very difficult time concentrating, and would lose her train of thought relatively quickly. In addition, she frequently complained of fatigue and indicated that she was too tired to go on. Despite her attention and fatigue concerns, it became quite evident that Sophia was a very bright and cognitively capable student. When the time came to assess her mathematical skills, however, a dramatic difference in performance was observed. Sophia struggled with even basic-level mathematical questions. Such challenges with the math component of the assessment led Sophia into an extreme emotional state. She began softly crying, and within a few seconds, her crying turned into an inconsolable sobbing. I tried my best to soothe and calm her, but this was not an easy feat. When she eventually calmed down, we were able to talk about her feelings, and she confided to me that math had always been a struggle for her and that her inability to answer even basic questions caused her to feel hopeless and stupid. Despite my repeated attempts to convince Sophia that she was a capable young lady with a lot of strengths and potential, she would not listen. She had an extremely pessimistic view of herself—and her future.

Following the completion of Sophia's learning assessment, I discovered that she had a math disorder. Given the data that I had collected, arriving at this conclusion was fairly straightforward and easy. Devising a plan to present these results to this already fragile girl, however, was not. I worried how she would respond to hearing the news that, in addition to her depression, she also had a learning disorder. I did not want to cause Sophia anymore pain than she was already experiencing. When the time came to share the results of the assessment with Sophia, her parents, and the school, I decided to be as up front as possible with her. Just as I feared, Sophia took the news harshly. I knew that she and I had a long road ahead of us in order to develop her academic self-confidence.

During my next few meetings with Sophia, I felt like we accomplished very little. Although we spent a great deal of time looking at and identifying Sophia's strengths, the problem was that for every strength that was acknowledged, she would also self-identify two weaknesses. My frustration increased and I worried that, despite my efforts, Sophia would continue to view herself as "stupid and hopeless." One day, after a particularly negative meeting with Sophia, I thought of Emmitt. I remembered the challenges I had faced while working with him, and recalled the importance of celebrating all of his accomplishments, regardless of size. I decided to try a similar approach with Sophia. Although this strategy took some time to get off the ground, with Sophia's pessimism often detracting from the celebration of her achievements, eventually she began to acknowledge her strengths, and focus less on her weaknesses. Sophia and I continued to work in this manner, and gradually her academic self-esteem improved. Although Sophia still at times continued to dwell upon the negative and focus upon her weaknesses, this pattern decreased. By the end of the school year, if one were to ask Sophia what kind of student she was, I honestly believe that for every weakness she identified, she would be able to tell you two strengths.

As I reflect upon my experiences with both Emmitt and Sophia, it is unquestionable just how momentous and meaningful they were to me. Although there were similarities in working with both children, there were also vast differences. Such differences led me to challenge my preconceived notion that depression was depression, regardless of whether a child, adolescent, or adult was experiencing it. My experience has taught me that this is not the case. Symptoms that present in one child may not be the same as symptoms that present in another child. As such, we must remember that no matter what label we give a child, or what category we place them in, they are first and foremost an individual, and therefore unique. We must always put the child first, ahead of the label. In addition, my experiences have caused me to change the perspective I held previously toward learning and what it means to become a competent, capable, and effective psychologist. Although I once saw learning as a means to an end, I now see it as an integral and continuing part of my role as a psychologist—and as a human being. If I were to offer any sort of advice to others, it would be that no matter how much education one gains, there is always room for more learning.

Meaning Is All You Need; Relationship Is All You Have

I have worked with children and adolescents for many years, as school psychologist, psycho-educational consultant, and psychotherapist, and for the past 10 years, as a counseling psychology professor. At the present, I am in private practice, and giving lectures and workshops on play therapy. My cowriter is a clinical psychologist and originator of the meaning therapy. In this report, we applied meaning-centered play therapy in treating a girl suffering from depression.

When Helen was referred to me by her mother's therapist, she was 10 years and 10 months old in fifth grade. The mother's referral concerns included noncompliance with school expectations, verbal aggression toward peers, poor eating habits, defiant behaviors at home, and extreme mood swings. Helen's only sibling, John, 2 years younger, has been diagnosed with autism spectrum disorder (ASD) since 18 months old. Her mother is a social worker, and her father is an engineer working for a large firm. Helen's counseling was funded by an Employee Assistance Program (EAP) through her father's company.

In a school system, it typically takes a long time and many referral processes before the children are assessed, diagnosed, and given appropriate interventions. More often than not, the identified children with serious clinical problems would receive an inadequate amount of psychotherapy to resolve their deep-seated issues because of lack of funding and qualified clinical psychologists in the school system. Helen had not been diagnosed or seen by a school psychologist before she was referred to me. She attended a private parochial elementary school with her brother, because her parents considered a private school setting would give their son more care and attention. During school weeks, Helen had to stay with the homework club, while John attended an after-school care program. She was rather resentful of the fact that she was not able to socialize with her peers after school. At home, John received intensive behavioral intervention (IBI) after supper for two evenings a week, funded by the ministry of education, because John was 3 years old. In a small house, his training required a specially equipped room, and a trained educational therapist to carry out the IBI procedures. Helen complained that her bedroom was much smaller and that she did not even have a proper desk or chair to do her homework. She felt shortchanged and neglected because so much of her parents' time, money, and energy were devoted to John. Her resentment and anger toward John were further aggravated by John's frequent physical assault.

Another major source of stress stemmed from her parents' marital problems. Due to the recent economic downturn, her father constantly worried about losing his job, as many of his colleagues had been laid off. Her mother worked part-time to supplement the funding for John's education, in addition to tirelessly working the government funding agencies to secure ongoing funding for John's IBI training. The financial stress, coupled with the difficulties in raising an autistic son, led to frequent flare-ups and fights. The situation got so bad that the parents were contemplating a trial separation. However, for the sake of the children, they stayed together, but the marital conflicts continued, making the home situation toxic and traumatic. Helen felt completely trapped in a helpless and hopeless situation.

At home, Helen had to help out with household chores, look after her brother, and try to de-escalate the conflicts between her parents. At school, she was very conscious of her family financial situation as compared to that of the rich kids in the private school. She was also very sensitive about her own physical stature—she was noticeably smaller than other girls of the same age. Consequently, she felt frustrated, isolated, and resentful. Prolonged stress on so many fronts eventually overwhelmed Helen, leading to episodes of major depression and acting out. Her inner tumult and poor self-esteem simply amplified the negativity of her circumstances. My counseling focus with Helen was two-fold: relief of her symptoms as well as enhancing her self-esteem and well-being so that she could become what she was meant to be.

During the intake session, her parents were concerned with Helen's weight loss due to poor eating habits (i.e., often throwing away sandwiches prepared at home, skipping breakfast, being a picky eater at dinner), with her apparent underachievement (e.g., incomplete assignments and lack of concentration), nonparticipation in class activities and school socials, and mood swings, in addition to her defiant attitudes. They also described her as accident-prone. Their concerns also included possible depression, noncompliance with school requirements, and habitual lying and talking back with angry words. They found it difficult to communicate with her.

As the counseling progressed, it soon became clear to me that the problems of noncompliance at school and defiant ways of talking to her parents were just a few symptoms of a very depressed child with a great deal of anger as a result of growing up in a dysfunctional family. The challenge was how to help the child cope and grow without first fixing the dysfunctional family system.

Over the past three decades, I have worked with children and adolescents with a variety of mental disorders, from autism to depression. In addition to employing appropriate psycho-educational assessment tools, I adopt an integrative approach to counseling, which consists of active listening, cognitive-behavior, and narrative therapy, as well as play therapy and art therapy. Understanding the context is also important. I communicate with parents on family issues, and have minisessions with them, which may include playing "Ungame" together.

Recently, my approach has become more meaning centered, focusing on the in-session relationship and client-centered meaning systems. Simply put, meaning therapy emphasizes a two-pronged strategy of facilitating personal growth and transforming negativity through meaning. Meaning is defined as PURE (Purpose, Understanding, Responsibility, and Enjoyment), which serves as a conceptual framework to clarify what really matters and how to achieve a more desirable future. Meaning transformation is accomplished primarily through ABCDE (Acceptance, Belief, Commitment, Discovery, and Evaluation), which allows the client to manage life stress in a more positive and adaptive manner.

To fully understand Helen's problem, we have to situate it in her family, which is like a pressure cooker. Apart from financial stress and the demands of raising an autistic child, there were also subtle cultural differences that complicate an already very tense situation. Different habits and expectations often led to conflicts and quarrels. The mother was raised in an upper-middle-class family in Hong Kong, while the father was born to a working-class family in mainland China, and the children were born and raised in Canada. The intergenerational gap was compounded by intercultural conflicts in terms of child rearing. Helen's parents practice the traditional authoritarian parenting style, while Helen desires to be treated like her peers. Consequently, her reaction used to alternate between open rebellion and feelings of helplessness and depression.

It remains difficult to have a clear diagnosis of childhood depression for a number of reasons, such as dynamic maturation process, immature cognitive development, and the complexity of the family system. However, my ongoing assessment revealed these themes: insecurity (e.g., wanting acceptance from parents and not accepting self and constantly feeling the need to be perfect), high anxiety (e.g., always biting fingernails, high frequency of constipation, and high-strung), difficulties focusing on tasks (e.g., parents' reports of being accident-prone and lack of concentration), eating problems, and frequently disobedient and lying. It appeared that Helen had many of the symptoms of adolescent depression. Based on all the symptoms reported and observed, Helen suffered from major depression disorder resulting primarily from chronic and extreme adjustment problems.

A meaning therapist believes that positive relationships have curative benefits, because they meet the clients' basic needs for understanding and connectivity. A therapeutic relationship also models for the clients' new ways of communication and relating. More importantly, the dynamic and fluid in-session interactions often open critical moments for diagnosis and intervention. During the early part of the therapeutic journey, my focus was on building rapport and trust with Helen. When we first met, she began by declaring, "Mom has a therapist, mom and dad go to counseling, and now I have my therapist." She was open and eager to unload her negative feelings regarding her relationships with each member of her family. She vividly described some of the scenarios at home: She had to hide her spaghetti, because mom controlled her intake of food; she was coerced to eat some food that she was not familiar with, such as some Chinese cuisine in the restaurant. Her main complaint was about unfair treatment she had received as compared to her brother: "He got away with everything, and I got yelled at. Mom lectured me even it was not my fault. Everything is about John. Dad was impatient with my piano practice, telling me how to play when he really didn't know anything."

Helen was articulate and insightful. She sighed often when she related her experiences, as she poured out her sadness, frustration, and a sense of helplessness. She clearly revealed maturity and intelligence in spite of her young age. My counseling with Helen can be regarded as a journey together into her world of meanings and beyond; it was a journey of healing and discovery of a brave new world in the future. She revealed her inner conflicts and feelings through both words and play. In one of the sessions, she went straight to the sand tray to play and self-talked. Putting two cars into the sand tray, on two separate tracks, she said, "Mom drives to work. Dad drives one hour to work." She paused, and then said, "Mom got up at 5:00 a.m., freaked, got breakfast, freaked, came home after work, freaked, put John to bed, and freaked." She illustrated a dysfunctional family with two parents going their separate ways, and each stressed out by the demands of life.

The next session, she was sitting on the floor, playing with a few toys, reflecting her feelings, "I am being ignored; my needs were rarely met." Then she picked up a small soft plastic monster head, saying, "This is John," and picked up a plastic fork, jabbing into it, showing more negative emotions. She whispered, "So annoying, a brother is a pain. I get into trouble if I do anything to him. Life sucks." She changed the topic and told me that her classmates were talking about getting menstrual periods and wearing bras. She was really worried about her own lack of development. She appeared to have low self-esteem, but she was also confident about her own inner goodness—a point that was often reinforced in the journey of self-discovery. I often pointed out to her that in spite of all her negative feelings toward her parents and brother, the ways she played with her brother, her sand tray displays, and musings often revealed her concerns and caring attitudes toward all her family members.

After five sessions, she showed clear improvement in managing her school assignments and homework. She was excited about her class presentations. Her expressed purpose in life of getting a good profession, and her understanding of her own academic abilities, especially in Mathematics and Science, provided the intrinsic motivation to do well in school. She also realized that noncompliance was a counterproductive and self-handicapping strategy to express her unhappiness. She was willing to assume the responsibility for creating her own happiness and her own future. As a result, she found schoolwork much more enjoyable. Thus, the PURE framework worked well for Helen, because it is all very intuitive and practical. It makes good sense to her. Thus, I could get across the essence of living a meaningful life without getting into philosophical issues or existential musings.

Another milestone of her progress was her understanding in the adaptive value of ABCDE. Acceptance is one of the key steps in meaning therapy. She came to accept the fact that she would not be able to change her parents, but she said, "I can change myself!" Almost in the same breath, "I wish there are no fights between my parents. They fought on Thanksgiving because dad was tired and mom was very emotional.

I was in the middle of it." Then she sighed, "It was so bad, I'm used to it, I don't have feelings anymore, just hang in there."

After 7 months, I witnessed clear changes in her outlook on life, and coping strategies. For an 11-year-old young girl, Helen learned to accept the ups and downs of life, the imperfect family, the extra burden of having an only sibling with autism. Once she said to me, "I really don't know what it's like to have a normal sister or brother," but she had accepted her lot. Her views of her future and her family also changed. She had also learned the second step of meaning therapy, which was belief. She believed that she could make a difference in her own life, if she kept thinking positive thoughts and pushing away negative emotions. She became less preoccupied with her present worries, and more optimistic about creating her future.

In a later session, her sand picture depicted a large bedroom with sitting and dining rooms occupying half of the sand tray, with barracks and cannons along the length of the home. She labeled the picture, "Overprotected family." At the same time, she told me that she had As and B-pluses on her report card. She began to feel more positive about home and school. At this time, her highlights were small group activities at her church where she felt belonging. She was happy with her academic achievement. Her lowlights were being bullied and teased by the boys at school. She said, "God is in charge of my life, but sometimes it seems that mom is in charge of my life." Her belief in God was helpful in strengthening her belief in future change—God is bigger than her parents, and God is bigger than all her problems.

She also learned the third step of meaning therapy—commitment. In addition to her commitment to excel at school, she was also committed to improving her coping strategies. To cope with her mother's yelling and nagging, she said, "I do my things and go to bed." Or "I don't think about it, focus on candy or computer." I posted this question to her: "This is the second last session before the New Year; what may be some differences in you between May when I first saw you and now?" She responded, "I used to carry many worries on my shoulders, but now I don't care. The other day, they had a big fight. I know if the fight stops in a few hours, that's not serious, but if it lasts for a day or more, that's serious." I asked, "What do you mean?" She answered, "They are ... my mom is going to divorce him." I responded, "How do you feel?" She said, "I now take the positive view ... I joked, 'Could I still have my room and take the bigger bed?' To cope, I change myself and used humor."

Her report card showed four As and seven Bs. She really discovered that hard work paid off, and she really enjoyed her new status at school. However, her problems at home proved to be more intractable, and she depended heavily on acceptance to make the situation more bearable.

In the following session, she revealed, "There isn't one weekend that they don't fight. If dad gets angry, mom yelled at me, and John gets angry. I go to my room, and close the door and read." An example of her outlet for her anger was: "Scream into the pillow, sometimes cry myself to sleep." Again she expressed her coping skills, "When I am in pain, go to my room, shut the door, and scream. Right now it's bad enough, one more bad thing happens, our family will be toast. Nothing I can do, I have to go to orphanage, or foster home." She expressed the fear of the uncertain future. However, I observed that Helen, in spite of her fears, has become more resilient and self-understanding. The home situation did not change much but her relationships with her parents and brother improved. She gained self-confidence and control. She drew up her ideal bedroom on a piece of paper, and was hopeful that her dad would fulfill her wishes. She claimed that her father began to trust her more, but complained about her mother's unpredictable behaviors. Her parents reported to me that the family as a whole had planned fun activities together; her father rewarded her for completing chores, and her mother gave her validations. She has demonstrated a number of strengths, such as intelligence, leadership ability, a sense of humor, openness to new ideas, and having a caring heart in spite of her being treated unfairly.

Working with the child without involving the family can be very frustrating and challenging. Whatever progress the child makes can be wiped out by a sudden worsening of the family circumstances. The prospect of domestic violence and divorce always hangs over a child's head like a threatening storm. Helen expressed her sadness and hopelessness that her home situation would ever get better. She perceived herself as "a punching bag" when her parents argued and fought. In a session, she confessed, "I am a worry head. Dad has hives that may cause sudden death, may lose his job. Grandpa died, he had diabetes and leukemia. Hope it will not happen to me, like skips a generation. Mom's cousin died of liver cancer. Our other grandpa died when dad was only 2." It is one of the signs of an adolescent with depressive moods to think about death of self and loved ones. In the playroom, she was in control of her environment and felt safe to share her death anxiety; at the same time, she gained some understanding that family connections and her faith in God could help her deal with her fear of death.

Although the home front has not changed much, Helen was able to recognize some positive changes in her parents. She stated, "Things are getting better, when dad's depressed, he buys things, and bought me a guitar. When mom is depressed, she eats; she's an emotional eater." As Helen's therapist, I also made a point of talking to the parents either separately or together. I had a couple of minisessions with the parents. I kept them updated about the progress without violating confidentiality. I shared with them the strategies of meaning-centered play therapy.

The psycho-educational component of meaning therapy works. Teaching Helen the importance of PURE facilitated goal setting and the development of a positive orientation. For example, after a few months, she wanted to set new goals for herself: At school, she wanted to strive to get on the honor roll; at home, she said that she would try not to let negative thoughts and feelings get the better of her. It was also helpful not to label her. By adopting a holistic approach, the focus is no longer on the symptoms of depression. Even though she used the word *depression* several times, she did not see herself as a patient suffering from depression. Instead, she focused on depressive thoughts and behaviors as something that could be overcome.

There are many questions that beg for answers. For example, how may a meaning-centered approach be applied to the whole family as the best way to facilitate healing and wholeness for my client, her brother, and parents? Play therapy worked well for Helen. It allowed her personal space to reflect and express her thoughts and feelings freely. She experienced what it was like to be in control of her life in the therapeutic environment. The ABCDE intervention strategy also worked well, because it provided a logical and practical guide of how to deal with uncontrollable and chronic stresses. The discovery (D) component of ABCDE did not work as well in this case, because it would take a longer journey before she could fully discover the joy of recovery and fulfilling a new dream.

Working with Helen over 1 year, I have learned that the child's progress has its ups and downs, but the overall gains are obvious. Although it is absolutely essential to create a safe, accepting, and trusting environment and to earn the trust of the child with unconditional love and empathy, it is equally important to earn the parents' trust and try to connect with them to have a positive influence on them as well. The outcome of my journey with Helen was a positive and rewarding one. An unexpected reward is that Helen's parents are committed to working through their marital problems with a psychologist.

Epilogue

From our perspective, some themes within and across many of the stories in this section include the following.

- Children and youth suffering from depression often relate that they feel overwhelmed by feeling of sadness, frustration, and a sense of helplessness.
- Children and youth with depression often feel mentally exhausted.
- Children and youth with depression seem to be unable to get satisfaction for themselves from activities that most others would describe as fun or enjoyable.
- The social–emotional development of children and youth in schools is as important as the academic development of children and youth in schools.
- Children and youth with depression share many symptoms that are manifested in different ways.
- Some children who are depressed will be irritable, have behavioral outbursts, will withdraw from social situations, and express body discomfort and fatigue.

Specific thoughts professionals and parents expressed in the stories that we thought were insightful and have heard expressed in similar ways over the years include the following.

- Depressed youth seem to overly perseverate or are perhaps hyperfocused on thoughts or feelings that generate a negative outlook such as the feeling that they have little or nothing to contribute in any situation.
- Sometimes we feel frustrated and helpless in our attempts to influence the feelings of our child.
- There is a feeling of relief from getting understanding and help from professionals.
- Building a meaningful bond with my depressed child is very important, as is making sure my family stays connected to and loving with each other.

Part III

Developmental and Learning Disorders

5 Learning Disabilities

Prologue

Children and youth who have learning disabilities have average to above-average intelligence but have lifetime processing problems due to identifiable or inferred central nervous system dysfunction. These processing problems cause a pattern of uneven abilities and observable weaknesses particularly in reading, writing, and/or math. Typically, children with a reading disorder, math disorder, or disorder of written expression have abilities in these areas that are significantly below that expected given the individual's age, intelligence, and age-appropriate education. Their disabilities (learning problems) significantly interfere with their academic achievement and/or daily living activities that require reading, mathematical, and/or writing skills, and are considered not to be due to environmental disadvantage; visual, hearing, or motor handicaps; emotional disturbance; or physical and/or health impairments, although they might occur concurrently with one or more of these.

In this section are five stories: one personal story, one parental story, and three professional stories pertaining to learning disabilities.

The personal story, entitled "From Confusion to Understanding of Learning Disabilities and the Journey Along the Way," is written by a young woman who has a learning disability. In her story, she shares her experiences, thinking, and feelings as she reflects on her progress through school with a learning disability and on how she becomes increasingly aware of herself as a learner. The parental story, entitled "One Step Forward, Two Steps Back, and Another Step Forward," is written by a mother who shares her experiences parenting her child who has a learning disability. In her story, she talks to us about the confusion, frustration, tension, and anxiety, as well as the resolutions she experienced in raising her son and the steps she undertook to maximize his development. The first professional story, entitled "From Experiencing and Understanding to Being Understood," is written by a psychologist and is about his experiences in working with a 12-year-old boy who has a learning disability. He shares with us the importance of understanding the struggles of this child while finding the best ways to address them. The second professional story, entitled "My Best Year of Teaching," is written by a teacher who relates her experience of teaching within an inclusive classroom and in particular her experience in teaching a boy who has a learning disability. This story speaks to the effectiveness of utilizing numerous resources and teaching learning strategies to foster academic development, and reminds us how having individual differences in learning is not really being different. The last professional story in this section, entitled "Keeping an Open Mind and Identifying the Primary Problem," is written by a psychologist who reveals his experiences related to working with a teenage boy whose learning disability was masked by other problems, circumstances, and issues but that led to a better understanding of this boy and his needs, as well as to more appropriate treatment and support.

Exceptional Life Journeys. DOI: 10.1016/B978-0-12-385216-8.00005-9

Personal Story

From Confusion to Understanding of Learning Disabilities and the Journey Along the Way

"I have a learning disability." The first time I said the words, it was like a sigh of relief. Finally, there was a reason why I was not able to perform like my friends in school. Finally, there was a reason why spelling, writing, and reading just weren't progressing, and finally, there was a reason why so many things were so hard that seemed so easy to others. However, at the time, what I didn't know was how big the impact of having a learning disability would be. The sigh of relief would not last, as I would soon learn that have a learning disability was lifelong, and that I was always going to have challenges and obstacles to overcome.

What I also didn't know at the time was the perception some people held of individuals with learning disabilities. It was years after I was formally diagnosed that my parents shared with me that the educational psychologist had told them that they would have to "downgrade" their expectations. That I would be "lucky to finish high school and that postsecondary was out of the question." The belief that the psychologist held haunted me for years, leaving me never quite sure if I would be successful, always trying to do the best I could but always having that fear in the back of my mind that she might be right. Nonetheless, I am happy to report that I am currently in my final year as a B.A. Honors psychology student and looking forward to graduate studies.

The journey to where I am now really began for me back in first grade. I still remember the classroom so clearly. It was a typical day, nothing too significant or memorable. The class was reading a book as a group. The teacher was calling on students one at a time to read aloud, and once they read a paragraph, the teacher asked some foreshadowing and critical thinking questions and then called on another student and so on. Although this would not seem like a significant day to remember for most students, it was for me. I remember taking one look at the book in front of me, trying to read the first sentence, and then thinking: How is everyone reading this? What am I missing? I had no idea what the words on the page said. I was lost in a sea of language, and I couldn't figure out why. Then I heard my teacher calling a name and I thought, "PLEASE don't call MY name." What was I going to do if I had to read next? I had no idea what the words said, let alone what the story was about. If she came to me, I would be so embarrassed and would probably be picked on at recess by the other students for not being able to do what they could. I tried to disappear into my chair while staring at the clock, hoping that if I watched it, it would move faster. The whole time, I didn't understand why I was so different from everyone else. It felt like it was some horrible nightmare that I would wake up from at any moment, but this was very real and I couldn't pinch myself awake.

The challenges I was facing continued for a few years before I had a name for what I was experiencing. In that time, I became incredibly shy and introverted; my whole personality imploded on itself. I perfected the art of "looking busy" as the class size was so large, that those kids that acted out got the attention and quiet students like myself were assumed to be working away just fine. But everything wasn't "just fine." I was confused and worried. I didn't understand what was going on. Why was I so different from my peers? Would things always be this difficult? Would it ever get better?

Every week was like a rollercoaster of emotion for me. My heart would sink Monday morning as I would have to start the new week at school, and my emotions would get progressively better as Friday approached, knowing that the weekend was coming. Saturdays and Sundays were my favorite days as that meant no school, and I could not have been happier or in a better mood: no reading, no spelling, and no writing. This break from school was great, but before long, it would be Sunday night and my rollercoaster would soon begin again.

Even before I was diagnosed, I knew that something was wrong or different but I didn't know what it was. Teachers and professionals knew too, but they focused on the fact that I was in French immersion at the time and suggested that my challenges were the result of problems I was having learning the two languages at one time and that with some phonics training I would be fine. Little did everyone know that it was much more complicated than that, and the phonics training wasn't helping at all. I sank further into isolation and worry, not wanting to stand out as being different from my peers, or for others to find out that I was not able to achieve the basics that everyone else could. I just wanted to be like everyone else.

I would soon have answers for what I was experiencing. In fourth grade, I was finally diagnosed with a learning disability. I remember having the assessment done. It consisted of a number of tasks, or what I thought were games at the time, that I was asked to complete. Some of them I was particularly good at, like the block design task where you have x number of boxes and have to re-create the image, increasing the number of blocks as the task got more challenging, or the picture completion, where you have to figure out what is missing in the picture. Those tasks were easy for me, and I enjoyed trying to figure them out. On the other hand, the digit-span task was a nightmare. I did so poorly on this activity that I only did a few trials; I could not remember any of the digits after only a few had been added. The worst activity, however, was the word-reasoning task. Here, fake words were written on a page, and I had to try and pronounce them. Normally with words in class, I remembered them in isolation and not the individual sounds, as I didn't process the phonics involved. Now I had to try and sound out words I'd never seen. I knew that wasn't going to happen. I didn't know the sounds in real words, so fake words weren't going to be any easier, and I really struggled with this.

The assessment was able to point out the areas that I was struggling with and those I was good at. I had excelled in reasoning tasks such as the block rotations and picture completion. These strengths could assist me in school as it was suggested that I rely on picture and context clues to decode information. My grasp of abstract concepts and reasoning was above average, and those skills would assist me in many learning tasks. When it came to the areas that I was struggling with, there were no surprises. I had difficulties with sound–symbol associations and underlying automaticity in reading and spelling. By fourth grade, I had a very limited sight vocabulary and inefficient word-attack strategies, as I could not segment words or understand underlying phonics. When trying to sound out words, I would frequently miss the sequence, and add or omit sounds, further supporting the finding that I did not understand phonics or segmenting of words. Therefore, I had to learn words in isolation, which would be time-consuming, but necessary if I wanted to achieve academically. Additionally, it was noted that I had a very limited short-term memory, which was evident in tasks such as the digit-span task, but that I had excellent long-term memory. This would result in my reliance on one type of memory over the other when it came to remembering information, which would affect my learning, understanding, and studying skills in the future.

Although all of that makes sense to me now, at the time, I did not fully understand the importance of the label of "learning disability" and what it all meant. But I understood that I felt better because of it. I knew I was different already; I just didn't know how or why. I felt better after having the diagnosis because, without it, what was I left to infer? I didn't want to believe I was dumb or stupid. If I didn't know why I was having troubles, what other conclusions was I left with? In that moment, I truly felt better. I had a name for what was different about me and with that, I could learn ways to make up for the areas I was struggling with.

The only downside to the diagnosis was the recommendations of the school psychologist. Not only were my parents told to "downgrade their expectations," but I was also going to be immediately removed from my class and placed into an English-only classroom. At the time, it felt like a punishment. I didn't want to leave the friends I had there as I had been in the same classroom with some of these peers for 3 years. I didn't want to go somewhere new where I didn't know anyone, and these other students might tease me when they realized I couldn't read like them. My friends knew I was good at other things, but these new students wouldn't know me at all. Why couldn't I stay in French immersion? It was part of who I was. But I tried to stay positive, being told that this was the best option for me to be successful at school, so I put on a brave face and entered the new classroom with as much optimism as I could muster. Looking back now, I still don't think that this recommendation was necessary, and today, I don't think an educational psychologist would suggest the same.

The next few years were a time of trial and error. I was trying to find my footing and learning what skills I needed to be successful. Also at this time, I learned that not all teachers were created equal. Some were eager to help me achieve and others were not. My fourth-grade math teacher realized that I didn't understand what I was being asked to do on the word problems, and she tried reading the questions to me. When I tried to read the word problems, it became more of an English test than a math test. I didn't know what the question said, and thus I never actually got to the math problem. Through reading the questions to me, we learned that having the auditory as well as the visual input helped me to understand what was being asked.

On the other hand, my homeroom teacher in fifth grade, who was also my math teacher that year, ignored my requests for help. Being a very shy and introverted student at the time, I was uncomfortable asking for help in front of everyone. Therefore, a system was developed where if I needed assistance from him, I would cough and he would come over and answer my question. Well I would cough and cough in his class and nothing would happen; he ignored my need for help even though he had promised my mother that if I used our signal, he would come over to my desk and answer my question. I was left to struggle the whole year.

My parents realized early on that self-advocacy was going to be important for the future. They felt it was imperative for me to begin to develop these skills early. Therefore, they encouraged me to answer or ask one question a week in class. Although it didn't seem like a lot, it was to me. By this point, I was uncomfortable talking in class, and my doing so up until this point was extremely rare. But I did as they suggested. At the time, I hated it, and it would usually be Friday before I would put up my hand, but now, I realize the importance of self-advocacy and I am grateful that they pushed me to develop these skills early.

It wasn't until seventh grade that I really felt that everything was going to get better. This was the year I firmly believe that everything changed for me, and was the turning point in my life. I started at a new school that year, and everything seemed to be just like all the other years. Then, a couple of days into the semester, I was called down to the Learning Center. Not too sure what that was, I proceeded down with apprehension. I learned that this was a place I could go whenever I needed assistance, and one of the resource teachers would help me. This seemed like an amazing idea, and I wondered why none of the other schools I had been to had such a place. Then the resource room teacher asked me what I wanted to accomplish that year. I thought about it for a moment or two and then said, "Well, my sister is on the honor roll, so I want to be on the honor roll too." Her response to me was perfect. She said, "We will work together and see what we can do." The reason this response was perfect in my mind was that she didn't say that, of course, that was possible, and then if it didn't happen, I would be crushed with false hope. She also didn't say that would never happen, even though at the time, I was a C or D student, and the idea of my making the honor roll probably didn't seem very likely. But she promised me that she would help me try to achieve that goal, and the honesty in that statement was the best response I could have received.

That year, I came to understand a lot about myself as a learner. I learned that when it came to exams, I needed the questions read to me, to help me understand the meaning of the words. The written words were not enough if I was to answer the questions properly. I also learned that I needed extra time to allow myself to deconstruct the information presented to understand what was being asked in the question and to pull the necessary information from my long-term memory to answer it. That additional time was essential to be able to express what I knew. I also understood by this point that spelling was never going to be my strength, and when possible, I would need to either have access to a computer or not be penalized for spelling (such as in instances where spelling was not the main concern of the assignment). Year after year had gone by where phonics had been taught to me, and now professionals recognized that this remediation was not going to work. My poor word-attack strategies and sight vocabulary were going to be a lifelong challenge for me. That year, I began to develop skills and strategies to become a successful learner, and through my hard work and determination, I was able to make the honor roll. It was a great feeling. I felt like I had climbed the highest mountain. Every school night, I would stay up late to study, and what took some students only a half hour to complete would take me hours, but I persisted with it. I wanted to be successful. I wanted to be able to achieve just as much as my peers. I didn't want to be limited by my learning disability.

The significance of this year was not found in this moment; it was found in the second term. I didn't make the honor roll this time around; I was off by just over 1%. This was when I was faced with the turning point in my life. I had two options: I could accept that I had achieved the honor roll that one time, and that it was a fluke, and be happy with that single accomplishment, or I could keep trying to be the successful student I wanted to be. And with my accommodations and supports, I was able to achieve the honor roll again and again for the remaining two terms that year. This gave me hope that things could and would get better, that I wasn't destined to struggle at school, that I could be successful.

The only downside of seventh grade was my math teacher. Although others would allow me to go down to the Learning Center whenever I needed extra help with an assignment or have the questions read to me for a test, he would not. After being questioned about why he had such apprehension about my visits to the Learning Center, his only explanation was that I did better there, and therefore I must be cheating. There were so many things wrong with that statement. It was heartbreaking to hear that he didn't believe that I could be successful in his class, considering all the extra effort I was putting into his class to understand the concepts and get the work done on time. He believed I would cheat, which I would have never considered as I wanted to learn and wanted to be successful at school. Furthermore, the ability to cheat was highly reduced in the Learning Center, as students from all grades and courses would work down there and often no one was working on the same assignment. In the Learning Center, I was given more one-on-one instruction, which meant more eyes on what I was doing, reducing the ability to cheat even more. It was insulting that he believed that was the reason I was doing so well. In my mind, he clearly did not understand what a learning disability was, and because of that, he couldn't understand how I could do better with some modification to how I was learning the material. Once again, all teachers were not created equal.

A problem I would soon face was that there was no consistency between schools in terms of supports and that there was a significant variety in teachers' attitudes toward special learners. The following year, we moved to a new house and therefore had yet another new school. Here, there was no Learning Center and only one resource teacher that was going to be there half of the day. I knew immediately that this year was not off to a good start. I needed the support of a Learning Center to help me develop the skills I needed to be successful, and to provide me with appropriate accommodations when necessary. I felt very alone this year. What made it even worse was my homeroom teacher. She was one of the worst teachers I ever encountered. Her signature lines she used when I didn't know the answer—"I'll come back to you" and "Just think harder"—were always like nails on a chalkboard when she said them to me.

First, "I'll come back to you." For what purpose? If I didn't know the answer then, my worrying about the answer for the next 5–10 min wasn't going to make the situation any better. If anything, having me worry about figuring out an answer was distracting me from following along with the lesson at hand. The only redeeming factor about that statement was that my friends in the class got wise to it, and when she had moved on, they would sometimes give me the answer so that I wouldn't have to sit there in panic for the next little while. And that is when she developed her second signature line: "Just think harder." Now instead of coming back to me so that one of my friends could help me when I didn't understand what was being asked, she made me sit there with all eyes on me while I "thought harder." I didn't even know what that meant. Was it really possible to think harder? She was one of the worst teachers, and just confirmed in my head what I had learned in previous years, which was that all teachers are not created equal.

I was lucky that I was only there for 1 year, as I don't think I could have survived the unsupportive nature of the school for long. I had to work so hard on my own to try and maintain good marks, and there was little support at school to help me achieve my best. With what I had learned the previous year, I knew that I had some skills I could use, but they were not developed enough to support my independent efforts in the classroom (and language arts was a nightmare every week). I began hating school again and wishing for Saturdays. My parents tried to help me as best they could, but there was only so much they could do when working with such an unsupportive and uncooperative school.

The high school the next year had to be better, and lucky for me, it was. Here, there was a Learning Center and very supportive resource room teachers. When I started there in ninth grade, a resource teacher went through my IPP with me, and I was able to provide input as to what I was going to work on that year. Furthermore, through the Learning Center, I was able to take a special course called "Learning Strategies," where I was able to apply different tools to accomplishing my homework tasks. This course was perfect for two reasons: The first was that it reduced the course load for me as I was already beginning to feel the pressure of all the additional time that was required to complete the same assignments as my peers. When tasks were given for homework, my classmates usually took a half hour to an hour to complete them. For me, it took so much more than that, and some nights I couldn't even finish all the assignments given. I needed to reduce my course load so that I could accomplish all the tasks. This course allowed me to work through my assignments during these hours, which allowed me to complete all my work on time.

The second reason this course was perfect was that it allowed me to try out different learning tools, depending on the task I had to complete. When it came to essays, for instance, we would try creating web diagrams, flow charts, and so on, to help me organize the information. When I had to take notes from the textbook, we tried color-coding, webs, chunking, and so on, to help me better understand and remember the information, as well as to see what worked and what didn't. So, when I didn't have the Learning Center, I could use these strategies on my own. This was extremely helpful and gave me the tools I needed to complete assignments and essays in ways that worked with my abilities. This course was perfect in helping me develop skills and expertise in these areas, and important for my future schooling. I realized from previous years that there was not always going to be a Learning Center, so when it was available, I had to take full advantage and develop what I would need.

Another great thing about the Learning Strategies course was that it was undetectable to other students. When I had this course, it was at the same time as a number of other classes and no one realized where other students went. I didn't have to worry about being pulled out of class for all my peers to see. I was able to go down during the break between classes, and my friends and peers never questioned it. They just assumed that I was in a different course during that time block. It was nice not having to self-disclose to them. It wasn't that I didn't want to tell my friends; it was just easier not to. I had learned over the years that the general population is not very familiar with what learning disabilities are, and often times people begin to look at you funny. The big problem I find is that having a learning disability is hidden. It's not like a blind student who has a cane or a seeing-eye dog, which are indications this person has visual impairments, or a deaf person who has hearing aids. There is no defining feature that would allow someone to deduce from my appearance that I have a learning disability. Therefore, when people don't understand what a learning disability is, they often question it or look at you funny while trying to find out what it is all about.

The great thing about being involved in the IPP was that I understood what was in there. Not only that, but once the IPP was finished, it was my job to take it to all of my teachers and have them read and sign it. This was an important step in developing self-advocacy skills, and it helped the teachers be able to identify who in their class had special needs and would need what accommodations. Having to take my IPP from teacher to teacher and year after year made me more confident in talking about my learning disability and advocating for myself. It was a great extension to the first self-advocacy skills I learned back in earlier grades.

These years I learned a lot about myself and strategies that would help me in the future. I learned how to balance my schedule. In high school, I did this by taking English and social studies in separate semesters to reduce the amount of reading and writing I would be required to do in a semester. Also, I learned to take math and science in alternate semesters so that I didn't have three core subjects at once, and this better balanced the workload. I also learned to select my instructors. For instance, if there were two teachers instructing the same course, I would talk to other students and learn as best I could about their teaching styles to see which one was a better fit for me. Furthermore, I was constantly learning skills and strategies through the Learning Center to help me with my studies. These years helped me develop confidence as a learner and believe that I could achieve success academically. By the end of twelfth grade, I had proven that the educational psychologist from back in fourth grade was wrong as I completed high school while being on the honor roll every semester and by being accepted into university the following year.

Throughout university, I learned that there were always going to be new challenges and obstacles to face. But after 4 years of developing the strategies I needed in high school, I felt confident moving forward. There were always going to be professors and peers that didn't understand what a learning disability was or what I would need to be successful, and I learned to take everything one day at a time. Through the years, I have continued to develop my self-advocacy skills and they have served me well and I am grateful that I was pushed so hard to develop them.

Although I know that school will take me longer to complete, I am determined to be successful at it.

In the end, there are a number of important things I have learned. The first is how grateful I am that I was labeled as a student with a learning disability. Without such a distinction, doors would not have been opened for me to receive the accommodations and proper supports I needed. I have never understood why people don't want to be identified as having a learning disability or when parents don't want to tell their children about such a diagnosis. Not having the label doesn't mean it is going away. No one can help you in the areas you are struggling with if they don't understand why you are struggling. The label opens doors to accommodations and supports that one needs to have to be successful. Why wouldn't you want that?

I learned that various people in the community are not as supportive and understanding as to what I needed to be successful, and I hope for an increase in community awareness and knowledge with respect to these things, for both myself and others in the future. Awareness about learning disabilities is difficult to promote because there are different types that affect people differently. Not only that, but learning disabilities are not typically visible to people, which makes it difficult for them to understand and appreciate what it entails for me. At times, I wish I didn't have a learning disability when things get so hard, but I realize that it has helped me develop into the person I am today and therefore I am grateful for it and everything I have learned because of it.

I have also learned that "fair" is not that everyone receives the same but that everyone receives what he or she needs to be successful. So many times I would hear that it wasn't fair that I got extra time or that I was able to use a computer to write my exams. To those individuals, I would ask if it is fair that I struggle so much with reading, writing, and math, that I can't rely on my short-term memory, or that I can't sound out words I have never seen? These things do not seem fair to me but it is what it is and I can't change that. I can however, receive the accommodations I need to try to put me at a level playing field compared to others, and that sounds fair to me.

I learned that not all teachers are created equal, and I hope that teacher training provides more of a focus on students with special needs in the classroom. There should not be such a difference among teachers with respect to their level of understanding or willingness to help students with learning disabilities. Supports for students should be uniform across schools.

I also learned that having a learning disability is only part of who I am. Yes, I will struggle in some areas, but these challenges are only a part of what makes me the person I am. I also have many gifts and talents. I have learned to be a confident person with a learning disability, and how to cope with my challenges. In this regard, having a learning disability has taught me how to be resilient and overcome many of my obstacles.

Looking to the future, I am now in the final year of my undergraduate degree in honors psychology. I hope to go to graduate school in the area of school psychology. In this field, I hope to help other students who struggle like I did and assist them in their journey. Through my work in this area, I hope I can influence people's understanding as to what a learning disability is and how individuals can help those with learning disabilities reach their full potential.

Parental Story

One Step Forward, Two Steps Back, and Another Step Forward

I am the mother of a son with a learning disability. To me, the word *disability* conjures up negative connotations immediately. It was so hard for some people to understand that he was very capable of learning; he just needed to learn differently. He was born in late October, into a loving, well-adjusted family consisting of a mom, dad, and two sisters. Our oldest daughter did extremely well in school and was always at the top of her class. My younger daughter struggled a little with reading and math but overall did very well in school too. Our son, though, was always a little different. Long before he entered the school system, I knew he showed signs of "something different," but I just didn't know what that was. He resisted doing anything new or unusual because he was unsure about how to do it. He stayed within his comfort zone of only doing things that he had mastered. He desperately wanted to play Nintendo with his sisters but absolutely refused to try. He sat and watched intently and finally, one day, he picked up the controller and played masterfully. I knew he was bright and that there was nothing he couldn't do age-appropriately, but there was always a nagging feeling that he needed more help than either of my girls did to make sense of the world.

He entered the public school system early, being an October baby, at the age of 4, almost 5. He didn't take well to the everyday business of going to school. He knew he had to, so he went, but he really didn't want to. He displayed extreme separation anxiety and became fearful of some of the most normal occurrences that happen in schools. For example, he refused to use the washroom at school in case an alarm would go off while he was in there, or whenever the alarm did sound, he would run into the coatroom and plug his ears. He never did these things at home because he felt safe here. I think that he always depended on me to make his world understandable; I was like his interpreter, his liaison to the world. When I was not with him, he panicked. Fortunately, he grew to adore his kindergarten teacher and after several months, he settled in to the routine of daily school. He was always a very polite and sociable child and had no problem making friends and following rules. When it came to learning his letters and numbers, his teacher and I could see that he became uneasy and unsure of himself, which in turn increased his anxiety again. This was the first time that I had any inkling that he may have some trouble learning. Because he was still very young and only in kindergarten, I didn't give it too much thought but I spoke regularly with his teacher about how to help him overcome his fears and anxiety. He had two kindergarten teachers: one was a kind, smiley, gentle person, whereas the other was brusquer and didn't seem to care that my son was struggling to try to keep up with the rest of his class. He dealt so much better with the gentle teacher who seemed to understand him and make him feel safe. He hid behind the most charming smile, everybody commented on it, but it was a facade protecting him from others getting close enough to notice that he couldn't do many things. He did nothing to draw attention to himself.

It was in first grade that my fear for my son really began to take form. His kindergarten teacher had made sure that he was with the most gentle first-grade teacher, who was a lovely and patient young woman. He still cried every morning because he didn't want to go to school and again when he returned home for lunch and had to go back for the afternoon. In retrospect, I think that the whole idea of learning was confusing and terrifying for him. Even at that young age, he could see that the other children could understand what was going on in class far better than him. His teacher and I tried in vain to teach him his numbers from 1 to 100, but even with a lot of extra work at home, he just didn't seem to be able to retain the information and he didn't seem to see the logic in the patterns. I have to say that, at this point, I felt very afraid for him, but I didn't know what to do except carry on trying to teach him at home myself so that he could catch up to the rest of the children in class who seemed to be sucking up knowledge like sponges.

His learning progress was slow but he was progressing fairly well because of his good attitude. He never stopped trying to please both his parents and his teachers. I'm sure he could sense that I was concerned for him, so he tried and tried to work hard for me. He accomplished enough to be sent on to the second grade. I regretted starting him in school at such a young age but felt that holding him back, after all his hard work and effort, would cause more harm than good. He had made friends and, by the end of first grade, he was more comfortable in class. His reading and math skills were still weak and he was definitely below grade level, but I still clung to the hope that he would have a mental growth spurt and catch up to the other students. He could articulate very well.

Second grade continued much the same as first grade. It wasn't that he couldn't learn; he just needed more time and a different way of being taught. He actually worked harder that most of his friends, but didn't get the results that they did. The thought of pushing him on to another grade and falling even further behind began to worry us. I was trying so hard to help him at home, and because no one had any good advice to offer me, we thought about holding him back in the same grade. I mentioned this to his teacher and I received a letter from the principal of the school in March stating that he would be held in second grade for another year. I had only put that suggestion out there looking for some help and advice, and now they were telling me he should be kept back. Upon reading this letter, the reality of the situation slapped me in the face and I decided that he needed a more appropriate school setting. I did not really believe that keeping him back a year would solve our problems. It was obvious that this school could only do so much. I was not offered any suggestions from either the resource teacher or school administration as to where I might look for such a setting. I remember spending hours on the phone trying to find someone who could help me. I felt incredibly alone and feared for my son's future in a world that placed academic learning at such utmost importance. He had so many strengths and I knew he was intelligent. While I was searching for a new school for him, I continued to work hard with him and with other professionals that I conferred with.

I brought an occupational therapist into the picture at this time to assess his visual perception skills and motor skills. I was trying to eliminate all possibilities that he might have some physical reason for his learning difference. I also took him to get his eyes checked and his hearing. I was actually praying that someone would find something that could be fixed easily with glasses or whatever. No such luck. His eyesight and hearing were good. He was far above average when it came to hand–eye coordination; his fine motor skills were great, large motor top-notch. The next method I tried was to bring a tutor, who had received her training working with children with learning disabilities, into his classroom. We received permission from the principal to allow her to come in to his school and work with him there. He received extra tutoring during school hours for the last 2 months of his second-grade year. This was hard on him because the only class he could afford to take time out of was physical education, which he excelled in and truly felt confident in what he was doing. She tested him to see if his eyes were tracking properly when he read and taught him to watch the mouth movements of each vowel. He worked so hard and put up with so much. All the other students knew that he was "different," but he never seemed to get hassled about it.

One of my many phone inquiries led me to the public schools only special setting school for children with learning disabilities. By this time, we had seen a psychologist and had an assessment done on him. He was officially coded LD. To attend this school, he would have to leave his home school and attend a completely different school for 1 or 2 years, leaving behind the familiar surroundings and all his friends and acquaintances. The decision was tough, but we knew he had to get some additional help. It seemed like a good place to start.

Starting up in a brand new school, taking the bus, and eating lunch in a loud, crowded lunchroom instead of a quiet comfortable home was one of the first things he had to overcome. The school was old and not modern like his community school had been. He was placed on an IPP and had a low student–teacher ratio. He progressed steadily, and I finally took a break from the heavy responsibility that I had been carrying for 3 years. It felt like heaven knowing that he was enveloped in a school where he would get the attention he needed and he would learn some strategies to help him cope with the rest of his schooling. He was no longer swimming upstream all the time and so he was more able to step out of his shell and risk trying some new things, such as acting on stage and doing a comedy act on his own.

All good things must come to an end, it seems, and after his second year there, he needed to transition back to his home school. There isn't enough funding for children with learning disabilities, and although they would benefit greatly from staying in a school such as this, he was only afforded a maximum of 2 years. In comparison, when children are considered gifted, they are moved to a more fulfilling environment and allowed the privilege of continuing there until twelfth grade, regardless of how many years that may be. I wonder about that one. What else are we to think but that those children with learning differences are not as important? He was well accepted back at his home school and settled in quite nicely with his old friends. He was older now and had built up some confidence in himself from his years at the other school. He was in fifth grade now, and it wasn't long before some of the curriculum he had missed at the special setting school, due to their heavy concentration on reading and writing, began to show up as failures. I had even taken him to a math tutor over the preceding summer to help him catch up. He didn't have the background knowledge of the other students in social studies, math, and science, and health class was something he had never experienced. Suddenly, courses that would have been difficult for him had he been taught them year by year were even harder for him now. One step forward and two steps back.

He enjoyed being back at his home school and being able to just walk and come home for lunch, but by now, his sisters had both moved into high school and I was working and couldn't be home at lunchtime. At school, his teachers tried as best they could to continue to support him. He worked his way through fifth and sixth grades diligently, with many nights of tiring homework help, a few tantrums, and immense frustration on both him and me. Again, both of us were back to our nightly stressful rituals. By the end of sixth grade, when students are invited to their new junior high school, I got the opportunity to speak with the resource teacher who would be responsible for all the students with learning disabilities in the school, those new to the school, and those already attending. That number was 90; 90 kids all on IPPs! She told me that she was totally overwhelmed and just barely able to create 90 IPPs, let alone offer the help that any of these kids needed. She said the resource room was crowded, noisy, and that sometimes there would be dozens of students from all three grades there. I mentioned that I had been throwing around the possibility of private schools, and she unabashedly told me that she highly recommended it if we could find any way to afford it. I left the interview feeling depressed and frustrated. It seemed to me that, just when we got settled somewhere and things were beginning to run smoothly, it got totally messed up again. We couldn't afford it but we felt that we absolutely had to place him somewhere he would not just be overlooked. The financial strain added even more anxiety to everyone in the family. I felt guilty for taking so much of the family budget to pay for an expensive private school but I knew if I left him in the public junior high school, he would fall further and further behind. It was obvious that they did not have the manpower to offer a good education to these children with learning disabilities who were attending their school. I felt like taking on the school board at this point. I sat down and composed a long, poignant letter, pouring out my heart and my great disappointment with our situation. I truly felt that our education system owed all children a good publicly funded education, especially those with special needs.

I never sent the letter because I knew that their answer would be to put our son in an LD III division, stuffed to the brim with a hundred or more students, all with varying issues, including bad attitudes and behavioral problems. I had worked so hard to help him keep up his confidence that I just couldn't risk throwing it all away now. My husband found a way to put aside all my paychecks for his $10,000 tuition plus $800 a year for the busing fee. Our oldest daughter's first-year tuition at university, which happened to be the same year our son started at private school, was nowhere near that amount. Fortunately, she had managed to obtain many scholarships to pay her way. That's the only way we could manage it.

My son suffered very much the first few months of school. There were new rules and regulations that he had not been subjected to before. He was very alone and out of sync with the school. He needed to ride a bus everyday for at least an hour and a half. When I think back on how awful those beginning weeks were for him, I still feel like crying. He had already gone through so much to obtain an education, and he was only halfway through. The school had two streams of learners, one comprehensive/remedial and one regular stream. When I first heard of this private school, I incorrectly believed that this was a school for special learners only. The school was originally founded as a school created to provide education for children not benefiting from the conventional system because of their learning difficulties. When we were being shown around the school at the open house, one of the students leading our group referred to the kids with learning disabilities as "those who just don't get it." My heart just fell. What were we getting into now and at such a financial cost? I approached one of the deans with my concerns, and he tried to put my mind at ease but there was always the nagging thought in the back of my head that our son would be treated by some of the students as "those students, the ones who just don't get it." In the 3 years he was there, he definitely felt the segregation between the two tiers of students. He remained in private school for his 3 years of junior high. We didn't want to disrupt him anymore than we had to, and he was experiencing some success (not nearly as much as we expected, but he did learn more strategies). His math marks were way below grade level when he left ninth grade. I knew that once he reached high school, he would be able to choose a course of studies more appropriate for him, classes that he could do well in and feel a sense of accomplishment, and that this would ensure he received his diploma. So it was back to the public system.

Coming from a small private school did not prepare him for the large, noisy crowds of students in a regular high school but, to my surprise, after the first few weeks, he felt comfortable and settled. He was assigned a resource teacher to report to and was given credit for attending. She worked with him on subjects that he needed more help in, allowed time for homework and help if needed, and prepared him for tests. He was allowed more test-writing time and was able to take all his exams in the resource room where he was more at ease. He finally attained a success level he had never experienced in any of the schools before. His high school years were the best 3 years of his whole educational experience. He was able to get resource help all six semesters of high school. Most students coded with learning disabilities just got one semester per year of resource, but our son got it everyday for the whole 3 years.

At no time had I ever held his teachers responsible for his difficulties. I was never demanding or rude and, in return, my son and I received nothing but help and kindness from every teacher he ever had. I think that because he was never a discipline problem and always worked hard with a good attitude, he found his way through the maze. With his newfound confidence, he became a most enjoyable, humorous student who got on very well with all of his teachers. He passed twelfth grade with more than enough credits and enough courses to get his diploma. It was a very happy day, absolutely the high point of this whole journey for me. But he didn't quit there. He had only been working for a half year when he realized he needed more education. To get into college, he needed to upgrade his math and English. He set out to do just that. He attended night school while he was working and attended summer school to get the rest. He had to repeat Math 20 two times before he could move on to Math 30, but he persisted. He also needed to upgrade English 30-2 to English 30-1. He passed the diploma tests, and this was no small feat. Two years after graduating twelfth grade, he applied to college and was accepted. The colleges and universities have departments set up to help students who have various learning issues, and I found that very impressive. He has been in university for 2 years now and is only able to handle a workload of three courses a semester. If he wants to work part-time, he can only handle two courses. This makes getting a degree a long process. Our son has opted for a 2-year diploma and will work toward a degree in social work from there. At least he has options. I know that he will be excellent at working with troubled children and teens; his own life's struggle has made him a caring and sensitive young man. He will never outgrow his learning disability, and he will always have to work very hard for what he wants academically. I know he will make it.

Professional Stories

From Experiencing and Understanding to Being Understood

I first met Brandon in my office when he was 12 years old. He was just finishing the first half of sixth grade "in unspectacular fashion," as his father put it. I was working as a registered psychologist in private practice in a large urban setting. I worked with many students like Brandon in the past. His parents had called with a story I had heard many times before: Their son appeared to be a bright kid who could hold a good conversation with his friends and with adults, but when it came to school work, he struggled to perform at even the lowest end of what was determined to be the "acceptable" level. In fact, by sixth grade, Brandon was technically "failing" at school. However, his school's policy was not to retain students, but to find "appropriate placements" that would "best serve the student's needs." To date, no such placements had been identified for Brandon, as he did not "fit" the criteria necessary to access such services. Specifically, he had not been diagnosed with a disorder significant enough to warrant placement in such a program. As such, by February of his sixth-grade year, he was facing the probability that he would be entering a regular junior high program with little to no support. His parents, both professionals successful in their fields, were understandably worried about what would happen to their son. They were concerned specifically about his potential to become "a behavior problem" should he continue to struggle with meeting success at school. His parents wanted to know if I would be able to help. I hoped that my experience would be an asset in helping their son, but was cautious. "Expect the best, prepare for the worst" remains my cynical optimist creed.

My own experience with learning disabilities extends back to my own educational experiences. I have often noted to clients that if I, as a third-grade student, walked into my own office, I would likely have diagnosed myself with a learning disability in the area of written expression. Specifically, I struggled with getting my ideas on paper, not only due to fine motor control issues (which continue to remain problematic for me to this day), but also with organizational skills. I always felt that I had good ideas, but what appeared on the page was so far below what I felt my ideas deserved, that I had the tendency to either rush through work so I could get the ideas out as quickly as possible, or do no work at all. In fact, by third grade, I was one of "those" students who got pulled out of physical education and other "fun" classes to work with the resource teacher on my handwriting skills. As one can imagine, the idea of trading in a game of parachute in the gym for a 40-min appointment with the school's resource teacher in which I would write and rewrite sentences over and over was not particularly appealing. I wonder if the resource teacher ever wondered why I always became so thirsty at the start of our sessions (as I was forever asking to go get a drink of water), and then had to leave early to go to the washroom toward the end of each session. Anything to avoid more lines of repetitive, boring, and tedious work that, even as a 7-year-old, I could recognize as not being effective. I continued to experience challenges with handwriting through my elementary school years, but became much more effective at dodging the resource teacher. Although I'm sure she was a wonderful person, my relationship with her was one that could best be described as being mutually nonempathetic. We engaged in a ritualized and, in hindsight, quite ridiculous game of cat and mouse each time we were to have a session. Ten minutes before physical education was to start (which was also my appointment time for what was euphemistically called "Writing Sciences"), I would ask my teacher if I could use the washroom. My teachers, perhaps being empathetic to my reluctance to go to the sessions, would let me go. Twenty minutes later, I could be found wandering through the stacks of books in the library, most frequently finding a comfortable corner in some of the deep stacks toward the back of the library, reading a book and quite enjoying myself.

I continued to experience difficulties with writing through my junior high school years, and in seventh grade, much to my horror, I was again placed in a writing remediation group with two other students. One of my fellow students and I had negotiated an escape based on promises to both sets of our parents to improve our writing greatly without further assistance, or pledging to return to the resource group. A combination of charm, dedication, and moderate focus on handwriting allowed us never to return to Writing Sciences. In fact, we became very close friends and remain in contact to this day, occasionally looking back with a sense of wonder at our escape.

I believe that these experiences, in no small way, contributed ultimately to my decision to work with students with learning disabilities. Since the start of my career in education, I have worked as a teacher, a guidance counselor, an associate principal, and as a psychologist. Although my day-to-day work is characterized in my role as a psychologist, I consider myself an educator above all else. And through these experiences, I have felt prepared to work with students such as Brandon.

Brandon entered my office a defeated boy. In hopes of limiting any potential anxiety, his parents had not told him about our appointment or about the rationale for our meeting until they picked him up from school that morning. He was not happy about this decision, saying later that he felt that "only weirdos need psychologists." However, upon entering my office and seeing an array of toys, most specifically a Rubik's Cube, he seemed to relax somewhat. In interviewing his parents, it became clear that Brandon's issues with reading and writing were not new and had been simply accelerating over the past 4 years. He had not identified himself as an individual who enjoyed reading, and would read only if absolutely necessary, primarily for school or to appease his parents' wishes. When asked what kinds of materials he enjoyed reading, Brandon replied flatly, "I don't like to read." I picked up a copy of a *Calvin and Hobbes* anthology from the table next to me. "Have you read any *Calvin and Hobbes*?" I asked.

Brandon rolled his eyes: "That's just a comic book. ... That's not reading."

"Okay," I said. "It is a comic book, but it's still reading. It's just not a novel. There are all sorts of things that involve reading that are not novels or textbooks."

Brandon seemed to warm up to this idea, but was still cautious and protective. His parents again told a familiar story. A child who was born following a healthy pregnancy and delivery, a child who met his developmental milestones at appropriate ages, a child who was active and engaged and enjoyed the company of others and playing with friends. Nothing could be more normal, more typical.

Preschool and kindergarten were "fun" for Brandon. He enjoyed the activities, the games, and meeting new friends. He appeared to be a bit slow in picking up basic reading skills in first grade, and struggled with basics of writing, including writing his own name. His teachers, although concerned, were simultaneously cheerfully dismissive, indicating that his challenges did not appear to be developmentally significant and that he would likely "get caught up" as he grew in maturity.

Second grade presented further issues for Brandon. His reading and writing skills developed very slowly, particularly compared to his classmates. He was assigned extra worksheets to complete at home to enhance his written performance, and his parents were provided with a list of texts that might engage him in the reading process more effectively. Moderate success with the strategies was met and moderate success in relation to reading and writing was noted; however, significant problems related to frustration and anxiety were developing. He started to have stomachaches in the morning before school (although he was rarely sick on holidays or weekends). For the first time in his life, he became argumentative with his parents, typically right after supper, which was when the family worked on doing homework. The experiences of frustration, anxiety, and somatic complaints continued through his later elementary school years, culminating in sixth grade.

Throughout this time, Brandon continued to be a polite, respectful, and well-liked student at school. His teachers, according to his parents, tended to view him as a hardworking but struggling student. However, his struggles were seen as being less complex than those of many other students in his classes and, as such, resource assistance and extra help were generally limited to occasional support from his homeroom teacher or from a literacy specialist, who was at the school 2 days per week and had a caseload that rarely permitted her to work with Brandon.

By February of his sixth-grade year, Brandon's struggles were becoming more and more noticeable, not only to his teachers and his parents, but also to his classmates, who had taken to avoiding working with Brandon in group situations as they feared that his challenges would result in an overall lower grade on their work. Brandon became more sullen at home and withdrawn at school. His midterm report card, for the first time, did not contain the phrase "hardworking," and had been replaced with "shut down."

Brandon's parents were heartbroken and fearful. They met with his homeroom teacher, who had noted that Brandon's struggles with reading and writing were not being alleviated and were, in fact, becoming more pronounced. She clearly indicated that she wished that there was more that she could do for him, but that other students in her class had more severe special needs than Brandon did, and that is where she needed to place her resources.

An Internet search conducted by Brandon's parents provided them with some understanding of what may have been happening with their child. Descriptions of children who were bright, articulate, outgoing, and enthusiastic learners, but who struggled with basics of reading and writing were not unusual. However, such challenges appeared to be most common among a group of students who were identified as having "learning disabilities." Based on their Internet research, Brandon's parents felt that further investigation into his challenges and learning profile was warranted.

My work with Brandon began with very basically breaking the ice. He was initially reluctant to engage in the assessment process, until it was explained to him what was required. Specifically, he was quite pleased with the fact that conducting the assessment would mean that he would miss school. By the end of the first session, he asked, "How many times can I come back?"

We began our second session, following a first session in which a comprehensive family and developmental history had been obtained, with a standardized measure of intellectual functioning. Brandon quite enjoyed most of the activities throughout the session, as they did not seem to be very "school-like." Instead, he approached many of the activities as he would to games. He became more and more "chatty" throughout the testing sessions, ultimately to the point in which we had to create a "code word" so that we would not get too far off track as we chatted about hockey, movies, and skiing. Throughout the testing sessions, which occurred over three separate occasions, each lasting approximately 60–80 min, Brandon was generally enthusiastic and positive. He even enjoyed math-based subtests, commenting that math was his favorite subject at school "because you could always know when you have the right answer." His enthusiasm faltered, however, when engaging in tasks requiring reading or writing skills. His oral reading of passages was, for lack of a better phrase, agonizing for both of us. This chatty and articulate boy stuttered, stammered, and muttered to himself as he struggled to read a passage; sitting across the table from Brandon, my stomach turned thinking of his experiences in the classroom in which he had to read aloud. I asked Brandon how he felt about reading aloud at school. "Well, my teacher doesn't ask me to do it very much because she knows I get embarrassed. ... But sometimes she forgets. I hate it. I feel stupid, and I know the other kids don't mean to, but they laugh at me."

Upon completing our assessment, reviewing the data from all of the tests that we had used, along with information from Brandon's report cards and interviews with his parents and teachers, it was determined that he met the criteria as a student with a learning disability, with specific challenges in reading–decoding skills and written expression. However, we also found that Brandon, as measured on a standardized intellectual measure, was in the high average range of verbal skills. I sat down to meet with his parents to review these results. "I knew it!" his father explained, excitedly looking at his wife, beaming with pride. Brandon's father was a man with a tremendous sense of humor, and had a strong and positive relationship with both his son and with his wife. He had always tried to be optimistic about his son's academic future, but had been worn down after years of struggles, frustrations, and well-intentioned but meaningless platitudes referring to his son's "hard work" at school. He was of the belief that hard work, although valuable, was not as viable as "smart work." He wanted his son to get the best benefit for all of his efforts, and was concerned that his son's efforts were beginning to diminish due to ongoing frustration. We reviewed the assessment results, and then moved on to all the recommendations arising from the assessment data. We started by looking at the assets at our disposal in developing Brandon's reading efficacy and writing performance, along with some of the barriers that may have proven to be inhibitory factors. Among our assets were a desire on Brandon's part to develop his skills effectively; committed parents; a teacher who, although overworked, appeared to have Brandon's best interests in mind; and time. Inhibitory factors included limited resources at school, frustrations on Brandon's behalf (along with an emergent sense of learned helplessness in relation, ironically, to learning), and again, time. We felt that the window was closing on our range of opportunities to develop effective strategies, implement them, and alter them as necessary, with Brandon approaching junior high school. Obviously, we felt that a collaborative approach would work best, starting with Brandon himself.

I met with Brandon to review his results in somewhat general terms. His initial response, interestingly, was a negative echo of his father's. "I don't believe it," he muttered, looking down at the table. After chatting for a while about the nature of learning disabilities, particularly the fact that the diagnosis generally requires the individual to have average to above-average intellectual functioning, Brandon slowly started to identify specific areas on the tests where he felt confident in which he had performed strongly. Ultimately, he was enthusiastic to learn that he wasn't "stupid," and that, in fact, having a learning disability meant (in his mind) that he technically *couldn't* be stupid, as the definition indicated that he had to have at least average to above-average intelligence. It was an interesting meeting—I saw a student mature in front of my eyes.

In consultation with Brandon's teacher, we developed a strategy plan that, in the jurisdiction in which Brandon went to school, involved him being identified formally through the school system as a student with a specific special need. An IPP was developed with his teacher, his resource teacher (upon whose radar Brandon now appeared), Brandon's parents, and myself. Later, we involved Brandon in the process to fine-tune the plan. We identified three specific areas to focus upon: finding genuine and interesting texts in which Brandon would be interested; using a brief and focused phonics approach to reading these texts that could be implemented at the school and pursued at home; and using technology to assist Brandon with both reading and writing skills. This latter goal involved the use of basic technology available to him both at home and at school (including word processing), along with somewhat more advanced technology available to him primarily through the school due to financial constraints at home (including voice recognition software to enhance his written output). My own experience guided this latter recommendation. I had undergone 4 years of intermittent but frequent remediation on writing skills with limited results. Once I discovered word-processing software and, eventually, voice recognition software, my difficulties with handwriting were largely alleviated. (In fact, I should note that this entire story has been written using voice recognition software. I have been writing while pacing about my office, listening to music, and only infrequently using the keyboard to replace the cursor as needed.)

I have maintained contact with Brandon and his family over the past 5 years. He is currently in senior matriculation classes at his local high school, doing quite well, and has become an active member of the school of fine arts program, in which he has been identified on a regular basis as a leader, particularly in relation to performance. He has aspirations of university, and his grades make him a competitive applicant. Interestingly, but perhaps not surprisingly, he wants to be a teacher.

I believe it important for us to understand not only the daily struggles of children with learning disabilities in academic environments, in which they are asked to perform tasks that require them to perform with skills that they do not have, but that they do this day after day, week after week, month after month, year after year. I often comment to parents who are struggling or frustrated with their children who have learning disabilities to imagine themselves working for 8 h a day at a job for which they are not qualified, requiring skills they do not have, and being evaluated on an ongoing basis, only to go home and have to do even more of the same thing in the evening. Most parents, happily, understand this analogy and, more often than not, give their child a hug once they hear it.

My Best Year of Teaching

A couple of years ago, I taught a rowdy, talkative bunch of fifth-grade students. In a class of 27, 2 students were diagnosed with autism, 3 were English as a Second Language Learners (ESL), 1 had severe behavior problems, and 3 were diagnosed with learning disabilities. Additionally, with only 7 girls in the class, the 20 boys often directed class discussion and activities. Although this was quite a handful, it also became my favorite year of teaching. Twenty years ago, I would not have been prepared to handle this class because I had little experience or understanding of learning disabilities. Through the years, working with great mentors and developing personal strategies to meet the diverse needs of students had prepared me for this challenging group of students.

Joey was one of my students who had been diagnosed with a learning disability in reading and writing when he was in third grade. To say he hated reading and writing would be an understatement. He did everything possible to avoid it, from needing to go to the washroom to playing with objects in his desk to poking the boy next to him. However, Joey hated being left out more than he hated reading and writing. This particular group of fifth-grade students had developed a strong sense of community, and Joey would do anything to belong. When I discovered this, I used it to my advantage.

First, I had to encourage a love of reading in Joey, knowing that I had to promote this love of reading with all the other students as well. Because the class was primarily made up of boys, I chose to study the novel *Holes* by Louis Sachar. I knew that it would appeal to their sense of adventure, justice, and friendship. I relieved the stress of reading the novel aloud by having them follow along to either me reading it aloud or a prerecorded CD. Discussions around the themes in the novel were lively, and many debates ensued. One such theme that came up was illiteracy and how illiteracy could hinder one's success in life. We discussed how not knowing how to read could make it more difficult for one to get a driver's license. Seeing the look in Joey's eyes when he realized that reading was necessary for him to get a driver's license was priceless. It was an "aha" moment for him. For book reports, books on CD were provided and he was allowed to choose from a variety of high-interest, low-vocabulary books. Reading in small groups or pairs was common in all subjects, not just language arts. I made sure to not call on students to read aloud, but if they offered to, I would encourage them. Basically, I wanted to create a safe and supportive environment for all students, including Joey, to take risks.

Next, I had to encourage the importance of writing. It was during writing activities where Joey would most often go absence without leave (AWOL), running off to get a drink of water from the fountain or to the washroom. We needed to start small, so one of the first activities was to apply writing to real-life situations such as writing notes in our agendas. While teaching how to take notes, students designed abbreviations and acronyms for common words, making note taking easier, a lot more fun, and a great deal like the text messaging they were used to. Another activity was having the students draw or map out what they were going to write before actually writing. Graphic organizers and mind mapping were frequently used for planning and organizing our writing activities. Clear expectations and steps for writing, along with rubrics for evaluating student writing, were always provided, both visually and orally. Just like reading, writing assignments were sometimes completed in small groups or pairs, allowing Joey the opportunity to work with and learn from more competent writers in the class. Finally, when producing a final written product, all students were given the opportunity to use word-processing programs on a computer to complete their assignments. Using assistive technology such as laptops, AlphaSmarts, PowerPoint, and voice-activated software inspired Joey and other reluctant writers in the class to become more fluent writers.

Like many students with learning disabilities, Joey didn't want to be seen as being different from others in the class. If all the students in a class were to learn and progress, then they would all need somewhat different amounts of attention and different resources. Class meetings and discussions held where students would talk about their differences and how being different is not bad, just different, wasn't enough for Joey. Not until hearing about my own struggles with spelling and writing in elementary school and having his peers share their strengths and weaknesses did Joey finally realize that he wasn't that much different from everyone else and that each one of us learned differently. By providing choices and options for completing class assignments, tests, and homework, along with flexible instruction, Joey was able to draw on his strengths and employ a number of strategies to deal with his weaknesses. Additionally, a variety of learning materials and differentiated instruction were employed for the benefit of all students. Learning was a goal that Joey needed to buy into, and he did, wholeheartedly. Watching Joey and the rest of the students in the class learn, progress, and succeed made this my best year of teaching yet!

Keeping an Open Mind and Identifying the Primary Problem

Mike was 14 when I first met him. He looked to be more like 17 or 18 years of age, particularly due to his height and facial expressions. Mike was referred to me by an uncle, Henry, with the consent of his parents and to some extent with consent from Mike. Although Mike may have agreed to see me, it was a rather reluctant agreement, and in our first meeting, I wanted to ensure that I respected his wishes and that he did have a say within the context of legally being considered a minor. During our initial meeting, Mike stated that "psychologists are all nuts and they just mess with you enough to get money and feel good." However, he did agree to work with me.

The primary motive for the uncle being involved was that he believed that Mike had potential and that things could work out. However, Mike was out of step with his home and argued constantly with his mother (Rita) and, to a lesser extent, with his older brother Allen. The situation was disintegrating rapidly, with the mother threatening to put Mike out on the street or into custody of children's services. In addition, Mike had been on a long-term suspension from school and had several outstanding charges related to theft, vandalism, and loitering. The uncle and Mike described Mike's father Reg as a nonentity in the home who opted out of family responsibilities. He regularly deferred to his wife, who took an uncompromising stance on every issue. Although Mike's dad was effective in his work, he drank heavily on the weekends. Mike also had drinking problems, drinking frequently, drinking to excess, and being drunk once or twice a week. As well, he regularly used mild forms of other drugs and experimented with a variety of drugs. With regard to the report that his mom took an uncompromising stance on every issue, I was about to learn how uncompromising his mom was.

Our first meeting was at noon on a cloudy Wednesday. Mike and his mom and dad arrived looking stressed and angry. My normal practice in situations like this is to welcome them as a family and suggest I meet with the youth first to validate him in the process and get his perspective. This is followed after about 15 min with a similar meeting with the parents for about another 15 min, and then yet another 15-min meeting with all of them together. This first session is intended to review the process, obtain or confirm consents to proceed, and most importantly for the youth member, to know that his story will be heard from his perspective and to determine what, if any, goals he might have for the process. Naturally, parental goals are also important. However, balancing the goals of all parties and knowing whether or not the goals of each person are consistent is an important aspect to this initial meeting. In a sense, this initial meeting is setting out the logistics and expectations for all of us as well as establishing at least an initial timeline to begin the process.

Mike entered my office with his ball cap pulled low over his eyes. He didn't make eye contact. He slumped in the chair and folded his arms. He was dressed in brand label clothing, including a Tommy jacket, Tommy jeans, expensive high-top sneakers, and what looked like a huge Guess watch. This is significant, as his parents were in the lower middle-income range from what I knew of the father's occupation. I started the session by asking Mike what he understood as the reasons for being here. His reply was brief: "Dunno." I waited about 30 s and told him I was interested in his perceptions, as I had his uncle's perceptions and would soon get his parents' perceptions. I reinforced that his perceptions were perhaps the most important. That opened the door a little, and he revealed some things about his mom, school, and past psychologists he had seen. I didn't defend or take issue with his comments but spoke next about what I did and how I worked. I asked if we could focus on the issues that were most important to him. He indicated that "getting mom off his back" was most important and that he needed to stay at home.

We clarified a little on what he meant about his mom and the importance of being able to stay at home. As well, we discussed what roles each party played in accomplishing these goals. We finished with me asking if we could work together on a better home relationship over the next six to eight sessions. He tentatively agreed. I indicated that I'd chat with his parents next and then all three of us would come together, or we might go back to another individual session before breaking this first time. I also asked him to push the hat back or remove it, as I preferred that both of us could have eye contact when we talked. He sort of sneered but complied.

Next I met with his mom and dad. This session was challenging in that his dad said little and his mom tended to dominate. She outlined a long list of items that she had written down prior to the meeting. They were items that had to change in the home, with the most focus on a clean room, no back talk to her when she gave a command, and attendance at school. The list was very detailed and covered individual pieces of clothing in the room that had to be in certain spots. After the 15 min were up, we agreed that their top goal and focus for the initial sessions would be on working together so Mike could stay at home. We discussed his goal of "getting mom off his back" and came to an understanding of how we would proceed when all three came together. Mike's mom indicated that she was ready to discuss a compromised list and a negotiated living-together arrangement. We decided we wouldn't meet as a family today and that I would work with Mike for several sessions to bring them together.

I went outside to discuss this with Mike in another office. He agreed but warned me that things would break down pretty quick as his mom only said she would try and that she didn't really mean it. This was interesting, and we discussed how to deal with breaches of the tentative arrangement until the next visit.

Following my conversation with Mike, I went back to the parents for a similar discussion, which went well except that his mom said Mike would be a problem within minutes of leaving. I then called Mike in and reviewed the tentative arrangement and booked an appointment with Mike for the next day so as to keep things rolling. All agreed with the initial goal and they left the office. As they were walking through the waiting room, I heard his mom say to Mike, "Now you remember you have to listen to me and follow what I say. I want that room cleaned and no more back talk when I tell you things." Mike exploded and told his mom to "**** ***" as he left the room, slamming the door behind him. His mom turned to me and said, "See what I have to put up with?"

After some discussion and follow-up with Mike's mom and dad and then with Mike, I decided that his mom would have to be a key part of this journey. I also wondered about the prior diagnosis of Mike, having seen the way this event unfolded. The manner in which his mom appeared to be at the center of this incident was revealing to me. However, many questions entered my mind, and I set about answering them.

Mike came with a long history of past psychological intervention. He had been hospitalized for a 90-day mental health observation when in fourth grade. He had been in and out of day treatment programs for the past 4 years, none of which were successful and none of which went full term due to his being asked to leave or his quitting. He had seen two other private psychologists, two psychiatrists, and two school psychologists. As well, there had been one attempt at family intervention. He had been placed on a variety of medications over the course of his life, including medications for attention problems and depression. I reviewed all these files and, with parental consent, consulted with many of the past people who were still available. In summary, the professionals consulted, and their files notes indicated that the best hypothesis by all concerned was that Mike had a conduct disorder and that his family situation exacerbated attempts at better functioning. However, I was uncomfortable with the diagnosis of conduct disorder.

Over the course of the next six months, Mike and I met at least weekly and sometimes more frequently, especially if there was a crisis. While I follow a brief therapy model and while Mike continued to work with me, we were constantly updating our goals and revising strategies. Because Mike was unwilling to abstain from using alcohol and other drugs, I followed a harm reduction model in terms of his use of alcohol and soft drugs. His mom, dad, and I also met concurrently, and I held several sessions with his older brother. Despite the use of best evidence-based strategies, consultation with a new psychiatrist and a resultant change in medications, reintegration into the school, and strategies for improved home management, his situation was not significantly improving in any independently sustainable manner. At selected times, especially during longer home stays, the situation would deteriorate and he would either run away briefly, be asked to leave the home, be suspended from school, or have an in-school suspension. His drinking was still present, as well as some moderate use of mild drugs. However, this was more on weekends with friends and less of an ongoing usage as in the past. Nevertheless, all reported that his life was moving forward in a slightly more positive manner. As well, no new legal charges were added to his profile. I felt as though we had a huge bandage on the issue but that we were not coming to terms with sustainable change.

My analysis of his functioning, supported by various assessments that he agreed to complete with me, revealed low-to-average overall ability, which was being depressed or pulled down by significant learning disabilities in the areas of expressive language and social perception. He had attention challenges that impacted his recall, and he was mildly hyperactive. I did not concur with the finding that he had a conduct disorder.

One of the key aspects of a conduct disorder involves aggression to people and animals, typically but not singularly doing physical harm to others and to animals. In all his history, there was clear evidence that he never hit anyone other than when engaged in the typical fights that exist in the youth culture. Evidence by all concerned indicated that although he would slam doors, kick garbage cans, and break off rulers or pencils, he never hit or attempted to hit a teacher, his parents, his brother, or any other school or community member, even when being restrained or having his path blocked as he attempted to leave a situation. Furthermore, his father had a dog that Mike would take care of, pet, and play with but never abuse, tease, or threaten in any way. This type of behavior would be atypical of youth with conduct disorder. It was my opinion that, although he would meet criteria for a conduct disorder and for an ODD, his behaviors were more a manifestation of his frustration with his situation and not an underlying disorder.

This perception was soon to be challenged as he was arrested for being involved in a swarming incident. After his release, he and his mother got into a serious argument, with the result that she threw his clothes out of the house, telling him to leave and never come back. No one informed me of these incidents nor was I called until I heard from him via my pager at 4 a.m. on a cold and stormy night preceding our next visit. We arranged to meet.

I went to a designated place and picked him up. After talking for a while, I took him to a youth shelter and we arranged for him to become a ward of the province as his parents stated he couldn't return. He remained there for several days, being assessed by the in-house psychologist and social worker. All of us consulted with one another and, although we all agreed he didn't fit with the group, it was his only option for the moment. As well, his court date on a range of outstanding charges was within a week. I suggested he obtain legal aid prior to that court appearance. This would prove to be rather significant in the outcome of his situation.

Mike was at the center 3 days when I was called by the on-duty social worker to come to the group home to see if I could help settle him down as he had become quite agitated. He had been having a difficult time with the routines, as well as with one of the other youth. By the time I had arrived, six patrol cars were at the house, and he was inside destroying the kitchen area. The officer in charge of the situation didn't want to enter the house until Mike had settled, as they feared he had a knife. Everyone else had been evacuated. The officer asked if I could talk to him. After about an hour of talking, Mike had calmed down and surrendered to the police. He was taken to the hospital for treatment of some cuts he had received in breaking things, and from there, to the police station for booking. He spent the night in one of their cells.

These were clearly another series of critical incidents. It appeared that Mike had growing pressures with the pending court appearance. Although we were working on these, the unfolding crisis involving the swarming, at-home confrontations, and the most recent incident at the group home, resulting in further charges that had moved the pending court date to a less prominent position. As well, as previously noted, Mike continued to drink but only on weekends and continued to associate with his old friends who tended to have regular conflicts with social norms. Furthermore, his mom was very reluctant, unwilling, and/or unable to abide by any of our agreements, strategies, or de-escalation techniques, with the result that home life continued to be confrontational and problematic. Mike's dad continued to opt out.

With regard to his association with old friends, this unfortunately led to a very serious incident. Specifically, his group of old friends joined with another group and together they attacked a youth with whom one of the members had a long-time grudge.

Important to the diagnosis, Mike told me he never hit the victim nor did he participate in the beating in any way other than to stand back from the action pretending to watch for police and/or witnesses. At the end, he was asked to kick the person to prove he was with them. He did so with a light kick to the hip area. When asked for his reasons for agreeing to kick the victim, Mike responded that it was kick or become another victim. Shortly after, the group ran off, with all being arrested within 24 h, and he was subsequently asked to leave his home. It would become my argument that there was no intent to harm and that this incident did not compete with my original diagnosis of his not having a conduct disorder. Again, I argued it was a manifestation of poor decision making, associating with an inappropriate group, and being in the wrong place at the wrong time. All of these factors pertain to poor or impulsive decision making, use of alcohol or other drugs, and failing to consider consequences of one's actions. As a result, we began to review our approach to the harm reduction model relative to drinking and other drugs. As well, I set up some new strategies to avoid contact with his old group of friends and he and I agreed he was moving on, despite these recent setbacks.

I invited mom and dad to a meeting to discuss Mike. My goal was to move the father into action and to use the crisis as a stepping-stone to some real resolve. In our meeting, I reviewed the long history and used some direct quotes and examples of how Mike was looking for his father to take an active role in his life. I also summarized some painful examples of how his mom was not living up to her agreements in our negotiations. His mom began to argue with me. At that moment, his dad spoke up with the comment that he had been silent for too long and that he needed to take some ownership of what was happening. He went on to talk about Mike and his good qualities, and finished by stating that if he and his wife didn't get their act together, Mike would end up in jail or worse. Surprisingly, Mike's mom said little, but she did state that Reg, the dad, could take the lead on managing things and she'd stay out of them. I arranged for a follow-up meeting with Mike and his family. Mike was back at another group home, a more secure treatment facility. Here, he had severe restrictions on his movement and on his association with others. He was being treated as a criminal with explosive tendencies and a severe conduct disorder. Although I had arranged for a consult with his psychiatrist, this had yet to take place.

As we chatted while waiting to admit his parents, he told me he had met with his legal aide lawyer. I was amazed that it was one of the top criminal lawyers in the area who regularly volunteered his services to legal aide. I obtained consent from Mike to talk with his lawyer, as I was optimistic that some accommodations may be possible in sentencing. Given the nature of his charges, it was clear he would have some type of detention-imposed sentence.

The meeting with his parents went extremely well, with his dad taking the lead. In fact, things went so well, they agreed to take Mike home once the family court judge agreed to a change of living arrangements and set forth the conditions of his house arrest.

Consultation with Mike's psychiatrist resulted in our agreement on a diagnosis. Mike had a significant learning disability, particularly in expressive language and in attention problems. These manifested in sufficient diagnostic richness to sustain a secondary diagnosis of ADHD of the combined type. Together, these manifested in inability to cope with the regular classroom, particularly with tasks involving reading and writing, but also impacted his math and other subject areas. His challenges also manifested in frustration, and poor problem solving by virtue of the fact that he frequently made impulsive decisions without full consideration of the consequences. Together, the frustration led to behaviors that were in competition with his own basic principles of right and wrong. In effect, he was troubled by his behavior but couldn't control or escape the situation in which he found himself, although it was in part of his own making. His behavior did meet criteria for being labeled oppositional defiant but it was held that this was a manifestation of his underlying attention and learning disabilities and not in and of themselves a causal factor. It was argued that there was no sustainable evidence to support the diagnosis of a conduct disorder, even of a mild type.

Given agreement on this diagnosis and the presenting conditions of how Mike came to be involved in the swarming and other incidents, Mike's lawyer asked for a delay in the initial trial and worked to have all the charges considered in one session at a subsequent date. The lawyer then worked out an arrangement with the prosecution regarding an agreed-upon statement of facts, that many of the minor charges involving issues such as loitering would be dropped and that Mike would be tried only on the swarming charge and five of the other outstanding charges, albeit the most serious being the damage charge at the group home. In each case, Mike was part of a group, and his lawyer introduced evidence that demonstrated that Mike was a follower and not a leader in these events, and that he was often intoxicated during these events. Coupled with the psychiatric and psychological reports speaking to a positive prognosis, and the detrimental outcome for placement in a detention facility, along with confirmation from Mike's dad regarding his willingness to provide active ongoing supervision and that Mike would continue in regular treatment, the court decided that Mike would be placed on probation for 3 years with a variety of conditions, but released to his family. Although this was a very positive outcome, it was only the beginning of the next phase of intervention for and with Mike and his family. First, long periods of counseling regarding self-management and strategies to manage frustration followed.

Secondly, Mike applied to and was recommended for acceptance at a private school for youth with learning disabilities. The local public school jurisdiction agreed with this and was part of the recommendation for his placement in the special private school. Indeed, they covered the cost for Mike under a provincial government funding arrangement that permitted public schools to contract out services for youth with significant challenges to private school operators with the specialized personnel and focus to more fully respond to the challenges and strengths of such learners.

His admission to private school proved very valuable to Mike, as he needed individualized and supportive attention beyond the capability of the public system. Although the expertise actually existed in the public system to address his needs, many of his past associates also were readily available to him in the public system. As well, the ratio of staff to students simply did not enable the type of individualization, attention, and ongoing monitoring possible at a private school whose single purpose was to work with individuals with the strengths and challenges Mike exhibited.

The third piece of the action plan was the family. Here again, long sessions involving Mike, his mom and dad—and his brother to a more limited extent—followed with the result that his dad took on more of a leadership role and his mom backed away from many of the confrontational approaches she had used previously. Although Mike's mom did not pursue therapy with another psychologist to help address her own issues as I recommended to her, she did manage the manifestation of those issues much better and appeared comfortable with supporting Reg in his work with both boys.

The journey continued for another 3 years in this fashion, not without critical incidents and not without its challenges. Staff at the private school was fantastic and provided Mike with many accommodations, adaptations, and a great deal of personal and professional support. One of the actions resulting from the private school program that proved most valuable was the opportunity for community service and for career mentoring. Mike had expressed interest in plumbing, and he entered a co-op program with a local plumber.

At home, issues continued but were approached more from a problem-solving paradigm. As we had often discussed, life brings problems and challenges. It's not the problem but how one deals with the problem that makes the difference. Mom, dad, and Mike now had a process for problem-solving issues that they used with more and more frequency, thereby de-escalating the problems as they arose. Mike's mom was at least comfortable with letting go of management issues now that his dad was taking a more active role. For his part, Mike's dad readily took to the role and even reduced his drinking. Mike also made changes. He reduced his intake of alcohol, ceased using drugs as these seemed to interact with his meds, and began using many of the strategies for self-management. He was not able to manage the writing act nor did he cease having trouble with frustration tolerance. However, he did find ways of managing his frustration, and he began to celebrate more of who he was and how he functioned, with an emphasis on what he did well. Eventually, Mike managed to obtain the equivalency of tenth grade and entered an apprentice program, subsequently taking courses that would enable him to become a plumber. He is now working for a viable local company and doing well. He now has a steady girlfriend, an apartment that he manages, and his own truck, as well as shared ownership of a small cottage. The cottage is very important to him as he claims it is a refuge for him from the day-to-day struggles.

For me, this case reinforced the importance of keeping an open mind when working with a person. Mike had a learning disability with attention problems complicated by a challenging home and behavioral component. The behavior took precedence for many of the individuals who worked with Mike. This resulted in the underlying learning disability taking on a less significant role in the way Mike was approached, with the emphasis being placed on treatments for his behavior and for what was believed to be a conduct disorder. We have to keep an open mind that sometimes what is presented as the background information of a case and how others interpret that information is not necessarily all there is to know about a situation. It is important to appreciate how various pieces of information can be taken together but result in multiple interpretations. Once an interpretation is made, much of the subsequent analysis and intervention follows based on that initial interpretation. As in this case, alternative approaches to intervention can be missed, and assumptions about how a person is responding to or failing to respond to intervention can result. In this case, the underlying issues were those pertaining to a learning disability and attention problems that tended to raise issues of self-concept and frustration tolerance, resulting in avoidance and anger. Treating the anger without also treating the underlying conditions was only providing a partial answer to all that was needed.

A second aspect this case reinforced for me was the importance of looking at the entire context of family, friends, community, and the social environs of the person. As one example, the issues that Mike's mom was experiencing and her response to Mike as a part of those issues—coupled with the nonresponsiveness of the dad—were critical ingredients to how the situation was being exacerbated. Although this was reported in the past and although attempts were made to address it in family intervention, a major incident and a readiness to act were also needed to help gain perspective and for action to result.

A third (although sad) finding from this case is the role that critical incidents play in the outcome of cases. Fortunately, in this case, the outcome was one of enabling action and a positive resolve. However, critical paths emerge all the way through personal situations, and the path that is chosen or happened upon can have multiple outcomes, some of which can be very devastating.

Fourth, the role of good fortune can never be forgotten as exemplified in the present case. Gaining access to a lawyer with experience and credibility as well as having psychiatry and psychology work together in such unison were good fortune as to the selection of the individuals and the way they bonded to act in the case. True, one can arrange for colleagues to work together and one can network in the system to ensure such possibilities, as was the case in selection of the psychiatrist. However, this can result in an uncritical analysis of issues sometimes. As well, "luck" also is a factor at times, despite the best planning and intention. One has to recall that chance can also undo the best-laid plans, as in the case of Mike choosing to join his friends at the time that they were planning revenge on another young man.

A fifth lesson that was reinforced for me was the need to focus on who is the client and how far one can go in dealing with other issues that warrant attention but are beyond the challenges the person who is defined as the client faces. For example, the issues that Mike's mom had and the issues that his dad had were in and of themselves important for action. However, those issues needed to be the focus of another professional and could only detract from my work with Mike or at least risk Mike thinking the parents once again were the most important focus and not him.

With regard to the barriers and challenges that exist in any situation, this journey totally reinforced for me that the challenges aren't the issue. Barriers and challenges exist everywhere and to varying degrees. Indeed, barriers often reveal much about the person and their situation if we are open to hearing/seeing what is really being said or displayed. It is how one addresses and solves the arising challenges that are important.

Having a process is critical. Key guiding points to that process include focusing on who the client is, being willing to explore more alternative interpretations and actions as one option/interpretation proves incorrect or incomplete, and being aware of one's ethical decision-making process on behalf of the individual with whom one is working.

Epilogue

From our perspective, some themes within and across many of the stories in this section include the following.

- Children and youth with learning disabilities have average to above-average intelligence but also have significant learning difficulties.
- Sometimes the learning disabilities of children and youth are masked by their acts of aggression, defiance, and intolerance, as well as by their distractibility and feelings of anxiety and depression.
- Many children and youth with learning disabilities suffer from the anxiety, tension, and frustration that they experience from not being able to do in school what they think others can do so easily.
- One of the keys to success for children and youth with learning disabilities in school is to have supportive parents and teachers who provide strategies and accommodations for the children's particular learning needs and for the children to become empowered and skillful in applying these things independently.

Specific thoughts children and youth expressed in the stories that we thought were insightful and we have heard expressed in similar ways over the years include the following.

- It is difficult to know that everyone in class can understand and do the things that are required of them and that I am not able to achieve what they can in the same way.
- It is a relief to find out that the reason for my academic struggles and underachievement is due to a learning disability and not because I am not smart enough to succeed in school.

Specific thoughts parents and teachers expressed in the stories that we thought were insightful and we have heard expressed in similar ways over the years include the following.

- Sometimes parents of children and youth with learning disabilities feel they have to fight the school system and have to be in competition with other parents to get what is needed for their children.
- It is not about children and youth with learning disabilities "working harder" to be successful in school but about them "working smarter" through the use of personally effective strategies and skills.

6 Intellectual Disabilities

Prologue

Children and youth with intellectual disabilities are typically differentiated by the dysfunctional severity of their intellectual and adaptive behavior. The typical classifications are mild intellectual disability, moderate intellectual disability, severe intellectual disability, and profound intellectual disability, which typically require different levels of support (e.g., intermittent support, limited support, extensive support, and pervasive support, respectively). Children and youth with mild intellectual disability, many of whom come from minority backgrounds and/or have low socioeconomic status and disadvantaged backgrounds, are usually below average in their developmental and self-help needs. Typically, they are capable of achieving academic skills at advanced elementary grade levels and adaptive behavior skills at an independent level. Many children and youth with moderate intellectual disability have one or more organic causes for their disability (e.g., many children and youth with Down syndrome, genetic form: trisomy 21, and fragile X syndrome, genetic form: fragile site on X chromosome, have moderate intellectual disability). Although many of these children have marked developmental delays, often exhibiting physical and motor difficulties as well as communication weaknesses and requiring some help throughout their life, they can typically adjust reasonably well socially (e.g., learn how to share with others, respect others' property, and cooperate with others). Moreover, many of these children can achieve basic academic skills at lower elementary grade levels, and acquire semi-independent vocational skills. Children and youth who have severe intellectual disability typically have severe impairments and require lifelong care. Most of these children and youth have minimal communication and self-help skills, and some may be able to acquire low-level vocational skills and work under supervision in sheltered workshops or pre-workshop settings. Children and youth with profound intellectual disability typically have the most severe levels of intellectual and adaptive impairments. Most of these children have very limited motor development and communication ability and typically require lifelong care and assistance.

In this section are three stories: one personal story, one parental story, and one professional story pertaining to Down syndrome.

The personal story, entitled "You Go Girl," is written by a young woman with Down syndrome. In her "spirited" story, she recalls her childhood and youth with respect to some of her illnesses, interests, family life, challenges, highs and lows, and accomplishments. The parental story, entitled "Perfection in Imperfection," is written by a mother who reminds us that it is all about the child with Down syndrome, not about a Down syndrome child. In this regard, the message throughout her story is that you love your children for who they are, not for who you want them to be. The professional story, entitled "More the Same than Different," is written by a psychologist who reflects on her growing awareness and insight about children and youth with developmental challenges and leads us to her eventual realization that people are much more than what some people define them to be.

Exceptional Life Journeys. DOI: 10.1016/B978-0-12-385216-8.00006-0

Personal Story

You Go Girl

My parents call me an Olympic baby because I was born in February 1988, when the winter Olympics were being held in Calgary, Alberta, Canada. My name is Natalie and I weighed 7 pounds, 6 ounces. My mom said that she knew the minute she saw me that I had Down syndrome, but the doctor didn't believe her. My dad kept asking until they sent for a special doctor to do some blood tests on me. When the tests came back, the doctor told us that I had trisomy 21, which means I have Down syndrome. My mom told the nurse that she didn't know how to raise a baby with Down syndrome, that she only had experience with raising regular-type babies. The nurse told her that is what she should do then: raise me like a regular baby. To the best of their ability, that is what my parents have done.

One of the first things they did when I was a baby was to join a group called Ups and Downs. There, they met other families with babies with Down syndrome, and that helped a lot. We are still friends with those families because we have done many fun things together. Now, as an adult, I love to go to the Ups and Downs teen/young adult parties and dances. My favorite is the Family Dinner and Dance that we all go to every year.

They were busy days for my whole family when I was born, because I had so many special doctors who wanted to check to make sure I was healthy. My mom took me to infant stimulation classes, and an early intervention worker came to our house sometimes.

I am 22 years old now and have had some of the problems that people with Down syndrome often have. I used to have lots of ear infections until I had to get tubes in my ears. I hear fine now. I had to go to the Children's Hospital three times to have surgery to have my tear ducts unplugged. I used to get strep throat a lot and often had to miss school. I've had bronchitis, bronchiolitis, and laryngitis. I have had pneumonia at least three times. I have alopecia, which means my hair falls out sometimes. I have acid reflux, I wear glasses, and I have weak ankles. I also used to have lactose intolerance, and I have had my tonsils taken out. Other than that, I'm pretty healthy.

I live at home with my mom and dad and brother, Cameron, who is 14. My sister Marie is 19, but she is away at university for part of the year. I have an older sister, Kirstin, and two older brothers, Jeremy and Wesley, who are married. That means I have two sisters-in-law, and now I have three nephews. I like being an aunt. They all live nearby, so I love to visit them and they often come here to visit. We live in the same house that we moved to when I was 3 months old, so I know lots of people in the neighborhood.

I like shopping, hanging out with friends, going to movies, hockey games, bowling with the Special Olympics, dancing, yoga, skating, listening to music, and writing stories, especially about the pirates of the Caribbean. I love being physically fit. These days, I'm taking karate, yoga, belly dancing, and piano lessons. I've been in two recitals so far. I love it when people clap for me! With my support worker, we are studying English, ancient history, and cooking. I work at Wendy's fast food place once a week, and I volunteer twice a week at the "Bishop's Storehouse," which is a food bank that my church runs. Once a week, I am a reading volunteer for the second grade, where I went to school.

I would like to share with you the story of my life and what it was like growing up with Down syndrome. Having Down syndrome makes it so my brain works more slowly than other people, and it takes me longer to learn things. Sometimes it bothers me that I can't do everything that other people can do, like drive a car, but then I think about all the things that I *can* do and it makes me feel better. I have a poster on my door at home and it says, "Just because something is difficult doesn't mean you shouldn't try it, it just means you should try *harder*." When I really want to learn something, I just keep trying until I can do it.

First, let me tell you about growing up in my family. Because there were lots of children in my family, I always tried to do everything they were doing. I remember playing dress-up with Marie a lot, and we put on lots of puppet shows. I love my sister, Kirstin. She played with me a lot and helped me do things. I remember her giving me piggyback rides. Now that she's a hair stylist, she always makes sure my hair is a nice color because I like to look beautiful.

My family enjoys hiking in the mountains in the summer. When I was little, they used to carry me in a backpack child carrier. When I got a little older, I wanted to hike, but I couldn't go very far. My brothers used to go on the trail ahead of me and Marie. All along the way, they put treats like cookies or Smarties, or gummy bears on the rocks at the side of the path. I used to run so fast to try to get there before Marie! We could hike a long, long way just trying to get those treats. Then one day I sprained my ankle. The doctor told me that I had "congenitally weak ligaments" in my ankles. I didn't hike for several years after that because I didn't want to sprain my ankles again. Now my parents got me some trekking poles and good hiking shoes. They help me feel safe, and I can be a good hiker again.

Learning to ride a bike was not easy for me. I find it hard to balance. When I was little, my family had a bike trailer that Marie and I could sit in and ride. When I got too big for that, they got a special kind of bike that attached to the back of my dad's bike, so that I could pedal and learn to balance. It had its own handlebars, so I felt like I was riding my own bike. That was fun for me, but hard for my dad because I always leaned over to one side. A couple summers ago, my parents hired a neighborhood teenage boy to try and teach me to ride a bike on my own. Rather than go up and down the sidewalk with me, like everyone else did, he took me to a park nearby with a small hill. I got on the bike and he pushed me down the hill. It was either pedal, or fall, so I pedaled! I could go eight or ten rounds before I started to tip, but I was getting it! If they hadn't moved that summer, I think I would have learned how. I'll try it again this summer. ...

When I was 2 years old, I started going to the PREP program. It's a place where children and teens with Down syndrome can go to get speech therapy and learn many other important things to be successful in school. I went there twice a week until first grade, then one afternoon a week until twelfth grade. I really looked forward to seeing my friends at PREP each week.

When I started school, there weren't very many students with special needs in regular classrooms. Lots of people thought that we didn't belong in regular schools. Because my brothers and sister went to our neighborhood school, of course, I wanted to go there, too. When my parents took me in to introduce me to the principal of the school and tell him that I was going to start kindergarten in the fall there, he said that he would never allow a student with Down syndrome to go to his school. My parents registered me anyway, and we later learned that the kindergarten teacher had a plan. At least once a day, the teacher sent me down to the principal's office to take a note, or to show him my artwork, or do some other job. He was over six feet tall, and he had to bend down really, really far to talk to me. At first, he didn't want to talk to me, but pretty soon he looked forward to seeing me each day and he became my friend. After that, he decided that I really *did* belong at his school!

Sadly, my favorite principal retired and we got a new one. This principal didn't understand about people with Down syndrome. He wrote lots of letters to the school board saying that he didn't want me in the school. He said that there was no money for an aide for me. My parents had to write lots of letters to ask for an aide for me so I could have some help. I had an aide for kindergarten and first grade. From second grade to sixth grade, I had a different aide. They are still my friends.

Nobody likes to go bald, but when I was 4, my hair started falling out. The doctor told me that I had alopecia. My eyebrows fell out, too. I had glasses and a bald head. I think I looked a little funny. There was a "no hats" rule in the school, so the kids had to get used to me. My kindergarten teacher told me later that I spent a lot of my free playtime at the dress-up center trying on hats. Outside of school, my mom bought me some pretty hats to wear. My parents even decided to buy me a wig. One day we were walking down the street and I was getting too hot wearing it, so I pulled off my hat and my hair came with it! There was a car going by right then and the driver nearly hit the curb from shock and surprise. My hair grew back in third grade, and it didn't start falling out again until tenth grade. I was not impressed, but it didn't all fall out, and grew back again soon. I now have beautiful hair and we celebrate it. It used to bother me when people stared at me, but now I just say to myself that it's because I'm so good-looking!

When I got to first grade, the resource teacher asked my mom and me what I hoped to learn in school. We told her that I wanted to learn to read and write and do math and all the other things all the other kids did. She told us that people with Down syndrome couldn't learn to read, write, or do math and that we needed to face the truth. She also said that I could never go anywhere alone and that I always had to have someone with me. She said that my mom had to drop me off and pick me up every day at the school doors. That was okay for kindergarten and first grade, but in second grade, I wanted to walk by myself. Sometimes I walked with my brother, Wesley, who is still my favorite brother.

One day, mom told me what to do if she ever came late to pick me up. She asked me to wait just outside the doors. Then she showed me how to wait at the playground gate for her. Then one day she showed me how to cross the street with the crossing guards. Then she pointed the way I should walk if I ever got tired of waiting. Every day, she came a little later and I had to walk a little farther to meet her. Finally,

I walked all the way home! I walked in the door and said, "Mom, did you forget about me?" She was very proud of me and said I was a good problem solver. Lots of times, if I have a problem, I like to solve it by myself. I took a note to the principal after that, saying that I was allowed to walk home by myself.

Because my brothers and sister could all read, I wanted to learn, too. My teacher didn't think I could learn to read, but my mom did. Every day, she showed me pictures and I had to match them with the words. I had to find each person's name in our family and match it with the person. Then I read little books that had only a word or two on each page. Finally one day, I could read a whole book by myself! I proudly took it to school and showed my first-grade teacher. She was very surprised, but happy. After that, they worked very hard to help my reading get better.

I got my first taste of dance when I was 6 years old. I took 2 years of highland dancing lessons, but after that, it got too hard for me. It was fun, though.

When I was in second grade, the principal said that I was not allowed to stay for any special hot dog lunch days unless my mom came in and stayed with me because "something might happen." He also said I couldn't go on any field trips unless my mom came. I liked it when my mom and sister Marie came into my class. My mom suggested that perhaps my sixth-grade "sun buddy" could come in and eat with me. Her name was Alicia. Each student in the younger grades had a sun buddy. Alicia came in and ate with me, but she told the principal that there was no point because everything was okay with me. Finally, Alicia's mom went in to see the principal and talked to him for a long time and told him that it wasn't fair that my mom had to go in every time and finally he let me stay for the hot dog days by myself.

When my brother Wesley was 10, he started karate. Marie and I used to watch at the side with the moms. I tried to do what the karate kids were doing and kept asking to join. I was 8 when I started, and I took karate until I earned my green belt when I was in junior high. I loved it! I enjoyed it that much because my brothers and sisters were all in it, too. When my mom took me to the doctor (pediatrician) for a checkup when I was 12, he said that karate was very good for training my brain. The reason I gave up karate for a while was because I started a swim program for teens with Down syndrome. A young woman was working on a Ph.D. by studying whether swimming would help young people with Down syndrome to speak better. We swam three times a week and learned how to be competitive swimmers. I was super fit. I liked swimming so much that I took synchronized swimming classes at the YMCA. I loved that, too!

Something that bothered me when I was 12 was that my sister Marie was growing taller than I was. I thought the older sister was supposed to be taller. When I told that to the doctor at the Children's Hospital, he told me something that I have always remembered. He said, "She may be taller, but you will *always* be the older sister." Now that Cameron is 14, he is taller than I am, but it doesn't bother me anymore.

Because Wesley was my hero, I wanted to be a school patroller like he was in fifth and sixth grade. I asked, and they let me take the training. The people at the school were very worried that I might not be able to do it. After a while, they quit worrying. NOBODY crossed the street when it was not allowed when I was on duty. Later, we found out that I was the first student in my city with Down syndrome to be a school patroller.

The other kids at school were really nice to me during school, but after about second grade, they quit inviting me to their birthday parties and to their houses to play. I think it was because I couldn't talk as well as they could, and I was slower. Every day, I asked my mom who I could play with. I was lonely and wanted my own friends, not Marie's. One day, mom started a friendship club for me. She phoned up the mothers of several children with Down syndrome not too far from us. We took turns going to each other's houses once a month. The group got too big and so we had to divide it into a girls' group and a boys' group. Now that I am an adult, I still get together with those close friends.

When it was time for me to go to junior high, the school told us that usually students go to a special program after elementary school. They told me that they thought I would do really well at junior high, so I went to our neighborhood junior high school. I loved it there, especially taking hip-hop dance classes. I went to classes with the other kids and sometimes I went to the resource room to get help. I loved my aide. At the end of ninth grade, I was invited into a meeting with my mom, the resource teacher, the high school principal, and some school board people. The junior high principal said to the high school principal, "I was scared to death about having Natalie in our school, because we never had a student with Down syndrome before and I didn't know what would happen. I can tell you now, that you will fall in love with her like we did." I knew he was my friend!

Next came high school. The school never had a student with Down syndrome in its regular classrooms before, either, but the principal wanted to try it. I had an aide who helped me. Sometimes I was in the resource room and sometimes I went to my option classes. When my aide saw that I was fine, she didn't come to class anymore. My favorite classes were musical theater, foods, science, cosmetology, and dance. My musical theater teacher believed in me, and I had parts in the two musicals we put on. One was *Hairspray*, and I also had a small solo in the part of Gretl in *The Sound of Music!* In twelfth grade, I took a dance class and, when my mom went to the parent/teacher interview, the teacher asked my mom how long I had been taking dance. My mom thought that perhaps she was going to tell me the dance class was too advanced and I shouldn't be in there. Then she said, "Natalie has a natural gift and talent for dance. I highly recommend that you put her in a community dance class." I joined a teen hip-hop class in a community nearby and loved it!

When I was 16, there was a popular TV show on that was called *Fear Factor.* I loved the way the people would be so brave and do really hard (and gross) things. My family noticed how much I liked the show and my brother, Wesley, decided to try something out on me. Whenever my mom asked me to do something I didn't want to do like set the table, or help with dishes, I would say, "I can't" or "I don't know how" or "I don't want to." One day, my mom asked me to take out the garbage. When I didn't want to do it, he said, "Natalie! Use your fear factor!" Then I felt like I was brave and I could do it. I liked it when they said that phrase. Even today, sometimes I'll say it to myself when I have to try something new and I'm really nervous.

While I was in high school, I was awarded the "Greg Boswell Outstanding Teenager of the Year" trophy from PREP. I was chosen because I was a good role model for being physically fit, a good student, and for doing volunteer work in my church and community. I had to give a speech at a PREP golf tournament banquet to accept the Greg Boswell Award. After that, I had to give a talk in front of 600 people at a gala to raise money for PREP. I love public speaking!

One day I got to meet Darryl Sutter who is the manager of the Calgary Flames hockey team. I told him that I was a serious Flames fan, and he told me that he was assigning me to keep a record of the Flames wins and losses. To this day, I do that. My family tells me that they probably have those scores recorded at the Flames office, but I tell them that Darryl is counting on me and I haven't missed a game yet, even if I have to stay up late to write down the score. My mom says she would like to have a conversation with Darryl about that some day. I like meeting famous people.

To help me be more independent, I learned to take the C-train so I could go to PREP by myself in high school. Learning to do that was easy for me. Once in a while, I get on a train going the wrong way, but I know how to get off and get on the train going the right way. I know that if I'm in trouble, I can press the help button. My parents always tell me not to talk to strangers when I'm on the C-train and not to give them money, and *never* to go anywhere with a stranger because some people are bad. Finally, one day, I asked my dad, "Just tell me what the bad guys look like so I'll know not to go with them!" One day there were some *real* bad guys. These men were drunk and they were asking people for money. They asked me for money. I didn't say anything and just looked down. They asked the lady beside me and when she said no, they stomped on her feet. Somebody pressed the help button and they stopped the train. The police came and took them away. I came home and told my family that I was calm on the outside but crying on the inside. I had to use my fear factor for a long time after that, but I'm okay now.

The summer after high school, my best friend Johanna and I flew to Vancouver to be representatives at the World Congress on Down syndrome. We had to take many classes with CDSS (Canadian Down Syndrome Society) ahead of time to prepare to be ambassadors. I still do volunteer work for CDSS whenever they ask me.

After high school, I told my parents that I wanted to go to university. There was a program at our local university where students with developmental disabilities could take classes that other university students take. In a kinesiology class, I learned to skate and I took wrestling, too. I loved my dance and history classes. At the end of the year, I got to perform in a dance routine for the other students in the varsity education program and the parents. That was fun. The university is a long way from my home and I took the C-train. I got pneumonia twice that winter, so we decided to study closer to home. I now have a tutor who helps me with some of my studies. We're working hard at improving my reading, writing, and spelling, so I can take some harder courses later.

Since high school, I started taking belly dancing classes at the YMCA, and I'm really good at that. I think taking karate when I was young has really helped me. I was asked a few months ago to perform a belly dance presentation for the Association for Community Living. My teacher came with me to dance with me. Some people came up to me afterward and told me they were going to put their daughters in belly dancing, too. My teacher is another friend who believes in me.

I would like to tell others that people with Down syndrome are good people. Not everyone with Down syndrome is the same. Some people have more health problems than I do, so they have a harder time learning things. You can't say you know all about Down syndrome if you only met one person with Down syndrome. I think when people hear I have Down syndrome before they meet me, they think I can't do things. I can; I sometimes need some help until I can do it on my own. I hate it when I have to put on an application form that I have Down syndrome because people don't often want to give me a try. They should meet me first before deciding!

I have a happy life. I love my family and friends. I am smart and I try my best. I like to learn new things every day, even if it's hard. I would like to get my driver's license, so I study the learner's book often. If I can't drive a car, at least my dad lets me drive our quad if it's in a big field with no trees to hit. Since I started karate again, I would like to earn my black belt. My sensei says only three more belts until black. She says I can do it if I keep practicing! I told my mom that after I get good at piano, I would like to take violin lessons. I want to get a job I enjoy and move out and live on my own. I would like to live with a roommate, or even get married. I want to learn to be a good cook. I want to travel and go to Disneyland and more Down syndrome conferences to meet other young people like me. I want to be a Hollywood actress. I know I need help with all those things, but my family loves me and helps me a lot. My support worker is the best! Most of all, I want to be the best person I can be.

Parental Story

Perfection in Imperfection

What you are about to read is my story based on my personal point of view on parenting a child with Down syndrome. Of course, it is one experience, strongly felt and embraced, and also fueled by my own background, upbringing, professional interests, and personal passions. In addition to being a mother, I am a psychologist, an author, an advocate, and a volunteer. It seems important for me to say that my story is by no means intended to "represent" the perspective of all parents who have a child with Down syndrome. It is but one story in a myriad of stories that could be told. In some ways, it may be rather different from that of others. In some ways, it may share some common themes. I do feel honored to be able to share it with you.

My particular journey of being a parent of a child with a disability started before our daughter Alexandra was even born. It seems, looking back, that I had an interest and affinity for Down syndrome from a young age. I was the child on the bus that enjoyed sitting beside a particular person with Down syndrome who always shared interesting stories with me. I was also the teenager who chose to select Down syndrome as my topic for a biology class presentation, and the young adult who paid particular attention to the genetics of trisomy 21, the most common chromosomal arrangement that results in Down syndrome.

In university about 30 years ago, I continued to be inexplicably focused on learning more about Down syndrome. In particular, I remember vividly a snowy, cold, and uncharacteristically cloudy day. I had just turned 18, and was in my first year of undergraduate studies majoring in psychology at the university. I was studying for an upcoming exam, and had decided to take a break. As it happened, I had found a spot to park myself, my always-too-full book bag, and my purse, in a section of the library that housed medical books. I was not studying Down syndrome—yet, during this study "break," I found myself browsing the books on the shelves, and searching for one about Down syndrome. It contained a graph depicting the increasing incidence of Down syndrome with maternal age, with a sudden and dramatic spike at age 35. Looking at it, I remember that day deciding unequivocally that I would have all my children before I turned 35. As much as I truly enjoyed the few people with Down syndrome that I had met through my life, my heart cringed at the thought of bringing a child into the world that would have extra challenges and who might not be accepted by others. The world was a hard enough place to live in, I thought.

I didn't realize at that time that more babies with Down syndrome are born to mothers under the age of 35. This is because, statistically, more babies are born to younger mothers in general. Despite my best-laid plans, Alexandra was welcomed into our world 16 years later. I conceived her at 35 years old and delivered her at 36. She had Down syndrome, and I loved her with all my heart.

Alexandra was brought into the world in February, and she joined a family that had valued traditional views of intelligence very highly. My husband has his Ph.D., I have a master's degree, and we found out later that our son is gifted. When Alexandra was placed into this mix, it was the beginning of an education that I wouldn't get in any textbook, or experience in any classroom. It was an education about what makes life worth living. She quickly showed me what really matters in life and taught me so many things that textbooks can't teach. Although I can't quite describe how, she has humbled me in a way that nothing else could have, and I am thankful for it. Alexandra has shown us what it means to be intelligent in other ways—especially emotionally.

Our daughter Alexandra is a beautiful, energetic, friendly, and kind 10-year-old girl, with a wide smile and beautiful almond-shaped eyes. Lately, she has taken to blasting her Hannah Montana/Miley Cyrus music at 6:15 in the morning, when she tends to wake up. Alexandra has very much taken to cooking and baking, and recently created her own delicious recipe that she calls "baked bananas." She makes this herself with bananas, brown sugar, maple syrup, and cinnamon, all gently warmed in the microwave for "3–0 (seconds)." Alexandra loves moving to music, going to the swimming pool, cooking, reading, and writing in her many notebooks. She is the first to notice if someone is not feeling well, and to run to get the person water or make a "get better" card. Alexandra loves playing on the beach at the lake. She adores her older brother Aiden, and can't wait for the hour or two when he babysits her once in a while. If you asked her, she'd definitely tell you her favorite food is pasta. Contrary to the popular (well-intentioned) stereotype about children with Down syndrome, Alexandra is *not* always "happy and loving." Like all children, she gets grumpy, sad, and frustrated, and she fights with her brother too. She doesn't always want to set the table when she is asked. She gets angry when she doesn't get what she wants. She becomes frustrated when she is not understood. Like every child, she needs boundaries and clear expectations.

Yes, Alexandra also has Down syndrome. We don't like this to be the first thing we tell people about her, because it's too easy to reduce Alexandra to a label when we do that. Down syndrome results from her having an extra 21st chromosome in every cell of her body. We don't consider her as having a "disorder" or a "defect." In fact, we believe that Down syndrome is a naturally occurring chromosomal arrangement that has always been part of the human condition. One result of this chromosomal anomaly is typically below-average intellectual functioning when measured by IQ tests. Another result is delays in developmental milestones, including walking, speech, and toileting. Finally, Down syndrome can also be accompanied by significant medical and health issues. However, this is a "medical view" of Down syndrome, and doesn't really describe what it's like to have Down syndrome or to parent a child who has Down syndrome. In other words, these things that can come along with having Down syndrome don't define her, or our experience.

Some parents use the metaphor of a "roller coaster" to describe the emotional highs and lows that come with parenting a child with a disability. The highs are pretty high (like when our daughter first walked), and the lows are pretty low (like when we discovered our daughter had a possibly progressive hearing loss). Yet, if I had to truly characterize my overall experience of parenting our daughter with a disability, I would say it has been more like crafting a quilt. The quilt is woven from experiences over time. Each experience you have becomes a block in the quilt. When you have only one or two blocks in that quilt (a few experiences), the one or two blocks truly stand out and you can't quite imagine the end product. However, as more and more blocks are added, the quilt becomes bigger, the pieces start to come together, and it begins to take form. Eventually, you are able to step back and see what the quilt may look like when it's finished. You take it all in and are able to appreciate its beauty.

In much the same way, when you have one or two experiences parenting your child with a disability, those experiences affect you very deeply when they happen, and you may have trouble making sense of them. Then, as you grow and learn with your child, the individual experiences become part of something bigger. You begin to see the common themes, and with the right support and environment, you begin to have a clearer vision for the future.

My husband and I both agree that it was the medical diagnoses that overshadowed the diagnosis of Down syndrome in the early years. Not many people realize that Down syndrome does often come with health challenges that can be very significant. I vividly recall when we were told, hours after her birth, that Alexandra had a life-threatening stomach defect and a serious heart defect as well. We loved our little baby, and the thought of her undergoing surgeries, and at such a young age, initially brought both my husband and I to the brink of despair. We learned later that an infant with Down syndrome with the medial issues Alexandra had might not have been operated on 30 years ago. This shocked us, as it never once crossed our minds to deny her medical care to save her life. She was, after all, our baby daughter.

The rare stomach defect Alexandra had is called duodenal atresia (still rare, but more common, with Down syndrome), and it required repair at 2 days old. Alexandra was also born with a significant heart defect called atrioventricular septal defect, which was eventually repaired at 20 months old. Thanks to early screening and detection, Alexandra was also diagnosed with having a neurological hearing impairment early in life. She also had a very serious acute illness at 10 months of age, when she acquired respiratory syncytial virus and was hospitalized for 3 weeks. The infection had caused her to go into heart failure, in which she literally began struggling to breathe. It was one of the scariest experiences of my life because there was no treatment for the viral infection that had caused it. At this time, she was put on 24-h oxygen, and she came home with portable oxygen. We were immediately wait-listed for the open-heart surgery they initially thought she wouldn't need until she was school-aged. After waiting 10 months, she had open-heart surgery at 20 months of age. Her heart surgery was 2 days before the twin towers collapsed in New York on September 11. I remember the incredible contrast of watching that mass destruction on the small television hanging above her hospital bed at the very same time as I witnessed a sort of "rebirth" of my daughter as she recovered from heart surgery. By the time Alexandra was 2, she had overcome more health issues than many of us see in a lifetime, and had shown us so many times that she would beat the odds and surprise us. It's something she keeps doing to this day.

In a personal journal entry dated December 23, 2001, months after her heart surgery, I wrote the following paragraphs. Alexandra was almost 2 years old:

> *"Our Christmas present came early this year on September 6th when Alexandra's heart was mended. She is sound asleep now—the quick, shallow, erratic breathing of before has been replaced by even, slow breathing. God I love her!*
>
> *... Now that Alex's health is more stable, there is an ability to think, now, about the aspects of her syndrome that will affect her emotional health. There's time now to dwell on the reality of being different in a world that is not at all comfortable with it. It's unsettling.*
>
> *At the very same time, it is impossible not to notice Alex's newfound energy and strength—allowing her will to show itself in what she does. I found her first doing this in her crib as she peered over the railing with those beautiful brown baby-doll eyes. I really never thought I would see that day—it has taken so long to arrive!! She is now able to stand while holding onto something. It's indescribable the feeling of having her stand at my knee like I remember Aiden doing. She's not yet crawling (and she may never crawl I'm told), but I've taken to calling her my little "energizer bunny" because she keeps on going."*

So, now Alexandra's medical concerns have fallen away and are very much a thing of the past. At 10 years old, she can not only walk, but she also runs very fast. She has begun to talk in sentences, read, and write. She can count to 10, and sometimes to 20 and can recognize most other numbers. Our challenges now are of a different kind.

Possibly our greatest struggles recently are related to helping Alexandra live as inclusive a life as possible. Our goal with both of our children is to help them reach their full potential. We hope to give them a childhood that will lay the foundation for them to be happy, active, contributing members of the community they live in.

We hope to see both of them develop friendships, and find love and a lifetime partner if they so choose. We want for them to find work that is meaningful, pays enough to allow them to buy the things they need and some of what they want, and that fits with each of their interests and abilities. We hope for them to live as independently as possible, in an environment that suits their personalities and preferences, with family always available when needed. These goals are not always easy to achieve for any child. However, in our experience, achieving them has taken much more persistence, patience, and advocacy for our daughter.

Sometimes, I have found, the path of least resistance can lead you astray in crafting a good life for your child with a disability. It's easy to be swept up in systems that seem to want what is best for your child and that have a predetermined way of responding to your child and your family. Sometimes, these systems provide exactly the right thing, but sometimes they don't. The challenge is knowing when you need to ask for something different, find your own path, or when to take the road less traveled. It can be exhausting just figuring this out. I find myself asking, "Is the short-term pain worth the long-term gain?" on a pretty regular basis, and doing this helps me make better decisions for me and for my family.

One system that is still learning how to respond to the needs of a full range of individuals is the school system. We strongly believe in the value and benefit for all, including Alexandra, in a regular classroom with her typical peers. So far, we have been able to achieve this for her. Reviewing the highs and lows of that part of our journey would be well beyond the scope of this short narrative. However, I will say two things about Alexandra's educational journey. First is that we have been fortunate to have encountered passionate visionary people in the school system who also valued inclusion in Alexandra's early years. Alexandra's kindergarten teacher embraced having her in her classroom, particularly when a government-funded aide accompanied her. She took the time to include a lesson plan on "valuing difference" to the children. The principal at the elementary school in our neighborhood found out about Alexandra's birth, and actually visited our home with a gift and told us she would always be welcome at the school. You can imagine the joy we felt in hearing those words. Her overture at that time paved the way for what ended up being an extremely challenging, but overall successful, period of inclusion in first through third grades. The special-needs strategist who is assigned to Alexandra's area is also a wonderful partner in helping make inclusion possible for Alexandra. Most recently, Alexandra is attending our community school. In this school, there are a host of learners with exceptional needs, including a community of students who are blind. Inclusion is a day-to-day reality there. I will say that, despite the good intentions and efforts of everyone involved, the educators who work with Alexandra are regularly challenged by limited resources, sometimes limited knowledge and skills, and multiple children with a wide variety of needs needing attention and time. Teachers are stretched to their own personal and professional limits at times, trying to make inclusion possible. Sometimes they struggle alongside us in learning and growing so that Alexandra can reach her full potential. At times, perhaps, some of them question if inclusion is the right thing for our daughter when all the challenges are considered. This doubt sometimes translates into frustration—with us as parents, with the system, and with the reality of the situation. We've experienced teachers becoming defensive and pushing us away as we fought for what we felt was best for Alexandra at the time. In this way, helping our daughter receive the education that she has the right to is one of the greatest sources of stress for us. It continually challenges us on many levels.

A great majority of Alexandra's challenges, both at school and at home, stem from her speech difficulties. Alexandra is able to speak in broken sentences, and over time, with therapy, she is beginning to be more easily understood by people who don't know her. We know that, despite her trouble with speaking and being understood by others, she finds ways to get her needs met, even if sometimes these are not socially acceptable. She asserts her will in every part of her life—at home, in the community, and at school. Alexandra can be incredibly persistent and determined, and others label all of this as "bad behavior" that needs "management." We realize now that, although her behavior can cause problems and needs to be addressed, sometimes it is her only way of telling us what she needs from us. The professionals who work with Alexandra know they can always count on me to ask, "What is she trying to tell us?" when she has behavioral issues. "Bandages" work, but they only go so far and last so long. You can send Alexandra to time-out in the office as often as you want, and it may stop the behavior in the moment. However, it won't prevent it from occurring over and over if the real issue is that she is bored with what she is being taught or how she is being asked to learn. Usually, asking questions about the root of the problem helps us find solutions that are more effective and last longer.

I have been forced to become a staunch advocate for my daughter. To be an advocate means that I "give voice" to what she may be thinking or feeling or needing, because she is not capable of doing it effectively herself. Representing her, translating for her, and helping others makes sense of some of the unusual things she does can be difficult tasks and don't always win me friends. I tell myself that's quite ok— if we're on the right track with her therapy, soon she will speak for herself instead. Over time, I've learned the label "unreasonable" is often applied to parents who advocate for their child with a disability—a label I find is much less often applied to parents of typical children who want what is best for them. It has not stopped me from speaking on Alexandra's behalf when it seemed necessary.

In order to be an effective advocate, I've learned that timing is everything, style counts, and knowing the system and how it works is critical. It's essential to make use of all the resources at one's disposal. Not surprisingly, I've found it's not always easy to harness the emotional connection you have to your vulnerable children. On the one hand, it's what keeps you going and gives you the drive to do things that are difficult, draining, or taxing. On the other hand, it can make a person less reasonable or more emotionally involved than is useful. Finding the right balance of passion and calm has been critical, and 10 years into this journey, I figure I've gotten pretty good at it most of the time. This wasn't always the case.

It can sometimes be surprising what touches me deeply as Alexandra grows. In my experience, these moments sneak up on you, and it is often the simplest things that strike me as remarkable. For example, I yearned for the day that Alexandra would simply hold my hand and walk beside me without "bolting." When that day came, I was so overwhelmed with happiness that I called a friend and burst into tears. I wondered if Alexandra would ever form a sentence, respond accurately to a question, or be understood by others, and since then have seen her act in a play, answer the telephone, and be able to describe a toy that she had lost. When she and I enjoy a hot chocolate at the local coffee shop, especially when the event lasts longer than 15 min, a smile crosses my face. The day that Alexandra (and I) were invited for her first sleepover, I was deeply touched, both by the hospitality of a close friend, and next by the success of the whole event. It is these moments that happen in an ordinary week, which seem to really hit home for me. It's things like these that I sometimes took for granted with my typical son.

In the past, Down syndrome was traditionally characterized by its physical and intellectual manifestations. Modern-day descriptions are beginning to focus less on the pejorative physical descriptions. Today, they rightly make note that people with Down syndrome look more like their brothers and sisters than each other. Alexandra's beautiful almond-shaped eyes are something to be admired. From the moment she was born, I thought that Alexandra was absolutely beautiful and I still do. When the nurses handed Alexandra to us, I looked at her in my arms and told her so. She looked so much like her brother did at birth! As she grew, she became even more beautiful. She is now a lean, beautiful teen.

Not everyone, I discovered, sees Alexandra through my eyes. One of my most significant memories was of an otherwise kind acquaintance who wanted very much to "help." She sold nutritional products and she really wanted us to consider purchasing some for Alexandra. With all the best intentions, she explained that taking these supplements had been shown to change the appearance of people with Down syndrome so that they would look more "normal." I was so shocked, because to me Alexandra is beautiful the way she is—she looks like herself! Why would I want to make her look "normal" when, to me, she was extraordinarily beautiful? The biggest compliment for me is to hear that Alexandra looks like me! After I composed myself in private, I decided to share my thoughts and feelings with this acquaintance. I told her that Alexandra was beautiful in my eyes, just like her children were to her, and that there is nothing I would want to change about the way she looked. This was a surprise to my acquaintance, and that realization had a profound impact on me. I hope our interaction had a similar impact on her.

We have been fortunate to always have very caring professionals working with Alexandra. I truly welcomed the people who I saw would help Alexandra reach her potential. They made us feel as if we were not alone in meeting the challenges we faced. Many of them hold a special place in my heart and I am truly grateful for their guidance and advice. They have helped Alexandra walk, talk, eat, dress, and manage her own behavior. At the same time, it's not always easy having so many people who want to "help" in our lives, for such an extended period. We sometimes feel we live in a fishbowl and wish for a little more privacy. With our son, we had the typical well-intentioned friends, family, and even professionals giving their advice over the years. However, with Alexandra, this advice giving has been multiplied, and continues well into her late childhood. Even at age 10, Alexandra has a speech therapist, an occupational therapist, a behaviorist, a teacher's aide, and several doctors, all of whom provide us with goals, strategies, and techniques to help Alexandra reach the next milestone or overcome the latest behavioral issue. With every goal comes an assessment. Assessments can be exciting as you notice the gains that have been made. But, sometimes, they feel like microscopes on Alexandra's shortcomings. Securing funding, renewing funding, and coordinating services takes time, energy, and persistence, and often leads to frustration. In these ways, the services and support Alexandra receives are both the reason she is reaching her potential and one of the biggest sources of stress for us as parents. It can be difficult to reconcile these two ends of a spectrum at times.

From the start, I have always tried to leave as much room as possible to just be "mom" in the midst of all the therapy and goal setting. I'm pretty sure that parents are not a child's best therapist. As much as we need to follow through on the goals and strategies set out for us by professionals, in the end, I do believe something is sacrificed when we lose sight of our need to just be parents. This means being emotionally close, nurturing, and being "in the moment" with our children, as much as it means setting boundaries and providing structure. I have been more successful at being Alexandra's "mom" when I live more in the moment and keep connected to friends, family, and community.

I've learned over the years that the needs of your family as a whole sometimes need to outweigh the needs of your child with special needs. If what we do is going to be sustainable, it has to take into account this bigger picture. Each of us needs to be healthy and happy and fulfilled, and sometimes this means that our other child, our marriage, or our careers, take the front seat for a while. We have found that when this happens, Alexandra also does better.

Everything has a season, and there are moments when the best thing to do is take a break. We only have so much energy. We know we will walk alongside Alexandra over the long haul. This awareness of our own limitations means that sometimes we choose not to take up an issue, ask for something different, or swim against the tide on Alexandra's behalf. These are usually times when we recognize a need to recharge and reenergize. There have been months when the little things that could do with our attention have to take a backseat because we see something more important that will need our effort and attention in the near future.

Parenting Alexandra has also given me an important sense of perspective. As time has passed, small things do not become huge issues too quickly for me now. I am much more likely to laugh these days. Things that used to seem like big issues now pale in comparison to some of the challenges that she has faced and overcome. I am more likely to see the lighter side of things. Recognizing my own fallibility has made me a better person. I've seen that there is perfection in imperfection.

I could write much more about this journey that I am still on. I could write about Alexandra's remarkable brother who has also become her advocate, and the story of his e-mail to the school principal. I could write about the ways in which having Alexandra in our lives began other members of our family on life-changing paths. No matter what I write, at the end, I can truly say that Alexandra has changed our lives—mostly for the better. One day, I will thank her properly for what she has taught me, sometimes in painful ways, but always in ways that make me more human. I can't wait to see what the future brings for her.

Professional Story

More the Same than Different

Growing up as I did in a small town in a small maritime province, my exposure to differences throughout my childhood was limited. Not only were the residents of my community all Caucasian of mainly Irish/Scottish descent, but also the people with whom I played, attended school, and grew up around were all developmentally typical. When I entered high school, and my school community grew larger, there were a few students with developmental delays of varying severity, but none of these students were in my classes and I did not get to know them.

As I entered university to complete my undergraduate degree in psychology at a small maritime university, my "community" expanded again. Once again, I found myself among people who looked and behaved much as I did. I did meet and befriend individuals of different races and cultures from my own, which I very much enjoyed as an opportunity to expand my experiences. However, none of these individuals included anyone with a visible disability or a cognitive delay.

Upon completion of my first degree, I chose to travel to gain world and job experience. For 2 years, I traveled across several continents, living and working in three different countries. Because I was seeking to not only see the world but also to build my resume, I sought jobs as closely related to the field of psychology as I could. Among the many teaching and "helping" positions I held, I worked for a short time involved in "community care" for individuals with physical and/or cognitive disabilities of sufficient severity to require around-the-clock care in a group home setting. To successfully secure this "relief" position, the only requirements were my degree and security clearance via appropriate record checks. My interview was held over the phone. I was provided no training or preparation. I was told that I would be called when a shift arose for which coverage was required, and that I would be provided the address of the group home where I would be providing relief at that time.

After a short wait, I did receive a phone call that a shift was available for me. When I arrived, I was greeted by the one other individual with whom I would share the responsibilities of the shift. I explained that I was new to the organization and to the job. She gave me a brief explanation of the tasks that needed to be accomplished for the three individuals with severe physical and cognitive delays living at the home, and then split the work between the two of us for the day.

I can remember very clearly that I was afraid as I worked throughout that day. Paramount in my mind was the knowledge that I did not understand the needs of these individuals. All I could focus on was how foreign this territory was to me. What could these individuals communicate and understand? What reactions could I anticipate? What would I do if something unexpected occurred? I completed each task with trepidation.

Looking back, I now realize that my anxiety and uncertainty were due to a combination of factors. First and most obvious was the concerning absence of training in how to properly meet the demands of the job. Second, I now understand myself as a person naturally bent toward anxiety and perfectionism who is uncomfortable when uncertain about my ability to meet expectations. But most of all, and most unfortunate, I now know that when I looked at these three individuals, I only saw the aspects that were different from me. I saw the disabilities, instead of the people, that were before me. Instead of approaching them as people that I could get to know, and assist in the best way I could, I focused on the degree of their unique needs and the discomfort that was provoked by their unfamiliar behavior. I only did two shifts with that organization before I secured full-time work running an after-school program for typical children.

Upon my return to Canada, I completed a degree in education. Subsequently, I was a substitute teacher for a short period. During this time, I was not placed in a position where I was asked to interact directly with any students who had developmental differences. In those classrooms where there were students with developmental challenges, their educational assistants largely managed their educational and behavioral needs on the days I substituted for their classroom teacher. None of these students had Down syndrome.

After a short time, I undertook a master's degree in school psychology so I could become a psychologist, which had been the larger goal toward which I had been working for some time. While there, I was drawn to studying about learning disabilities and internalizing disorders. Topics related to developmental disabilities did not hold my interest at that time. Looking back, I now realize that the main reason for this complacency was that I had no real personal experience or frame of reference on which to hang this knowledge. When developmental disabilities were discussed, the pervasiveness of the impact of these diagnoses daunted me and left me feeling intimidated. As a result, I only gave these topics brief attention.

It was not until I was approaching the end of this degree that I started to learn comprehensively about Down syndrome. And to be honest, it was not by personal choice that I began this journey. I was seeking a job at a private mental health clinic where I had decided I very much wanted to begin my career as a psychologist. However, because I had not yet graduated, I was applying for a position as a psychological assistant. At the interview, I was told that a large component of my job, should I be hired, would involve the provision of life skills coaching to individuals with autism spectrum disorders and Down syndrome. I can clearly remember the anxiety that gripped me when I heard this news. I "didn't know anything" about working with individuals with developmental disabilities. I had no experience in the area. This population was "so different" from those with which I felt skilled and knowledgeable. This was not the position for which I felt I was applying, or for which I felt I was qualified.

Because I very much wanted the opportunity to work at this clinic, and because I was promised extensive training and support, I agreed to take the position being offered. As a life skills coach, I began working frequently with a number of teens and young adults with Down syndrome and other developmental disorders, going out into community settings with them to learn life skills in "real-life" settings. I began this job with that same anxiety and uncertainty about the unknown, because again, this population of people felt "just so different" from what was familiar to me.

With excellent training and support, I gradually developed my understanding and skill set. I learned visual techniques to support my efforts to effectively convey complex information to my clients. I learned to slow my rate of speech and reduce the amount of words I used in my verbal communication. I learned strategies to teach social problem solving to individuals with cognitive disabilities. I learned how to step back and allow my clients to make their own choices and experience their own successes and mistakes in the context of a supportive and accepting relationship. However, most of all, I came to realize something that has changed me profoundly as a person. As I got to know my first few clients, I rapidly recognized that, although they had aspects of themselves that were different from me, we were largely the same. Although I had always held this philosophical belief, I had never before had a real chance to test my ideals. As I developed a relationship with my clients and they began to share their hopes and disappointments, and as I bore witness to their frustrations and successes in their daily lives, I focused less and less on their uniqueness. Instead, I increasingly focused on each person with whom I was developing a relationship, and on discovering what skills each person needed to be confident and successful in his or her daily life. My enjoyment of my job increased in leaps and bounds, going from feeling anxious and dubious to passionate and privileged to have such a rewarding and important experience. Never since that time have I looked at any person, no matter how unpredictable or unfamiliar their behavior, without recognizing them as people first. For that, I am deeply grateful.

With my interest in working with clients with Down syndrome growing alongside my skill and knowledge in how to do so effectively, I began my practice as a psychologist with a number of clients with Down syndrome on my caseload. I provided both assessment and counseling services to this group of clients whose functioning was impacted to varying degrees by their diagnosis. I also became the director of coaching programs for clients with developmental disabilities. Eventually, I was invited to become involved in developing and implementing a new program designed to develop employment skills for teens with Down syndrome. This program required both a classroom component and a supported employment placement. My memories of my time involved with the teens in this program are among the fondest of my career to date, and are what I will focus on for the remainder of this piece.

The students in this program were not a homogeneous group. Some were diagnosed with a mild cognitive disability, were psychologically well, and had great success functioning in the social world. Others had cognitive disabilities that were more moderate in nature, as well as some considerable challenges interacting successfully with others. Still other students had psychological or psychiatric concerns. However, the students were remarkably similar in that almost every single one (and there were many who moved through this program) was noticeably excited to learn about employment skills. Most had a job or jobs that they were interested in securing. For the most part, the students were highly engaged in the learning, and very motivated to become proficient in the area. They all understood at their individual levels that having a job was a privilege, and that holding a job in the community represented responsibility, freedom, independence, and "adulthood." For those that deeply desired to be seen as "typical," holding a job was one more way they could be like their typically developing peers. Because these students had grown up in community with each other, they were, for the most part, friends, and were supportive of each other. This dynamic seemed to strengthen their excitement and pride as they shared this new experience with one another.

I consider this program to have been a great success. I watched most of the students go from nervous (some significantly so) and self-conscious, to confident and self-assured in both their job tasks and their job setting. Their increased self-confidence was evident in the way they carried themselves, the vigor with which they attacked their job tasks, and the manner in which they interacted with coworkers and/ or customers. For some students, the change was quite profound. I can recall one student who went from refusing to move without being led by the hand to working completely independently. I can recall another who went from refusing to speak to initiating questions and responding to questions by others on a regular basis. Another went from showing up in sweat pants and a T-shirt to coming to work in a dress shirt and dress pants. Still another went from sitting down and crying when things became difficult to responding in a professional manner to frustration. Still another went from paralyzing fear when asked to interact with supervisors to confidently expressing needs and accessing assistance when required. One student who could not communicate verbally learned how to make a valued contribution to a workplace by successfully communicating with coworkers through other means. Students learned to take the lead in negotiating for themselves regarding hours, time off, and shift changes. Some got to know coworkers and developed relationships. For some students, their increased self-belief positively impacted other aspects of their daily lives in terms of the risks they were willing to take and the attitude with which they approached new challenges.

Changes occurred not only for students but also for their parents. I witnessed parents go from uncertain and anxious about allowing their teen to enter the work world to pleasantly surprised and proud about what their teen was able to accomplish. From this experience, some parents began to feel a hope not previously experienced about what the future might hold for their child. Some of the students were offered the opportunity to stay on at their placements for paid employment. Because both the students and the employers were appropriately supported on the job site, some employers came to understand ways individuals with disabilities could have an increased presence at their workplace. I believe this change in view on the part of these employers had its root in their opportunity to experience that their employee was a person first, who happened to also have Down syndrome, as opposed to seeing their employee as "Down syndrome." Furthermore, employers had been given the chance to appreciate what I had observed to be common among my clients with this diagnosis, which was that the nature of their functioning allowed them to display incredible diligence, and both the desire and ability to complete a job exactly as it "should" be done. Most students did not tend to assign the same meaning to repetitive, routine tasks as did typically functioning employees, and as such, took great pride and enjoyment in completing these tasks thoroughly each time.

However, working in this program certainly presented some unique challenges. Not all students had positions that could be designed uniquely for their personal strengths and challenges. For some, the nature of their functioning made it very diffi-cult for them to manage a change in routine, a new expectation, or other requirements for flexibility. These students required extensive coaching and practice of a script for "surprises" and how surprises could be handled, both outwardly and inwardly. Education was also offered to their employers, which was received with varying degrees of success. For other students, behaviors related to their primary diagnosis, as well as those related to psychological or psychiatric conditions, resulted in con-cerns from employers. These students had to seek specialized and modified job place-ments that allowed them to make an important contribution to their workplaces while respecting the needs of the work site.

Another challenge we encountered was related to the role the students were asked to play in their jobs. Although all employers appeared to receive the students with excellent intentions, and made commendable efforts to work effectively with us, some did not feel comfortable holding the students to the same standards expected of other employees in the same position. For example, if a given student did not feel like completing all his job tasks on a given day, some employers felt compelled to make allowances for that to be okay. Although being treated fairly does not always mean being treated equally, we worked with these employers to strongly advocate for our students to be held to a consistent standard that was fair based on their level of social and job skills, just as any other employee. Other employers found it more challeng-ing to implement the accommodations, or to design positions, that would allow the students to be truly independent in their jobs. These employers tended to struggle in a different way with the statement that being treated fairly does not always mean being treated equally. Our efforts to team with employers to find workable solutions were not always fruitful. There were times where the students needed to move on to differ-ent positions, for various reasons, essentially starting over in terms of their learning and skill development as it related to their particular job.

As well, the social nuances that exist in the work world are incredibly extensive, excruciatingly subtle, and highly complex for someone with a cognitive disability. The students were coached (e.g., through both education and practice) at length about the social skills, rules, and conventions about which they would need to be mind-ful while at work. They role-played in the classroom and gave each other feedback. However, this coaching could not provide the students the ability to successfully navigate all of the many novel situations into which they were thrown. The biggest predictor of their success in the social environment of their workplace, whether with customers, coworkers, or supervisors, was the ability of those groups to see beyond the inappropriate behaviors to the person behind those behaviors, and give construc-tive feedback.

Similarly, an unexpected complication for the students at their job sites was that the social mistakes that occurred were not always on their part. That is, the students were sometimes the people in the interaction who were behaving according to the socially acceptable standard, whereas their coworkers, supervisors, or customers were the ones breaking the social rules. These scenarios were particularly difficult for the students to manage, as they had difficulty both understanding how to cope with the breach of social convention, as well as why they needed to maintain a particular behavioral standard when others were making different choices. For me, the social misunderstandings that occurred for these students were the most painful to witness, as the embarrassment, horror, and confusion with which they reacted once they understood what had occurred was often very poignant.

After finishing the program, some students continued on in the work world. Others applied to and were accepted by colleges in the area, where they attended modified programs. Many of the students with whom I worked were "trailblazers" in their job setting and later in their college programs, in that they were the first people with Down syndrome to ever hold these positions. I am proud of the way in which they represented themselves in these settings, making the way clear for others with Down syndrome who might seek similar opportunities in the future.

Common to me in all of these successes and challenges is the theme with which I began this discussion. I like to believe that the majority of people we encounter today would endorse the statement that all people have value, that all people deserve equality, and that people with disabilities are "people too." However, I also believe that a great many of us have not ever had the opportunity to put that belief system into action by being faced with a person whose behavior or needs are significantly discrepant with what is expected or familiar. From personal experience, I can say that, without this opportunity, there is the risk of seeing people with visible differences as more different from us than the same. I am appreciative of the opportunities I have had to be in relationship with individuals with Down syndrome. Not only has it enriched my knowledge and practice as a psychologist, but it has also resulted in me working harder to find the common ground between myself and others in all aspects of my life.

In reflecting on my experiences in this program and the trails that were blazed by these students, I am mindful of the fact that these students accomplished much more than they could have known. Not only did they make incredible gains in their personal growth and development, but they also created a path for others to do the same. Furthermore, I believe they changed the viewpoint of many people with whom they came in contact regarding what it means to have Down syndrome. In doing so, I feel they made a contribution to the larger picture of which they are likely keenly aware but not always able to articulate, which is that through increased understanding and exposure, they will be seen, received, and treated as more the same than different.

Epilogue

From our perspective, some themes within and across many of the stories in this section include the following.

- Many children and youth with Down syndrome have difficulty with communication (particularly expressive language skills) but, as they develop, they typically acquire relatively good social skills, can profit from supportive and effective vocational training, adapt generally well to their community life, and are often able to undertake unskilled to semiskilled work, with the guidance and support of others.

- There are many *inaccurate* stereotypes about individuals with Down syndrome:
  - They are always easy going and fun loving.
  - They are incapable of academic tasks, such as reading and math.
  - They are completely dependent (e.g., can't walk to school, always need help with toileting).

Specific thoughts professionals and parents expressed in the narratives we thought were insightful and that we have heard expressed in similar ways over the years include the following.

- When you first begin working with children and youth with Down syndrome, you immediately connect to the person, not the disability.
- There are typically four crises that parents of Down syndrome children have to address that are qualitatively different from what most parents have to address: (1) when parents are told they have a child with Down syndrome, (2) when the child first enters school and transitions through school, (3) when the child leaves school, and (4) when the child is on his or her own without his or her parents.
- Like all other children, children with Down syndrome can be obstinate, moody, temperamental, and difficult to get along with, but, also like other children, they can be fun loving, spirited, respectful, caring, and full of wonder, and need boundaries and clear expectations.

7 Developmental Coordination Disorder

Prologue

Children and youth who have developmental coordination disorder (DCD) have significant impairment in their motor coordination development. Their coordination difficulties are to such a degree as to interfere with their daily performance activities and are not the result of a medical condition or other developmental disorders such as mental retardation. Children and youth who have this disorder have difficulties with movement-based skills and activities (e.g., they have difficulty throwing, kicking, hitting, or catching a ball, and they display clumsiness when walking, skipping, or jumping, demonstrate problems with hand–eye coordination tasks such as handwriting, tying shoelaces, putting together puzzles, writing, and drawing) compared to what would be expected of individuals of similar age and development. Many children and youth with this disorder develop low self-esteem and self-confidence, which interferes with their academic achievement and social development.

In this section are three stories: one personal story, one parental story, and one professional story pertaining to DCD.

The personal story, entitled "Coming Out from Behind the Mask," is written by a 16-year-old girl who shares her story about what it was like for her growing up before she was diagnosed with DCD and what her life has been like since she realized she had this disorder. In her story, she reveals her hardships as well as her triumphs with undertones that indicate her courage of conviction and strength of character. The parental story, entitled "Battling On with Love and Commitment," is written by a mother who talks about her family's journey with respect to her daughter who has DCD. She begins her story with what she noticed relative to her daughter's motor skills and then moves into a discussion about some of the issues she encountered within the educational system and how her family coped with and overcame some of these challenges. The professional story, entitled "There is Much More to Coordination than Meets the Eye," is written by a psychologist who shares her reflections, developing awareness, experiences, and passion relative to her work with children and youth who have DCD.

Personal Story

Coming Out from Behind the Mask

My name is Tabby. I'm a 16-year-old eleventh-grade student. I live with my mom, dad, older sister, and three pets. I can be really shy, especially when meeting new

Exceptional Life Journeys. DOI: 10.1016/B978-0-12-385216-8.00007-2

people. Yet, once I get to know someone, it's as if I'm a totally different girl because I'm more comfortable around him or her. The cause of that shyness probably has something to do with the fact that I didn't quite fit in with classmates and was tormented for 5 years when in elementary school.

A bit more about me is that I love music. I love to dance to the beat of the music and am very passionate about singing; I practice on a daily basis. Because I tend to be so shy, it's been difficult for me to build up the nerve to sing in front of people. But over the past year or so, I've been getting a lot better with that, especially in choir at high school. Another quirky thing about me is that I am good with words, and I can be a bit of a perfectionist when it comes to writing—especially grammar. My best subject in school is English; it's the one class I tend to really excel in and I actually intend to pursue a career in that field. After high school, I would like to get a degree in broadcast journalism and print journalism, and hopefully someday work in television news. I am quite excited that this is going to be my first publication.

Before I was diagnosed with DCD, I was really quite miserable. I didn't get along very well with my classmates—or my mother. You see, she didn't really understand me. She had no idea what I was going through at school or how unhappy and depressed I truly was. It's not that she didn't want or try to understand; it's that she literally couldn't comprehend what life was like for me. She was always just so busy. Even when she did try to talk to me about what was going on—so she could understand and try to help— she didn't really have enough time to truly find anything out. That's because for me to actually explain it all to her would take much more than just 20 min before dinner.

Ever since I was a little girl, I've pretty much been a klutz. I was always falling, tripping, stumbling, walking into things, or hitting my head. I had practically no coordination whatsoever—my mom and I would even joke about how I was so clumsy and accident-prone whenever I would fall on my butt. This klutzy trait I obviously possess is one of the many things I was ridiculed about through elementary school. Because I was so uncoordinated, it took me much longer to learn things—like riding a two-wheeled bike. But I never stopped trying. Every time I fell off, I would pick myself back up and get right back on—time and time again. Because I never gave up, I eventually did learn how to ride when I was in sixth grade. It just took much more time and many scraped elbows and knees. Once I learned, it also meant that I was no longer afraid to ride my bike to school, and classmates couldn't tease me about not being able to ride a bike.

Even my hands were clumsy; it always took a long time to write because I had poor printing. When in class, I had trouble putting my thoughts and ideas down on paper. When we had to write notes, I had problems keeping up with the class. Where the other kids would take one line to write their letters, I would need to use two because I wrote a bit bigger. Even when I tried to write small, my hands would shake and so I was unable to do it properly. Overall, my printing and writing were a lot messier than the other kids. I also had problems concentrating and was easily distracted; that meant I got confused easily with certain subjects—especially math. Things like that are what put a target on my back early in school.

Over time, that target just got bigger and bigger and I became more and more lonely. Somehow, my character, personality traits, and actions made me come across as being "slow." That is putting it nicely. I was constantly called names like "stupid, moronic, retarded, ditz," and many more; they even altered my real name at times. Classmates also used a hand signal to refer to someone as a "retard" (I won't explain it so others can't use it). The teachers didn't know what the gesture meant so those other kids never got in trouble. That's a sample of the treatment I got from my classmates multiple times a day throughout those 5 years of life.

None of my classmates took the chance to get to know me or really see the kind of kid I truly was. They all just listened to the rumors and lies that other students spread about me—or they judged me by the way I looked and moved. Because of that, I was very, very lonely. Every now and then, a couple of kids did give me a chance and discovered that I was actually a nice girl who was just an innocent victim. A few times, these students even befriended me. Unfortunately they would never stick up for me when I was being tormented by others—even when I requested their help.

Eventually, I asked them not to stick up for me because I knew that it would only make things worse—not just for me, but for them too. I didn't want anyone having to go through the same kind of stuff I went through, especially if it was because they were trying to *help* me. This made me feel very lonely and sad.

In addition to being hurt by words, my classmates also shoved me off the playground, and tripped, pushed, or did other things to hurt and cause me pain. But that usually happened outside. They were very careful not to hurt me too badly so nobody would suspect I was getting beaten up. That way no "sensation" would arise. I am literally covered with scars; most of them are from the physical abuse I received at school. Probably 75% of the scars on my body are ones I acquired from the kids I went to school with over those 5 dreadful years. (I'm sure the other 25% is my own doing because I am a klutz and I'm not afraid to admit it!)

Because I did not want to be a snitch (and give the kids *another* reason to target me), I never told others about some of the abuses I suffered. Upon occasion, I would tell a teacher—and in so doing, try to get help, but no one ever believed me. Report card comments indicated that I was a tattletale and too emotional when explaining things. People didn't seem to know it was my cry for help. So, even teachers made my life lonely.

I kept every feeling from those 5 years bottled up inside because that is all I really knew how to do. I couldn't vent to anybody because not only did I have no one to vent to, but also nobody would understand anyway. After a while, I gave up on myself; I put on a brave face and acted as if nothing was wrong, so as not to worry my parents. But my mother could tell something was wrong—she just didn't know *what*. I would confine myself to my room all the time and usually only come out to get food, go to the bathroom, or have a shower. Quite often she would see me come home with new scratches, cuts, and bruises, but I would pass it off as a result of my clumsiness. One part of me wanted to tell her what was really going on, to tell her *everything*, but I felt it wouldn't help. I knew that, if anything, it would have only made things worse. I did not want to worry her. I was also afraid that if I told my parents, they might go to the school staff and then I was scared that the other kids at school might think I was a wimp and could not handle myself. By the middle of third grade, I was crying myself to sleep every single night. I think I was forced to grow up too fast and too alone, and therefore, I never really had a childhood.

Over time, I changed from being confused, worried, and sad to being very lonely and angry. I was never invited out to other people's houses for birthday parties and, because I was not very good at sports, I seldom went outdoors to play. I just stayed home and watched television or went on the computer. I snacked a lot, started to grow bigger, and gained far too much weight. That was just another reason for schoolmates to tease, ridicule, and bully me. Life was really horrid.

The first time I was "attacked" and really injured badly was on my seventh birthday. I don't think it was a coincidence. During morning recess, I was struck on the back of my neck by a very big and hard snowball while trying to climb up the ladder of the slide. It hit with such force that I was knocked forward. My knees bumped up against a step, I lost my balance, and then fell off the playground equipment. A large sharp rock became lodged in my right knee when I hit the ground. Immediately my knee started gushing blood. A couple of sixth-grade girls saw what had happened and rushed over to help me. When they saw how badly my knee was injured, they acted quickly by picking me up and carrying me into the school as fast as possible. I got help right away. The cut was very deep because I remember once it *finally* stopped bleeding, you could slightly pinch together the skin on my knee, look into the hole made by the rock, and see all the way to the bone!

What I remember as most beneficial that day was that once the sixth-grade girls got me to the nurse, they didn't leave to go enjoy the rest of recess. They stayed with me and tried to help in any way possible while the nurse took the rock out of my knee. They asked me things like what my name is, how old I was, if I had any siblings and pets, and what my favorite things and activities were. They also told me about themselves. I remember one of the girls told me that she had a little sister about my age, which is partly why she felt compelled to help me. But the biggest reason they decided to help me was because they saw what had happened—not just that I had been injured, they saw *everything*. They saw who threw the snowball. They saw how the person had timed it—waited for the perfect moment to strike—and how he had aimed it right for me. They said it looked like the person was aiming specifically for my head! They were completely shocked when they found out it was my birthday.

Sadly though, when they explained that they had seen the whole thing take place and told exactly what happened, the principal didn't believe them. He said that they must have been seeing things because, "No 7-year-old child would do such a thing— try to purposely hurt another young child, especially a little girl on her birthday." The two older girls were really angry about what the principal said. In fact, they felt so bad for me that they apologized for the principal's behavior, comments, and denial. They even stayed with me for the rest of recess to make sure nothing else happened to me. I still have that scar on my knee, and I know it will *never* go away, because the injury was so bad.

That day, two things became very clear: one, this was only the beginning and my problems would only get worse; and two, nothing would be done to stop any misbehavior toward me because no teacher would believe that a young person could be so cruel to another child. I am telling you now that my seventh-birthday attack was my *second* worse experience.

The worst experience of my life took place in fourth grade. It was extremely traumatizing; to this day, I still have nightmares about it. As such, I will not go into great detail. However, I know that, in a very purposeful way, a fellow classmate tormented and planned an attack against me to the point of trying to strangle me by pulling on my neck with the hood of a sweatshirt. There was even a brief moment when I saw my life flashing before my eyes.

That schoolmate had carefully planned out a specific bullying action during school so I would need to spend time in the bathroom cleaning up before I went home. While I was cleaning up in the bathroom—after classes were done—he waited outside the building for me. Then he followed me across the play field on a very cold and blustery day; everyone else had already left the school grounds. It seemed very obvious that he had planned it out and was trying to really hurt or kill me. When his attack was over and I had the courage and strength to go home, I decided not to report it; if I did, it would have been my word against his. Like I said before, no adult would believe a kid could be so cruel that they would try to hurt another kid, so there was no way they would believe one child tried to kill another.

My life was miserable in every way. But hope was on its way. After a close friend of the family saw how "klutzy," sad, and lonely I was, he eventually spoke to my sister and she talked to my parents about having me tested for learning problems. My mom was especially happy to get some help because she did not know what to do with—or for—me anymore. Soon after, I was diagnosed with DCD and ADHD. Things became easier for me because I actually knew what was going on. It felt much better to know that there was a reason for all my difficulties—it made sense in my head. Once the psychological report was done and given to my mom and dad, as well as my teachers, they better understood what I was going through and how frustrating it can be when you are struggling with some of the simplest of motor tasks. Also, the strategies the psychologist gave me helped me do much better in school.

Things the psychologist suggested worked, which included getting the notes from class, accessing extra help for spelling and writing, having more time to complete assignments if needed, having more access to computers for completing tasks, and reducing the number of questions in homework and written assignments.

After I was diagnosed, I was also able to get some extra help in a specific class. I did *not* need help for reading, but I did need help for writing. While in that class, I got to know some other students and eventually learned that I did not need to be afraid to interact with those classmates. Now I am a little more outgoing when meeting new people. It also took about 1 year to start to trust that the teachers were really looking out for me.

I also participated in a sports day camp after I was diagnosed with DCD. There, I learned how to do different skills using the proper form, and I had a chance to practice the skills in a safe place. I got to learn and practice with other students who were about the same age and also had trouble with motor skills. After that camp, I was able to do a lot better in gym class and when playing sports because I could understand why I had to move my body a certain way. Because I learned how to utilize my motor skills to my advantage, I became good enough to fulfill one of my personal goals: to make at least one sports team before I reached the end of ninth grade.

To be exact, I made the senior girls basketball team in my final year of middle school; we ended up having a great season. And in the last few years I have also become really good at playing baseball.

There were a few other good things that happened after I was assessed and diagnosed. The psychologist convinced my parents to meet with a pediatrician and, soon after, I started medication for ADHD. I was able to focus much more at school, and that helped me do better in all subjects. It also helped me to lose some extra weight. That and help for my writing and other motor problems really started to make a big difference. But my troubles were not completely over. You see, although I think I hid it well, I did slip into depression for several years. I started to keep my emotions bottled up inside more than ever before—like a time bomb waiting to go off—I knew I was going to crack some time, I just didn't know when or where it would happen. That emotional time bomb finally went off when I went to summer camp for a week. At the camp, I met an amazing person who I instantly bonded with—as he was able to relate to my past. He was someone who had also endured much teasing and ridicule as a youngster. He was the first person I ever met who could see past my "happy mask." He saw how miserable I was and that I was still haunted by my suffering. I told him about my past, about the traumatizing experience I suffered in fourth grade, and the depression that followed. Then I had a very difficult and emotional night; I broke down and sobbed for several hours—something that *needed* to happen. I've felt immensely happier and much better since then—both physically and mentally.

I am working hard to pick up the broken pieces—trying hard to put myself back together again. I still have a long way to go before I am completely fixed and whole again. But, I have hope now. If I were to give one piece of advice to others going through similar stuff in their life, I would definitely tell them this: Don't make the same mistake I did—don't bottle up your emotions if you are left out of things and/or ridiculed at school. To be clumsy and unable to participate in sports is hard enough on yourself—it's even worse when others tease, bully, and ridicule you about it. Do talk to others about your problems, or all the stress and anxiety will build up on your shoulders; it will eventually come back and bite you in the butt, and you will crack. Oh yes, no matter how hard you try to fight, you *will* crack!

Too many things in this life are sugarcoated or dismissed as not being important enough to worry about. That doesn't change anything, and it certainly doesn't help you deal with problems. With difficulties like DCD and ADHD, people need to be told the cold hard truth so it will stick inside their brain—otherwise nothing will change and the vicious cycle of bullying and being left out of activities will never end. You see, when kids can't perform motor skills properly, it doesn't mean they are slow learners, and they should never be teased and ridiculed because they move slowly or look clumsy!

Don't judge a book by its cover. You need to get to know people before you can judge their character and personality. Then you can decide whether or not you like them. After you get to know them, if you decide you don't like them, then that's fine. You don't have to like every person you meet—but that does not mean you have to belittle them because they can't keep up with others or do things the same ways as you. If you want to meet someone just like you, join a club! That way you won't have to be with—or around—people you don't like.

Finally, don't ignore the problems of being clumsy and having DCD because, if you do, many people will continue to suffer—like I and many others already have.

Parental Story

Battling On with Love and Commitment

We are a fairly typical family. My husband (Tyler) is a social worker who has worked in child protection and as a behavior specialist within the school system. I have worked in the world of special needs—initially helping to support adults who lived in the community and now working with children who require assistance in the school system. We met and got married in our early twenties and, after feeling secure in our careers, began trying to start a family. It is through our family life—specifically our 10-year-old daughter Beth—that we are going to talk about DCD. As parents, we are going to share our story to help you understand our journey and learn how we have arrived at this point in life—more or less intact.

To start, we need to tell readers how we became a family. We had tried for 10 years to have a biological child but for reasons that no one was able to tell at the time, we were unable to conceive. Although we had thought about adopting for years—as both of our families have made numerous adoptions—it was an option that we believed was going to take many years. Because of that, we were also fully aware there was no guarantee that we would be at a place in our lives to adopt when a child eventually became available. Finally, in November 1999, we decided that it was time to seriously look at adoption.

Whether the stars were aligned or it was simply meant to be, it all worked out perfectly and quickly. The process started in March 2000, when we met our daughter's birth mother; she had chosen us to be the parents of the child she was carrying. While pregnant, she had taken good care of herself and our little bundle of joy. Our hearts still feel like they might burst whenever we think back to that early morning in that small rural hospital where and when Beth was born. She was placed in our arms and our family life began. We are now blessed to be the parents of an amazing child.

Shortly after Beth was born, we settled into a comfortable routine. Tyler continued to work outside of the home while I decided to stay home with Beth. As time passed, Beth grew and developed. Though we are certainly biased, she is one of a kind. For instance, even as an infant, Beth was even-tempered. She rarely cried where there was not an easily identified remedy. Feed her, change her diaper, snuggle her up, and the crying would be done. She also grew and developed as expected. Although some of her skill development was a little later than average, she was still meeting most developmental milestones in due time. She sat up, crawled, cruised, and walked. She squealed, giggled, laughed, and cried. It seemed that her early motor skills were totally on par with her peers.

As time went on, there were a few small clues that Beth might have some issues with her motor skills. We started to notice that she was very cautious when she moved. You could see her concentrate quite hard and analyze everything going on around her. Then she would tentatively take steps. This in and of itself caused us no concern whatsoever. We thought, "Hey, she just doesn't want to fall." As time went on though, we began to notice that her ability to plan and do different motor tasks and activities was different than most others the same age. She seemed to have trouble learning how to perform tasks properly. An example of this was her method of going up and down stairs. Initially, she began going up and down stairs using two feet on each tread; left foot up on a stair, then right foot up on the same stair. Later, we learned this is called "marking time." She would repeat this until she got all the way to the top and then do the same action when coming down—all the while holding on to the railing and taking extreme care with each move. This pattern of navigating the stairs continued from 3 years of age until she was approximately 8 years old. Though we would show her how to transfer and shift her weight from one foot to the other on alternating stairs, she was clearly not comfortable doing so. Even using instruction and modeling of hand over foot, Beth continued to use the stair-climbing technique that she was most comfortable with.

Other tasks were a challenge for her as well. We noticed that she had problems with things like doing buttons on pants and shirts, using utensils, brushing her teeth, climbing, and jumping. As she got older, we began to get more and more concerned. We started to worry about other kids teasing her—as we could see her peers' movements becoming more fluid and confident while hers remained tentative, clumsy, and awkward. Knowing that, as kids get older, they begin to see anything that is "different" from their skills, we knew that eventually someone was going to say something that might hurt Beth's feelings. We were glad that we had focused on developing a strong personal identity and self-esteem in her from an early age already. She had been taught and therefore knew that everyone was different—but each had strengths as well as areas that they were weaker in. Originally, we had done this type of "inoculation" to prepare her to deal with some of the issues related to being an adopted child—especially because she was from a different ethnic culture as us. That way, she could answer the question, "Why do you look different than your mom and dad?" We like to think that this early teaching about her identity helped Beth develop a strong sense of individuality and self-worth.

The biggest challenge for us as parents has been getting professionals to view our daughter objectively. She is quiet, though not shy, and she is not one who jumps around and makes a lot of noise. So people don't think she has problems. I recall the day when we took her to school for the first time. We had homeschooled Beth for her kindergarten year. At the end of that term, we also purposefully worked with the school to make sure that she was ready. All agreed, and so Beth was registered for first grade. On the first day of the new school term, we sent Beth into a different world of learning. Imagine then our reaction, when picking up our daughter after that first and important day, to be told by the teacher that Beth was in the wrong class. The teacher said she should not be in first grade because Beth could not throw a Frisbee and was still 5 years old. Never mind that she was reading at a second-grade level and doing simple math. We were totally taken aback! Due to staffing changes over the summer, no one had passed on the details about Beth's problems with motor skills. That was the beginning of our battles with the education system.

Over the next few months, we were also told that Beth was not as socially mature as her classmates. This child, our daughter, had been taught during that year at home to sit in her chair, raise her hand, listen to the teacher, and generally follow instructions in a classroom setting. These well-developed skills and behaviors—which would make her an absolute joy to teach in a few years—were then seen as being evidence of some kind of social dysfunction. The school staff informed us that, because her peers were not doing these things, there must be something wrong with Beth.

That was only the beginning; many other struggles with the education system have continued. We believe that many factors are at play when trying to understand the many difficulties parents and their children with learning disabilities encounter when trying to navigate their way through the school system. One factor is that the basic university training in education does not prepare many teachers to address disruptive behaviors and challenges in the classroom; this then creates a situation in which a teacher must learn about classroom management "on the fly," while delivering a curriculum. This is very difficult to do well—unless you are a phenomenal multitasker.

Another factor is what we call the "double message." Teachers say they want parent involvement in their child's education—studies even show that it is needed and desirable for a positive outcome for the child. That being said, having parents actively involved is a lot of work for teachers. It means a greater degree of communication, consultation, and collaboration than teachers have training for. It also means that teachers must allocate time and energy to respond properly to parents. Although we are great advocates for our child and her needs, many other parents also want to advocate for their child's needs as well. Our demands often compete for the teacher's time, energy, and other resources. That school staff member may become overwhelmed, and miscommunication may result; parents then feel like they are meddling in the school or classroom.

The final factor is what was mentioned previously; we are really only concerned about our own daughter. Though we can sympathize with the school and its need to balance competing demands, we can honestly say that we are more concerned about our daughter getting what she needs than other children. Beth has a learning disability in the motor area and she has the right to be educated with the supports she needs. Getting those needs met is the challenge we have.

Neither of us are timid people. We both have assertive "no-nonsense" personalities. This trait has been our greatest asset in navigating both the education and health systems. We have long known that we are Beth's best advocates. *No one* within either system is more equipped or knowledgeable about your child and yourself than you. If you do not respectfully push the issues, no one will take up a banner for your child and say, "This child may seem to function okay but she really needs some extra help or accommodation." To the contrary, in our experience, if you are not actively drawing attention to the needs of your child, the professionals in the systems are going to focus on children who appear to have more significant challenges—or they may focus on students that are gifted. From what we have seen, professionals do not have the time or choose to allocate resources for children in the middle, especially those who seem like capable learners and have no behavior problems, but still really struggle in other areas. We have not yet been able to overcome that challenge. But we did overcome another challenge—the challenge of getting a formal diagnosis.

Because of Beth's ongoing motor difficulties but her ability to read, focus, and do math when at school, no one thought she had a problem. We felt much resistance by educators and medical doctors when asking them about Beth's troubles with learning motor skills. But we were not convinced. Eventually, in 2008, I started to spend time on the Internet, searching for information on motor learning problems. I had no idea what to look for, so started typing in terms like "8-year-olds who can't button" or "8-year-olds who can't cut their food." I came to different parenting sites and eventually heard about something called DCD.

I then went to our local health center and asked the nurse about DCD. The nurse said that she had never heard about it, but she referred us to see a general medical practitioner. When we met with him, he also commented that he had never heard about DCD. That doctor also told us that we were being neurotic and that we should stop worrying, as not all parents should expect their child to be an Olympic athlete. I responded, "But we should be able to expect that our child can dress herself." That must have made sense to the doctor as he eventually agreed to make a referral to a pediatrician. Fortunately, that pediatrician knew about DCD—and even admitted that he suspected the presence of DCD in our daughter, but he stated that he did not feel qualified to make the formal diagnosis. So, he agreed to refer us to see a physiotherapist and occupational therapist for some testing of motor skills. They conducted formal assessments, but then informed us that they were also not qualified to make a formal diagnosis of DCD.

Next, the pediatric occupational therapist recommended a specific book on DCD that she had bought some time before. We followed up immediately and bought and read the book on DCD. Thinking a specific book on the topic would offer credibility, we then showed the book to the school staff, showed them a website on this topic, and asked to have Beth assessed for this motor learning disability. They did not seem to believe that such a condition existed, and they did not appear interested in learning more about DCD. So, we had to make our own arrangements to travel to a far location to eventually obtain the diagnosis for Beth. But it was worth it! The affirmation of Beth's motor learning problems (DCD) was a significant "win" for us because, for years already, we had known there was something different about her ability to learn motor skills. We now knew for sure what it was. "It" was DCD. Now we have strategies for her educational programming and want other parents to learn about this condition as well.

Some other battles won have been small but significant. Our connections with local occupational therapists have certainly helped overcome some of the motor problems that our daughter struggles with. The occupational therapists helped Beth learn skills, such as tying her shoes, riding her bike, and skipping rope, that we were not sure she would be able to master. After working with the therapist for a couple of weeks, Beth was finally able to use the skipping rope properly and jump over it twice in a row. We recall that moment, as the look of pride and joy on her face was amazing! Seeing her achieve a physical goal is very satisfying for a parent, as we know that she is proud and inspired in having achieved it. Working hard and helping Beth to learn to ride her bike was also a very significant event in her life—and ours.

At times, it seems like we are soldiers fighting a never-ending war against an implacable foe that has more resources than we do. We do win battles here and there—only to face new ones. But, we battle on—for our Beth. As parents, that is what we do. We have to. We see her fight daily with her body not being able to do what her brain is telling it to do. Thankfully, she doesn't give up. She is a trooper, a true warrior who is proud of who she is and what she has accomplished!

There are many small victories that keep the three of us going, but we all know there will be more to come. As of today, we are in the process of working with her school again. Now that Beth has been officially diagnosed with DCD, we need and have the right to have an IEP in place for her. However, it has been over a month since the report was given to the school staff, and, although the promises of creating a specific IEP for her have been made, and phone calls and other communications have been made—but not returned—and accommodations have been discussed—but not implemented, there is no plan yet. ... So, it is the old refrain, "Once more unto the breach, dear friends. ..." We battle on.

Professional Story

There Is Much More to Coordination than Meets the Eye

This story is going to be about DCD. "What on earth is this?" you ask. "I've never heard of coordination problems being a 'disorder' and affecting psychological health and/or functioning! When will they stop creating all these new conditions?" You may be surprised to know that you are not the first person who has made that comment, and you probably won't be the last.

I am honored to inform you about DCD because you will soon come to read that this condition is very credible, has been researched internationally for over 40 years, and is absolutely related to one's emotional health and well-being. In fact, research and case studies clearly show the interrelationships between motor problems and psychological issues; indeed, that is why DCD is included in the APA's *Diagnostic and Statistical Manual*. What may also surprise you is that a similar motor problem (using different terminology) is identified in the well-known *International Classification of Diseases*—which is published by the World Health Organization.

In addition to my life at home with my husband and our four biological children (who are all young adults at this point in life), I am a registered psychologist, teacher, and kinesiologist. I have studied and worked in education, psychology, and human movement sciences since the mid-1970s. In fact, it is my formal training, university degrees, employment opportunities, and deep interest in psychology, the human body, education, and learning processes that have made DCD a natural fit for me. But you will also learn that my current interest and fascination with DCD is rooted in my experiences and observations as a young student.

Let me explain. I am one of eight children born to a young couple who emigrated from the Netherlands to Canada after the end of the Second World War. Because my father was a potato farmer and landscaper, we needed to live in a setting that would enable him to grow the various trees, shrubs, plants, and potatoes. As a result, the 10 of us lived on acreage. Years ago, that meant we attended a one-room country school with other neighbor children. A few years after I started attending that school, the county trustees made a decision to close that building, so in the fall of third grade, my school-aged siblings, the other neighbor children, and I were bused to a bigger school in the nearby city. I recall very clearly that the size of the play field at the "big" city school was much larger than the grass area we played on when I attended the little country school. I also recalled that, at our little one-room school, all of the children explored different areas of the play area, ran around the perimeter of the school grounds, and played together in other ways. No one ever stood alone.

In contrast, at the bigger city school—which contained about 600 students from first to ninth grades—I clearly remember seeing certain students standing alone during the recess break or at lunchtime. Yet, when in class, I knew that these same-aged students were capable learners, as they answered the teacher's questions and contributed to class discussions. As such, I remember being confused and wondering why certain students did not get involved in regular child's play and physical activity during the unstructured play times of the school day. These children had been a puzzle for me for a long time. Thanks to many different learning opportunities and life experiences, I believe I now know much more about these youngsters and why they were all by themselves during recess and lunch breaks. Identifying and helping these individuals has become my passion in life.

Specifically, during the last few years, I have come to understand that these students probably had a learning disability in the motor domain. They probably had undiagnosed DCD. You see, a motor learning disability and DCD are one and the same. Now I have an unquenchable desire to educate others about DCD and support these students and their families. Specifically, I want parents, educators, therapists, and other professionals to discover how to identify these children, learn how to make the needed diagnosis, and then know what to suggest for proper intervention and support(s). I am passionate about this because I continue to hear and see that the psychological implications of clumsiness and motor learning problems are substantial. The movement difficulties impact mental health and even physical health within a very short time. Yet, once identified and supported properly, these students usually have a very positive response to direct instruction and intervention.

As a psychologist who works in this unique area of motor learning disabilities, I have been given the incredible privilege of helping parents understand that they are not making things up—they are absolutely correct in thinking/sensing that something is wrong with their child's motor actions. Unfortunately, most parents do not know what is causing the movement difficulties. I have also learned that parents do not know what to search for when looking on the Internet, what the exact problem is called, and/or what to do about it. Indeed, they do not even know how to find a clinician who might be able to help them.

Since learning about and specializing in DCD, I have had the chance to meet families, and then discuss and learn about their child's problems via formal testing. After parents fully comprehend my assessment results, it seems that they are totally relieved to hear how all the "dots line up." Everything starts to make sense, and the parents are truly comforted to learn that their gut feeling was right. They knew something was "off the mark" about their child's general motor skills, and now with a diagnosis of DCD for their child, their questions and queries are finally answered. They now know how to move forward. And they do move forward.

However, as gratifying as this career is now, there are still many challenges and barriers in this field. For instance, it is a constant source of frustration to me that so many of my professional peers do not know about—or understand how—children who have motor learning disabilities could end up with very significant psychosocial and emotional distress in a relatively short time. I think that is because issues related to motor learning never even crosses their mind. In fact, there seems to be a general resistance to acknowledging the importance of the motor domain in the world of psychology. That is in direct opposition to Bloom's taxonomy of educational objectives, in which he identified three main areas: the cognitive, affective, and psychomotor domains. Although his seminal work was generated in the mid-1950s, and he too focused much more on the cognitive and affective realms, I believe that his fundamental theories were sound. You see, the motor domain is fully associated with human learning, development, and behavior. Why, just think about how Piaget's theories of cognitive development help to substantiate the connections. And we regularly ask students to use motor skills to explain or show what they have learned.

In some sense, my colleagues and graduate students can't be blamed for their lack of knowledge because most psychologists have little to no experience or training related to motor development and/or learning. In fact, very few college or university professors in psychology even know about the psychomotor domain. They tend to leave that area to occupational therapists and physiotherapists. Psychologists prefer to focus on one's cognitive processes, emotions, and/or behaviors. Nevertheless, issues related to motor learning disabilities and DCD truly lie in the field of psychosocial health. In fact, some students I have diagnosed and worked with display features of significant trauma as a result of the bullying and teasing they endured because of their weak motor output and/or poor coordination.

My journey through graduate school was rather challenging. Most of my fellow classmates knew that I had a background in physical education/kinesiology. As a result, I endured some social "dissimilarity" from my classmates. That's because there tends to be a feeling in society that most athletes aren't very bright ("How did *she* get into grad school if she is a jock?"). Well, if my classmates only knew—I was not a natural athlete. And, even if I was a good athlete, it should not mean that I am not a good thinker. Indeed, my interest in this field was born out of a sense of wonder and a curious mind. I am completely awed by the human body and how it moves; I also have a zest to understand the factors involved in motor development, skill acquisition, and learning processes—and how human movement relates to psychosocial wellness. As a result of those interests, I often asked my university professors to consider how various learning principles, theories, and/or the models used in psychology applied to the motor domain. They seldom had an answer—or would simply state that the topic being studied did not suit the psychomotor area. However, I must say that when given the chance to apply various theories, learning models, and other higher-order concepts in psychology to the motor domain, this graduate student was always able to show my instructors that there was an amazing and close relationship—often to their complete astonishment!

During the last 10 years, I continue to feel like an outsider or oddball when asking questions or making comments about the psychomotor domain when attending workshops, meetings, and/or other such professional gatherings. I sometimes have the feeling that people are tired of my comments. In recent years, I have started to remain quiet—even though my natural desire to educate others about this topic makes me want to shout out. Fortunately, there has been a slight change in recent years—with more attention given to DCD now than ever before. Yet, I remain somewhat discouraged that most educational psychologists in North America regularly deal with learning problems in the areas of reading, math, writing, attention, and memory—but seldom consider issues related to motor learning. Usually the underlying relationships and patterns are already evident; it's just that the school psychologists, family doctors, teachers, and other professionals simply have not recognized motor learning problems as causing various emotional, behavioral, or other psychosocial problems.

You see, when uncoordinated but intelligent students are not receiving needed supports, they usually start by asking for help, eventually defend their self-worth, and/or then choose to deflect away from their areas of weakness by using emotional/behavioral actions. It is like a three-step recipe that creates the same product almost every time. To many people, it seems like these students often tattle on others, act too emotional, are too moody or lippy—or, they become withdrawn and quiet. Yes, they may act unmotivated, oppositional, and be resistant to writing or engaging in physical education, but not because they are bad; they want to cover up their poor motor skills.

I also feel very sad and frustrated when I hear about the excessive costs in time and health care dollars that have been spent by most families—while they try to determine the cause of their child's motor problems. Researchers have shown that parents of children with coordination problems often visit multiple professionals (up to 16 different visits) to try to get answers. Sometimes they receive partial solutions, a misdiagnosis, and/or the information that they "do not need to worry." There remains a general perception in society that people should not fret just because their child cannot use a fork or knife properly, ride a bike, print or write neatly, manage buttons or zippers or shoelaces, and/or perform other fundamental motor skills in the same way as their age-matched peers. Most people think that if you can read and do math, you are going to be okay in school/life. I have learned and am convinced that this is completely untrue.

Think back to junior high school when you waited for the "athletic captains" to pick their teams in physical education. If you were one of the last two or three students standing against the wall, you wanted to shrink away into nothingness while the athletes pointed at and picked classmates one at a time. When there were only a few students left, the captain would sometimes turn to other members on the team and start to plan their strategies for the game. It was as if they didn't even see the remaining people anymore. Sometimes the physical education teacher simply said, "You, on that team and you go there." Occasionally, the gym teacher did not even know your name. What is worse, when you were quickly assigned to a group, the teammates would often say something like, "Now we are going to lose because she or he is on our team." But the absolute worst was when the team LOST. They would actually blame it on you. It's these students—capable, but "klutzy" and uncoordinated—that I am referring to when I talk about students with DCD. Over a very short time, and without a strong personal identity or proper supports, these clumsy and sluggish movers tend to lose any shred of self-confidence and develop feelings of low self-worth. By the time these students are in the upper elementary grades, you may see them walking by themselves in the hallways, heads positioned downward or wearing a "hoodie" to cover their head. They are usually very sad and move in a slow and listless manner. Well, those are my kids! Although you may not recall their names (because they usually pulled back socially and no one knew much about them at all), you may still have a mental image about those students. Let me tell you a bit more about these children and teens.

Students with DCD are actually good learners but have trouble with the motor (or movement-based) skills required of them. For instance, many of these students struggle with penmanship, and they may not know how to throw, kick, catch, or hit a ball. They may struggle with balance-related tasks and look clumsy when trying to run or skip. Because they do not know the cause of their difficulties, these students frequently develop anxiety, and over time, may gain extra body weight and become angry, depressed, and/or discouraged. Furthermore, many children start to withdraw or isolate themselves because they cannot manage hand–eye coordination tasks and/or fundamental motor skills. Then, self-confidence, academic achievement, and school attendance also become affected in a very negative manner. What most professionals and parents fail to understand is that this is one of the most difficult types of learning disabilities because you cannot hide clumsiness. It is evident from a very young age and, in contrast to popular belief, DCD does not go away as you get older.

Please don't get me wrong: I am not saying one type of learning disability is not as serious as another. However, although you might be able to hide a reading, writing, or math learning disability for some time, you cannot hide poor motor skill proficiency. Everyone sees clumsiness. Individuals with DCD could and should have been identified many years *before* developing the secondary psychosocial issues.

Whenever colleagues, teachers, and/or parents ask for information about DCD, I provide them with checklists, assessment tools, intervention ideas, and other resources. I also provide workshops, seminars, and all manner of professional development on this topic—simply because I want professionals and members of the public to be educated about this important topic. Then I choose to explain things in a simple and straightforward manner when educating others. Another thing that has worked for me is to write and publish a book on this topic, and I have several other manuscripts on the go. I also write reader-friendly articles in parent magazines and educational newsletters, and have offered sports clinics and other such forms of intervention to teach students basic motor skills. What is amazing is that the motor learning strategies work—indeed, sometimes within one or more hours, a student can learn to skip rope, throw a Frisbee, ride a scooter, hit a baseball, catch a tennis ball, and/or ride a bike, and so on. What parent isn't absolutely thrilled to see the transformation in less than 1 h? Parents have told me that some of the learned motor skills actually open doors for the entire family to now enjoy new recreational pursuits. That brings me great joy. So far, these different strategies have been working, as I receive referrals from local medical professionals, via the Internet and via word of mouth. One of my most memorable situations was the meeting of a rather large and overweight 13-year-old boy who was so weak academically, socially, and motorically that most people in the school thought he was mentally handicapped.

He had recently immigrated to Canada after leaving his homeland in one of the Caribbean countries. It was his first year in Canada and he had trouble with all aspects of school. In addition to having problems with reading and writing, he was very shy, discouraged, and depressed. I also wondered if he was a "slow learner" after watching him walk into my office, sit down, and try to talk. Because of his very weak oral motor control (evident in 50% of students with DCD), he was very difficult to understand. This student's posture was slumped and he looked absolutely deflated in every way. However, as requested by school staff, I started the assessment process. I was not surprised that he made errors on the first few age-appropriate tasks. That meant I needed to reverse the expected order to make sure he understood the instructions on simpler questions. (Later, I discovered that he probably got the answers incorrect because he was so anxious.) On the fourth item of a test I was administering, he used a sentence that got me *very* excited. He had to create a specific visual pattern using blocks. In fact, he commented, "Dis wiw wequiwe a diagonaw" (meaning: "This will require a diagonal"). I became excited because that sentence showed me he understood *much* more than he could show. After much testing, I was able to show staff at his school (and eventually his parents) that this student was actually a capable learner who had many gaps in knowledge because people had misunderstood his abilities. I was also able to prove that he had very good math reasoning—but was unable to perform numerical operations on a worksheet because of his motor problems. During the last few months, this student learned to read, write (using speech-to-text software), and perform basic motor skills, and he developed some good friendships at school. Now he has a smile on his face every day. After his father was informed about the test results, he said he would be forever grateful for truly understanding his son's needs and getting the right supports in place to help him. This student made amazing progress in school. And almost everyone else thought he was mentally handicapped! Sadly, his motor issues, depression, and problems with attention made him appear that way. But he did not have a cognitive delay.

I have also experienced other low points in the field of DCD. A little while ago, I was involved with a teacher who was unwilling to hear that a student in his class had a motor learning disability affecting fine motor control only. He maintained that the student was lazy, stupid, and unmotivated. He would simply make him practice and practice and practice writing until his work was neat. I spent several hours in a meeting with school administration—who backed my findings—but this authoritative teacher was still completely unwilling to hear that this student had a disability. "He is lazy and stupid—a dumkoff. When he writes that way, he should be back in kindergarten!" Unfortunately, by the time I met this particular student (toward the end of his elementary years), he was very anxious, his sleep patterns had been affected, and he was absolutely fearful of going to school. In the end, because the student was ridiculed, demeaned, and penalized because of his inability to write neatly, I needed to report this teacher for being emotionally abusive to this learner. That was indeed a very low point for me, and it showed me that many more members of the public need to learn about DCD and the psychosocial implications of this condition.

Another low point for me in the field of DCD is when professionals use outdated terms, offer an incorrect diagnosis for a person, or misuse a diagnosis to cover up another problem (e.g., identifying someone as having ADHD rather than a motor learning disability, stating that a child has DCD rather than a cognitive delay, or, informing parents that their child has Asperger's or autism, rather than DCD). The differences between these conditions are substantial, and I continue to hope that professionals will seek out educational resources on these matters.

One of the most important things I have learned is that not all people share the same passion and/or interest in the topic I love. That is completely okay. Professionals who want to learn will make changes in their practice, and interested parents, teachers, students, and others will likely search until they obtain information about DCD and then access needed help. I've also learned that because they will always be the best advocates for their child, I usually focus the most on educating parents so they will better understand this topic. They usually spread the news to others—so the circle of knowledge grows.

Parents: Trust your instincts. Most mothers have a gut feeling when something is not right; they seem to have been designed to have an intuitive feeling—or sixth sense about their children. They may go in opposition to their spouse or friends who think they are "babying" a child by performing motor tasks for them (e.g., helping to shampoo the hair of a 12-year-old boy, or tie the shoelaces of a preteen, or cutting waffles for a 9-year-old). In fact, most mothers will *not* stop looking for answers about their child's difficulties until they find a solution that makes sense to them. Good for them. Dads: Trust the moms. If you don't trust her to move forward in seeking knowledge, she will do it behind your back. Teachers: Don't be scared to address your concerns with parents. Most will want to know that something is amiss and want to know what to do to help their child. Doctors and other medical health professionals: Allow parents to share with you what they have found. In their need to know, they will spend hours checking things out. They are usually not desirous to be in opposition with you—they want affirmation and knowledge—as they will keep looking until a diagnosis genuinely makes sense to them. Don't be scared to admit that you do not know much about this topic. Grad students: Be willing to learn and accept information about a topic that many others will dismiss.

It is my true hope and dream that all parents, professionals, and others will soon learn that DCD truly is credible and deserves the same attention as other psychological conditions. Indeed, because of the long-term problems related to this motor learning disability, DCD is a fascinating topic that deserves our very close and careful consideration—both now and in future years.

Epilogue

From our perspective, some themes within and across the stories in this section include the following.

- During the preschool years of children with DCD, parents sometimes begin to observe that their children might have motor issues; for example, tasks like fastening buttons on pants or shirts or using a toothbrush seem unusually difficult.
- Many school-aged children with DCD experience significant social discrimination from being excluded from play activities and celebrations like birthday parties.

Some thoughts children and/or their parents expressed in the narratives that we thought were insightful and have heard expressed in similar ways over the years include the following.

- Early on, kids and others begin judging children with DCD by the way they look and move.
- Children with DCD can be traumatized by the persistent teasing and bullying they endure because of their weak motor skills and poor coordination.
- Children with DCD can have great difficulty establishing and maintaining age-appropriate friendships.

8 Autistic Disorder

Prologue

Children and youth diagnosed with autistic disorder tend to have a wide range cognitive, social, affective, linguistic, and behavioral deficits and abilities in comparison to others at their developmental level. For example, these children and youth have differential intellectual (information-processing) abilities (from low functioning to high functioning) and have typically marked deficits in the development and regulation of social interaction (marked impairment in the use of nonverbal behaviors such as eye-to-eye gaze, body postures, gestures, and facial expression as well as an inability to imitate the movements of others, problems in the development of peer relationships, difficulties in understanding and being sensitive to the needs and emotions of others, problems in expressing emotions, and lack of interest in sharing enjoyments and achievements with others), difficulties with communication (delayed development or lack of spoken language, problems with language comprehension, inability to initiate and sustain conversation with others, use of irrelevant detail in conversations, and a tendency to repeat words, phrases, and sentences), and a restricted and ritualistic repertoire of behaviors and interests (repetitive and stereotyped movements like rocking or arm, hand, or finger flapping and whirling, elaborate and sequential dressing and eating routines, and circumscribed interests such as memorizing various facts related to different topics).

Children and youth with autistic disorder are identified prior to 3 years of age and will vary with respect to their development through childhood and adulthood. In this regard, some will improve in their cognitive, social, affective, linguistic, and behavioral functioning, and others may continue to struggle in these areas or deteriorate depending on their intellectual abilities and familial as well as educational and social supports.

In this section are five stories: one personal story, one parental story, and three professional stories pertaining to autistic disorder.

The personal story, entitled "From Good to Bad and Back to Good Again," is written by a young man in his early twenties who has autism and who shares his story of what it was like for him to go through his school years with autism. The parental story, entitled "A Road Traveled with Many Stops and Turns Along the Way," is written by a mother who shares her story about her familial as well as medical and educational community experiences as she was raising her autistic son. The first professional story, entitled "Patience and Appreciation," is written from the perspective of a behavioral therapist who shares his story about working with an autistic boy to improve his communication and social skills and to reduce or extinguish his aggressive behavior. The second professional story, entitled "Getting There, Step by Step," is written by a teacher who recalls her challenging yet fulfilling experiences while working as a behavior therapist with a young autistic boy. The final professional story, entitled "Working Beyond the Diagnosis: Respecting the Bigger Picture," is written by a psychologist who shares her story about working with an autistic child and her acquired understanding of the importance of the family context in the treatment of autistic children.

Exceptional Life Journeys. DOI: 10.1016/B978-0-12-385216-8.00008-4

Personal Story

From Good to Bad and Back to Good Again

My name is John and I have been diagnosed with an autism spectrum disorder since the age of 2. The name of this diagnosis always bothered me and I feel that the diagnoses of social anxiety and a language learning disorder fits me best and is easier for me to take. I am in my early twenties now and have benefitted from the love and support of my parents, sisters, and brother. I have also benefited from some of the friends I have had made throughout my childhood and from some teachers who, unlike other teachers I have had experience with, supported me and helped me to feel better about myself. As I think about my childhood journey, I have many fond memories but I also remember a number of hardships I experienced. I feel good about where I am now and about what might be in my future, but my life up until now has been filled with "ups and downs." As I think about it, many of my school experiences were very traumatic, leaving me at times feeling very depressed and lonely. However, I had many experiences that lifted my spirits and some experiences that allowed me to go forward with hope for the next day, the next week, and the next year.

In first and second grades, I learned how to read and write. My favorite pastime was to play with Ninja Turtles. I used to bring these to class to play with. During these grades, I remember I had a teacher who was in the class to help me with my learning. She was very nice because she used to take me and some other kids to McDonald's and buy us hamburgers and French fries. Besides this, I found school to be very difficult for me. I remember that, at times, I would run down the hallways of the school and yell out, "This place is a jail!" I felt that having an extra teacher to work with me was difficult for me to accept because I did not want to be seen as different from everyone else. In first grade, my teacher was very nice to me. She was great with me. She sometimes allowed us to draw pictures, and I always drew the Ninja Turtles, but after a while, she found it annoying. I found it funny the way she reacted after she got sick of the pictures. I had a couple of friends during first and second grades. There was one girl who became a great friend of mine and who I remained friends with for a long time. There was another boy who often had seizures. I remember a birthday party of mine where my friend had a brain seizure. I did my best to calm him down. I patted him on his back, and was there for him. My mom has told me a few times that my connection with other kids and my sensitivity to others was not like other kids who have autism. I am very happy that I have been a little different that way and can make friends. I feel like I am warmhearted guy. So, being treated differently at times made me feel bad.

For my third-grade school, I was transferred to another school because my family moved to another part of the city. This was a very bad time for me at this new school. I did not have friends at this new school. Nobody knew me, and I felt very isolated. I had another aide at this school who, like the teacher at my previous school, was suppose to help me fit into the new school and help me with my learning. I very much disliked the principal of this school. He seemed to like to "rattle" people like me. My support teacher seemed to isolate me rather than help me fit in with everyone. She used to follow me wherever I went. This was very embarrassing for me because all the kids would see me with this person everywhere I went, which made them think of me as strange. It also made me feel very strange. One of the kids would call me "John Germs," which was horrible for me. My aide would boss me around and tell me what to do. I never seemed to be able to do anything I wanted to do. When my parents would come to school at the end of the day, I remember that my aide would always say something bad about me to my parents. She never seemed to have anything nice to say about me. It was always about a problem with me. This was "rough" on me. I felt that no one liked me. I was never invited to anyone's birthday parties.

In fourth grade, I was on a new medication. The doctors gave me Dexedrine, which seemed to help me a lot. I liked what this drug did for me. It seemed to calm me and make me feel better. It would "take the edge" away. In fifth grade, I was transferred to another school. I was transferred to another school because the principal did not want me at his school anymore. He was awful to me. He traumatized me. He feared me and did not like to be with people like me. Before I was transferred from the school, I had a problem with another kid. His father apparently phoned the principal and area superintendent of the schools and complained about me. My parents got a call from them and said to them that they did not want me at the school anymore. The problem was that the kid called me "John Germs" and ran away from me and locked himself in his dad's car. I got in trouble because this kid bullied me and I was seen as the problem. I did not get any good support or guidance from people in this school. I did have some behavior problems while I was there. I was so tired of being called names and being picked on that I would sometimes yell at other kids and act out. But, I was just defending myself. I was acting out because of the way I was being treated by others. No one seemed to be "on my side."

During my fifth-grade year, I went to another school, a private school. In this school, I was put with sixth-grade kids, and I loved it there. I made lots of friends and had a great teacher. But, the next year people thought I needed to be put in a lower grade to catch up with my learning. But this was not good for me because I lost all of my friends and did not fit in with the kids in my new class. I was made fun of again, and I avoided going to school as much as possible. People were mean to me, and I just was not able to fit in like I had been able to the previous year. So, from sixth to eighth grades, I made lots of progress in my learning but I never liked being at the school because of my social life and because I never seemed to be able to fit in with everything and everyone. People like me have challenges. I had experiences with good people and bad people. The bad experiences gave me a bad attitude about school. Although I made a few friends at the private school, there were unfortunately bad experiences with other kids. I was bullied sometimes.

During my eighth-grade year, I had found out about another school where kids like me were more accepted. I went to the school with the help of my mom and met the principal there and said to her that I would like to come to her school. The principal of this school was wonderful and let me come to her school for ninth grade. I liked this school much better than my previous school but it was not perfect. I still had my struggles and still had difficulty fitting in. But the best thing about the school was all the options they had for students (art, music, outdoor education, gym, multimedia). So, ninth grade was much better for me. One of my best experiences in my life was at this school. I made a friend with someone there. He and I, as well as other kids, got to go on outdoor school trips where we got to go mountain climbing and camping. The outdoor teacher told my parents that he was worried about taking me on these trips. But my mom ensured him that if he took me on these trips, he would find out that I was OK and would not be a problem for him. So this teacher took a chance with me and took me on an outdoor school trip. On this trip, this teacher had some health issues. When we were finishing climbing the peak of the mountain, he was struggling to keep up with everyone else because of his condition. I initially volunteered to help him with his gear, even though I wasn't suppose to. He said he was OK. So I continued with the trip and enjoyed the camping experience. After the trip, he seemed to be very nice to me. During the next parent–teacher interview, he told my parents that I helped him "spiritually get up the mountain." He said that I got his "spirit going." This was the first time I felt proud of myself at school. At this school, I had a support teacher but it was different than my past experiences with support teachers. Here, the support teacher helped many students and not just me, but the entire school. So, I did not feel that I was special or different from other kids. That is what I think should happen in schools with students like me. They should get the help they need but the help should also be given to all other students as well, because I am not the only one who needs help and I should not feel that I am the only one who needs help.

One thing that happened at this school was very upsetting to me. While there, some student put on the Internet that I molest little girls. This shocked me and angered me and made me feel very bad because I did not do this and once again people were presenting me as someone who is not like other kids. I was now seen as not only different but someone who is weird and someone to fear and someone who did very bad things. But once the outdoor teacher, the support teacher, and the principal of this school found out about this, they met with every student in the school and let them know that the rumors that they heard were not true and could be very hurtful and damaging to the student (me). They did not name me but they totally supported me. These teachers trusted me, stood up for me, and cleared this up for me. This was the first time I had such support in my school experience, and it made me feel so good! Up until this time, most of the teachers in my previous schools would embarrass me by telling other kids about my diagnosis and telling them about how different I was from them. This was not OK by me because I never wanted to be thought of as different.

After completing ninth grade, I had to go to another school for my high school. Once I realized I wanted to go to high school, I wanted to start right away. So, before I finished ninth grade, I went to the high school I would be going to the following year (with my parents) to become familiar with the school and become as prepared as I could for this new school experience. I talked to the guidance counselor and principal and some of the teachers. Unfortunately, high school, by and large, was not a good experience for me. I was put in classes where students would be mean to me and bully me. The most hurtful experience for me was in tenth grade because of my social anxiety and my difficulty fitting in. During the beginning part of my high school experience, I was beaten up by three other students who took the same bus I did. The police laid charges on these students and I was no longer bullied by them after this incident, but I never was able to establish friendships and feel comfortable in the school. For example, at one time during my tenth-grade experience, I left the school at the end of the day to get my bus, but because it was late, I went back into the school to wait. When I entered the school, I still had my hat on but was not thinking about wearing my hat. I knew it was against school policy to wear my hat in school, but it just did not occur to me at this time that I had my hat on. As I entered the school, I heard someone yelling at someone behind me but did not think it was about me. A few seconds later, I felt my hat being swiped off my head by someone. I turned around and saw a teacher standing in front of me as his hand was swirling around my head and knocking my hat of my head. I took my hat away from him and he "dragged me" into the school office and blamed me once again for my behavior. No one seemed to care that a teacher had struck me.

After tenth grade, I once again switched schools. At this time, I went to an "alternative school" in my area that had more experience dealing with students like me. I found this school to be much more accepting of me, and much more inclusive with students who were a little different than average students. I made some friends at this school and fit in better. What I loved about this school is that teachers treated students more like equals than like people to "look down on." The school allowed students to be much more independent and seemed to care more for the students. However, I did not stay at the alternative school for very long because I had another breakdown. Some of the things that happened to me at this school that caused me to have a breakdown and for me to leave were some of my interactions with some strange people. For example, there was one teacher who counseled me out of some courses because he did not think I would be successful in the courses, even though I found them to be areas in which I had strength. So, I did not get great supervision and it did not seem that I was going to be able to reach my potential at this school. So, my parents and I decided that I should go back to a regular high school. In my next school, I had a great social studies teacher who I loved learning from. But I also had a science teacher who did not seem to care about me, and it was very difficult to learn from her. By this time, school was really wearing me down. So, I went back again to the alternative high school that was by and large more of a calming place for me. Eventually, I passed twelfth grade.

Since my time at high school a few years ago, I have spent my time traveling and doing odd jobs here and there. For example, I sometimes drive the ice cream truck around my neighborhood, and once in a while, help my dad with his work. I also took a 2-month course at a college, called the BEST program. It helped me figure out what career directions I wanted to pursue (like helping people). I really enjoyed the course and the school environment. More schools should be like that. I also take tae kwon do and am involved with Parkour training. This is a fun way to work out and seems to help me overcome my fears of obstacles. The people in charge of both of these activities are very good people and are very helpful and considerate of us. The group of people there are awesome as well.

As I think about my school history, I would like to recommend that the school system provide more options for students so kids like me can become more involved in school and have a better chance of fitting in. I would also recommend that schools consider having support teachers for all students and not just for individual students so that no one particular kid is considered strange or different from others.

It is important for me to say that, although school was difficult for me and even though I had some bad experiences with some kids and some teachers, I also had some good times. I had some very good teachers and I made some very good friends. I think school can be a very insecure place to be for many students and especially for kids like me, so I also think that schools need to find a way to make schools more of a secure place for everybody. I think this would allow more students to do much better in school and help them become more successful and have a more enjoyable experience.

So, it was not all bad and I am glad that I had a great mom and dad and great sisters and brother. I am also glad that I had some great teachers and friends while going to school.

Parental Story

A Road Traveled with Many Stops and Turns Along the Way

Thinking about writing about my experiences as a parent of a child with a development disability caused my thoughts to run in many directions because I have had so many experiences. Right now, my son is 19 years old and has just graduated from high school. I joke that I have also graduated from high school (for the second time), and it has been a hard-won battle. I have decided that I would talk about four of my experiences. The first is my experience with the medical community; second, my experience with the educational community; third, my experience with family, community, and social situations; and finally, our life after school.

My son was diagnosed with autism when he was 5 years, 2 months old. Alex was not like a typical child diagnosed with autism; he met early developmental milestones. He sat, crawled, and spoke on time. By the time he was two and half years, we began to notice some differences that, at first, we were unsure of. He didn't play the same way as his sisters; he didn't attempt to engage us the same way as his sisters; and his conversational skills were atypical. We decided that he needed to see his pediatrician just to see if we needed to be concerned. At the time, my husband was working out of town a fair amount, and I had the task of taking my child to the doctor on my own. My concerns were swept away by our pediatrician saying, "You can't compare boys with girls" and "If there was a problem, I would have picked up on it by now." Alex had been seen by the same pediatrician since his birth. I personally breathed a huge sigh of relief because, like most parents, I desperately wanted my child to be normal.

The next time I brought Alex to our pediatrician, he was 3 years old. Again, I indicated that his behavior was not typical. He still had no real play skills. We purchased many toys to attempt to engage him, but he would not engage with toys in a typical way. With Lego, he would scoop the Lego, bring his hands to eye level, and watch the Lego fall. He would spin the wheels of toy cars, and he was obsessed with Batman (he wore a Batman cape for 2 years). He also showed some very significant behaviors that, in hindsight, I now know were sensory issues. I struggled with him on a daily basis to get him dressed; he hated noise, and more and more, we were beginning to live in a very quiet household because of the aversion to sounds, music, or even people speaking. He became extremely upset and agitated if someone touched him, and he developed an unusual fear of spiders.

Still, I was sent away and told not to worry, that there was nothing wrong. In some ways, I felt that there wasn't really anything wrong. Alex was developing skills in many areas. He knew most of his letters, numbers, shapes, colors; he could complete puzzles, and had an extensive vocabulary. Alex started playschool at 3 years, 9 months, and within a 3-month time period, the playschool teacher was able to outline her concerns. At around the same time, the supervisor of the day home where Alex had child care also expressed concerns. Both indicated that Alex had no interest in the other children: He was aloof, refused to engage with his peers, and threw tantrums if other children attempted to engage him.

At age 4 years, I once again took Alex to our pediatrician, but this time, I went armed with letters from both his playschool teacher and the supervisor of his day home. If there was one piece of information I can pass on to parents is for them to ask for documentation from teachers, day care staff, or other professionals involved with their child. The documentation helped to get the services for my child, and it allowed our pediatrician to see that I wasn't the only one seeing issues. Our pediatrician referred Alex to the local rehabilitation hospital for a full assessment. I indicated earlier that Alex was diagnosed with autism when he was 5 years, 2 months old. We waited over a year for the referral to come to the top of the waiting list. We eventually were seen, but the process was extremely difficult. Alex was seen by five professionals on one day and three professionals on another day. Alex is a child who does not adapt very well to new people, new situations, or new environments. It was painful to watch him struggle through this assessment clinic. At the end of the assessment, we were then told he had both autism and ODD. They described him as so defiant that they felt he would benefit from a placement in the behavior unit at the hospital instead of placement in his community school.

At home, we didn't really see Alex as behavior disordered but as a child who didn't adapt well and one who had sensory issues. When dealing with Alex, we knew that we couldn't spring anything on him. We had to plan and it did not matter if the activity was his most favorite thing in the world; without proper notice, he would balk at the activity. We learned to adapt our lives so that behaviors were minimal. I was shocked by the ODD diagnosis but not by the autism diagnosis. We went along with the recommendation that Alex attend the behavior unit; however, within 2 weeks, we were having interviews with school staff, questioning the appropriateness of his placement within a behavior unit, and they were recommending he be placed into his community school with support. He was eventually transitioned to his school and began his kindergarten year with his age mates and in the same school as his two older sisters.

Alex certainly would not be considered an easy child but we felt he was adapting to his school environment and that the school staff were programming for him in an appropriate manner. His first year of school was a success. However, in the middle of June of his kindergarten year, my husband and I were called into a case conference and were asked to remove our child from his community school and have him placed at the district site for children with autism. We were asked to visit the school to look at the program, and we were assured that this would be best for our child because there were specialists and people who understood about autism. Well, again we followed the school's advice. We went to look at the program and the class Alex would attend, and felt "no way, under any circumstance" would my child attend that school or that program. Alex would have been the highest-functioning student with autism. What was the solution? We did have enough knowledge to understand that we could have fought our community school for placement but we really felt they did not want him, and we were fearful that our child would not get the necessary supports. Instead, we moved him, along with his two sisters, across the street to the Catholic school in our neighborhood. At that time, the Catholic schools had a more inclusive policy for children with exceptionalities.

We have had extensive experience with teachers, schools, special education teachers, administrators, and supports such as occupational therapists, speech therapists, and educational psychologists over the years that Alex was in school. I guess I would say we have had a continual fight for services and we had to always be watchful to ensure that school staff did what was necessary to meet the needs of my child. I think many parents have a naïve idea that everyone will do what they are supposed to do to provide for your child. This, after all, would be the job they were hired to do. I learned that, as a parent of a child with autism, you always have to be on top of the services and must never assume. I became known as a "red flag" parent with my child's school. From my perspective, this was positive but from the school's, I am sure they saw me as a thorn.

I have had a variety of other experiences over the past 13 years. We have experienced some of the most wonderful caring teachers that have worked hard and invested a great deal in the success of my child's program. We've met teachers who were dedicated, open, and flexible and didn't take my child's behavior as a personal affront. We've met teachers that worked hard and included me and Alex's dad in the process of educating our child. These were the situations that were so positive and welcoming. We have also met teachers who were inflexible and difficult to deal with. Once in a transition meeting between second and third grades, the receiving teacher kept saying things like, "In my class, we do things like this and that won't work for ..."; after about five or six instances of her stating what the classroom was like and how my child's current program would not fit in with her class, we started to realize that this was not likely the best fit. After the transition meeting, we set about investigating the other third-grade teacher. We asked questions of Alex's second-grade teacher and other parents. We requested a meeting with the principal and outlined the reason why we felt the designated classroom was a poor fit and requested a switch to the other third-grade class. It was a good decision.

In fourth grade, we had a special education teacher tell us that she felt that Alex had reached a plateau and that his skills were no longer developing. We were told that he had reached the highest reading level he would obtain and that his math skill would not likely progress any further. Can you imagine writing off a child's skills at 9 years of age? I had never heard of such a thing, and we simply carried on and insisted that the school continue to work with Alex and his program as if he would continue to success. We didn't believe that Alex would not learn. At his high school graduation, Alex won an award for the highest grade in English. My advice to parents is that you know your child best and push for what you believe and don't give up hope. For professionals, it is critical that you have high expectations for our children, and even if you think a child's progress has slowed, you need to continue to work with them because they may surprise you.

From a parent perspective, it is devastating to hear your child is not progressing and will not progress any further.

I want to talk about two significant instances that occurred with my child once he hit high school. The first was a programming issue and a problem we encountered with a high school subject-area teacher. The second is a peer issue. My child had the support of a teaching assistant from the time he entered school until he finished eighth grade. At the end of eighth grade, the board administration, in conjunction with school staff, determined my child no longer required teaching assistant support. Instead, they moved to a model where my child would have a reduced course load and would have support from a special education teacher one period a day. I admit this model was effective for the most part.

To provide you with a context, Alex had been assessed by an occupational therapist when he was in seventh grade, and the assessment results indicated that he had severe motor planning deficits. On the Beery VMI, Alex obtained a borderline score on visual motor integration, an average score on visual perceptual skills, and a well below-average score (or a score that is in the mildly mentally handicapped range) on motor planning. His difficulty with motor planning had created difficulty with writing throughout his school career.

In the second term of Alex's ninth-grade year, he started rumbling about a teacher who moved too quickly through his course notes and he was becoming very frustrated. One thing we have always done is try to avert problems before they occur. Alex's IEP indicated that he was supposed to receive photocopied course notes as an accommodation. So I asked for a meeting with the special education staff as well as the subject-area teacher. The subject-area teacher was polite and appropriate with us and said all the right things, and we came to an agreement that he would provide a copy of his PowerPoint notes to Alex on a memory stick and we would print them for Alex at home. The school reported that teachers had a photocopy budget, and copying information for Alex would eat into that budget. We thought it was settled; it was certainly an easy solution for Alex simply because he would no longer have to take handwritten notes. Three days after the meeting, Alex still wasn't receiving the notes, and he was continuing to rumble about taking notes. Four school days after the meeting, I got a call at work to come to the school to pick up Alex because he was reported to have attacked the teacher.

People who know my child understand that Alex doesn't lie so I asked him what happened. His story was as follows: The teacher was zooming through the

PowerPoint notes and Alex said, "Hey, put that back on I'm not done." He admits that he had a rude tone. Alex reported that the teacher ignored him and carried on giving his lecture. Alex left the classroom and sought out the school counselor because he was frustrated. The counselor calmed him down, and after Alex was calm, he realized he had forgotten his books, binder, and backpack in the classroom. The counselor allowed him to return to the classroom but, once there, Alex reported that the teacher shouted at him: "What do you think you are doing back in my class? Get out." Alex reported that he went from zero to one hundred in a matter of seconds, and he thought he would give the teacher a piece of his mind and stormed toward the teacher. He didn't manage to get close when the teacher pinned him, at which time Alex used his feet to kick the teacher. Alex was excluded (not suspended) from school for 10 days. No paperwork was completed and no record of the incident was placed on Alex's file.

In the intervening time, I started a paper trail with the teacher, school, administration, and counselors. I wrote about the meeting and our agreement to obtain the information on a memory stick and wrote about the incident from Alex's perspective. The teacher responded to my correspondence and copied it to everyone with a comment that made me realize that some teachers have very limited understanding of what it means to have autism and what it means to have motor planning deficits. The teacher asked, "How will Alex ever learn to write if he isn't made to write?" I found the question frustrating, but more frustrating in light of the fact he had already agreed to give my child the notes. My response to the teacher wasn't friendly, and I demanded and received a full-time scribe for my child for the remainder of the semester. I got it!

I know the next scenario is not one that people will expect a mother of a child with autism to report on. In Alex's third year of high school, he became the subject of a love interest by another student at his school. She decided that Alex should become her boyfriend but he wasn't interested. What ensued was nothing short of sexual harassment while he was at school. Alex was pursued in and out of school; he was followed around the school, not just by the student who wanted him as her boyfriend, but also by a group of her friends. She had male friends follow him to the bathroom to "spy" on him. The student and her mother drove around our neighborhood and parked near our house; she called him continually.

One evening Alex came home from school distressed, because the female student had exposed her breasts to him in the corridor outside the classroom Alex was waiting to enter. My husband and I asked Alex to speak with the school counselor; however, we are unsure of the outcome. Shortly after that incident, she came to school wearing shorts and told Alex that she was not wearing underwear and then proceeded to expose her genitals to him. Alex did go to the school counselor right away, but the counselor thought the whole process was "funny." The counselor actually laughed about it, which frustrated Alex. Of course, my husband and I did not see the humor in our child being harassed at school. I put it this way to the school: If my son took his penis out and showed it to a female student, the police would have been called. The whole situation made me aware of the double standard set by our society, and I asked that the school attempt to resolve this situation. My concerns were twofold: First, I was concerned for Alex's well-being and safety at school, and second, I was concerned about what might be happening in the life of a young woman who feels so desperate to have a boyfriend that she would risk doing those things. I am not exactly sure what process was used to handle the situation; however, the student was removed from Alex's classes for the remainder of the year.

The other arena where we have had experiences that have been upsetting has been in the community. We live in a family with three children and Alex is the youngest. Our older children were engaged in a number of activities as children such as gymnastics, karate, swimming lessons, synchronized swimming, piano lessons, and so on. Alex had no interest in engaging in activities that involved social or peer groups; however, that didn't stop us from trying. We tried several different types of activities such as swimming, bowling, and karate. Through the support of a handicapped children's support agreement, we were able to engage rehabilitation workers to assist him in the activities. However, it is incredible how people in general have very little tolerance for differences in children. After a particularly difficult swimming lesson, I removed Alex early because of his behavior (I actually had another parent advise me that "it is parents like you who create mass murderers."). I have been in restaurants where people have made comments about my child such as, "Now there is an example of a very spoiled little boy," and what do you do with this? Do you divulge your child's medical history or do you smile and move on? We are generally of the smile and move on variety, but I do think public education about what it is like to walk in the shoes of a parent of a child with autism is critical. It would keep people from jumping to conclusions and from making comments that are hurtful. My husband and I know our child is different and we don't need people who know nothing about us making comments.

What is it like to live in a family when a child is affected by autism? I believe that it affects every aspect of family life. A typical school day would start with routines; everything has to be done in a certain way and in a certain order. In Alex's early school days, this was a huge issue and would set the tone for his whole day. As he got older, we were able to switch his morning routine around (on purpose) once he settled into his school year. When Alex first started school, getting him dressed in the morning was a constant battle; he wouldn't want to get dressed (the issue was one of making the transition from pajamas to daytime clothing). Getting him to eat was also a battle, especially once we started on medicines (Ritalin) to assist him with attention, as was getting all his school supplies, a lunch he would eat, and then for the ultimate, getting his outside shoes and jacket on. We battled at every step. So what worked for us? I have a few tips. The first is that I started to bathe Alex at night (every night) and then I would dress him in the clothes he would wear to school. Usually it was sweatshirts and sweatpants. Once I started dressing him in the evening, it cut down on many tantrums and battles. The other thing I did was to write out schedules for him. I would review the schedule every morning and it actually seemed to work. We used charts and gave him stars for getting his "jobs" done in the morning. Once Alex got older, we started to purposely mix things up so that he could adapt.

Occasionally, I would forget the procedures of dealing with Alex and make comments like "let's go to McDonald's for dinner," and Alex would thrown major temper tantrums: "I'm not going and you can't make me," and so on. Of course, my two other children would end up being furious with me, because once Alex started down the tantrum arena, there was no turning back and no way we were getting him to go, even though he loved going to McDonald's for dinner. Our house also became the quietest house. We no longer play music, we watch television with headphones, we don't listen to the radio, and we do not have the type of atmosphere my husband or I grew up in. It makes life much easier to have quiet. Alex is now calmer, and we have adjusted.

Professional Stories

Patience and Appreciation

This story is about my interactions and experience with Billy, a young boy diagnosed with autism. This work, the beginning of my experience with autism spectrum disorders, occurred 12 years ago. I had the benefit of working with Billy during the last year of my undergraduate education. I was keen to explore a career in psychology, and asked for guidance from one of my course instructors regarding how I could improve my graduate school application to make it more competitive and appealing. His sole piece of advice, the best I have ever received, was to acquire "real-world" experience in working with individuals with special needs. To this end, he was willing and able to put me in contact with an organization that provided early behavioral intervention services to young children with an autism spectrum disorder. At that point, I had heard of autism, likely to the same extent that much of the population had—through movies and popular media at the time. The thought and suggestion intrigued me and so, nervously and with interest, I followed up on the offer. This advice began a long journey in working with individuals with autism spectrum disorder. To date, I have worked in many organizations that provide support and services to individuals with autism spectrum disorder in various capacities. And, to date, Billy remains one of my fondest memories and greatest challenges.

My first experience with Billy was difficult to say the least. At that time, Billy was 4 years old and displayed many problematic behaviors associated with autism. He was extremely interested in a restricted number of activities to the exclusion of much else, predominantly Thomas the Tank Engine and Disney movies. His parents had purchased an extensive collection of small Thomas toys and a train track set that was spread throughout their house. They owned a considerable library of Disney movies, which Billy insisted on watching throughout the day. He would typically react negatively and with some measure of aggression if either of these activities were interrupted. In addition to this, Billy had very poor verbal communication skills and struggled to express his needs and wants, often resulting in frustration and additional aggression toward caregivers. He had little understanding of social situations, and much preferred to play on his own and to be left to his own devices. Most attempts at cooperative play or social interaction were regarded as intrusions, and Billy reacted with his typical anger and frustration.

Being largely unprepared and inexperienced regarding the typical nature and behaviors of autism spectrum disorders, I left that initial meeting feeling quite overwhelmed. How could I learn what would be required of me to properly work with this child? How could I make a positive impact and be of benefit in this family's life? Is this sort of work what I really wanted to do? These and many other questions went through my mind over the next several weeks as I continued to spend time with the family and the other early intervention workers whom I continue to hold in the highest regard.

I began to learn more over the next few weeks. The nature of autism spectrum disorders and the specific type of behaviors that Billy presented with were explained in detail, and I did some additional reading to familiarize myself with this field. The most difficult aspect of the job, as I remember it now, was watching one of the experienced therapists engage him in an activity, understand what they were doing and why, and then engage in that same activity afterward. There was a steep learning curve to working with Billy, but small gains were seen!

My work with Billy initially focused on three areas of functioning. Billy was relatively nonverbal. A strong proportion of therapy time was dedicated to improving his communication skills, verbal or otherwise. He was taught to use simple signs for requests and slowly to use words simultaneously. As time passed, we began to focus more purely on verbal communication. At the same time, social skills and interaction were a second focus of intervention activities. We created a list of typical social activities in which Billy would engage (e.g., asking his sister to help him get a drink, asking his parents for help in dressing). Of the most concern was Billy's constant aggressive behavior. Whenever he was unsuccessful in any task, regardless of the amount of support provided, he would respond with some sort of aggressive act, varying from attempts at hitting us or his parents to throwing objects (such as television remote controls or, a few times, scissors) at us.

As therapy, and Billy's skills, progressed, we began to see improvements in his communication skills. His attempts at social communication were very stilted and structured. For example, when he wanted a drink, he would walk up to his mom, pull on her shirt three times, and say, "Hello, excuse me mom!" and stick his face in front of hers so that she had no choice to respond. Slowly, we began to teach him other ways to get attention. This improvement in communication, in addition to teaching him appropriate ways to refuse tasks that he did not want to engage in and improving his enjoyment of many tasks, eventually led to a noticeable reduction in his aggressive behaviors. Now, rather than feeling overwhelmed by certain tasks or upset about participating in certain activities, he was more capable of expressing himself and understanding of his own autonomy.

The best approach to working with Billy involved a combination of understanding his thoughts and desires, anticipating his objections, and simply being a fun and playful person to be around. Billy keenly desired to engage in specific activities or routines, and knowing what those were and when they would become problematic was key to working with him effectively. If Billy wanted to play trains or some other preferred activity, we had to anticipate his reaction to being requested to play a different game or activity and help him work through that frustration. Understanding who he was as a person, and respecting his thoughts and desires, became pivotal in him feeling comfortable and secure when working with us.

I remember Billy's significant challenges well. His frustration was always a problem, particularly at the beginning of our work with him. He became quickly fixated on preferred activities or objects (typically trains and Disney movies), and any attempts to divert his attention away from these and onto activities and programs that we wanted to work on with him resulted in rapid escalation and frequent outbursts. These became particularly problematic when his frustration was directed toward his sister, who was young herself at the time and did not understand the nature of her younger brother's difficulties, or his mother, who held herself personally accountable and responsible for her son's challenges. Over and above this difficulty, initial progress with Billy was slow. It felt that, day after day, we would work on the same skills and activities with little improvement. This, in turn, produced frustration on my part. However, Billy's communication and social skills did develop, albeit slowly at first. I believe this development resulted in large part from the consistent support we as a team provided Billy and his family, our understanding of Billy's needs and how to best meet them, and Billy's desire to interact with others. Although he presented with significant behavioral challenges, Billy is a loving individual who desires social contact. He simply needed the help of others to be able to show this need and to be able to meet it effectively. Indeed, by the time we stopped working with him, Billy had entered elementary school, was able to function well in an integrated typical classroom, and had made friends that he played with on a regular basis.

My fondest memories of Billy continue to be of the amazing successes and his tremendous leaps forward in terms of development. Billy loved trains. So, once a week he was offered the opportunity for a community outing to a local park where an old steam train is still in operation for passenger use on the grounds. If Billy cooperated and worked hard during the week, then Saturday was his "train day." He looked forward to this event each week, and so did I. I've never seen a child more excited than Billy during that car ride. As well, the trip allowed us the opportunity for Billy to practice his social and communication skills with people he hadn't met before, allowing for generalization of learning, which is so important with individuals on the autism spectrum. We would play with other children, work on eye contact and appropriate responding to children's overtures for joint play, and of course ride on the train. Several times, in fact.

In addition to "train day," going to the zoo was another of Billy's favorite activities. He loved to watch the animals, and this community outing afforded another opportunity for him to practice the skills we had worked on throughout the weeks and months and to get out and play with other children. Of most importance, these outings gave me the opportunity to visit the gift shop and purchase several animal masks (a new one each visit) and some other toys and props that we used to create new imaginative play scripts with. Billy loved putting on the masks and pretending he was a hunter, a bear, or a lion. And these imaginative games soon became the preferred mode of social interaction with his sister and peers at preschool. He would invite them over to play, and they would all get dressed up as different animals and pretend to be in the jungle. To watch Billy, a boy who had little to no communication and significant frustration/aggression difficulties, invite peers to play, dress up and effectively pretend with them, and enjoy himself the entire time continues to be a heartwarming memory.

Throughout my journey with Billy, and the dozens of other individuals with an autism spectrum disorder that I have worked with, I believe that the two greatest things that I have learned are patience and appreciation. These individuals process information differently than those who are typically developing. As a result, I expect that there will be differences in perceptions, communication style, and understanding of the world in general. I now appreciate these differences and understand that I cannot expect others, typically developing or otherwise, to perceive things as I do. Nor should I expect that supporting individuals with exceptional needs is an easy task. It is actually often quite difficult. However, doing so can be, and has been, the most rewarding experiences of my life.

The best advice I can give to individuals interested in supporting people in this population is to be patient and be prepared to invest a good portion of your time getting to know the family and the individual you are supporting. Dedicating that time, particularly at the outset, can enable you to become more aware of the unique needs and circumstances of each individual, and can help prepare and orient you to best meet those needs. As well, prepare to have fun. Many people begin this type of work with the intention of having a "job," though quickly realize that it is often much more. I often found myself waking up each morning eager to go to "work" and spend time with each child. They have each become a part of my life, and I theirs.

In looking toward the future, I no longer work with individuals with an autism spectrum disorder in the same way. However, I continue to look back on my experiences, and those with Billy in particular, as formative in my ongoing desire to continue supporting such individuals and their families. I now work in a clinical capacity, frequently conducting diagnostic assessments and providing oversight for intervention services for children in this population. Despite this change, I continue to feel excitement and interest each time I interact and work with these special individuals. They and their families continue to inspire and motivate me.

Getting There, Step by Step

Sometimes empathy is a struggle. It is one of the unique traits of human beings, but that does not make it easy. It becomes even more difficult when the individual you are trying to empathize with is nothing like you.

It is generally a requirement for a schoolteacher to work with a wide variety of individuals, many of whom have difficulties or disabilities in their daily academic or social tasks. One day, while feeling sorry for myself after putting in tedious hours of working with struggling students, I was given some genuine advice on how to better appreciate the feelings and ideas of others. It was something along the lines of this: "Imagine your weaknesses; whether it is math or public speaking or whatever it may be. Now, if your weakness is math, imagine that it came easily to everyone around you. What if the world functioned mainly in the area of your weakness? What if math problems became the most common source of communication? How trapped would you feel? How much anxiety would that cause?"

I've been a schoolteacher for 3 years, but that is not where my most meaningful personal experiences with autism have come from. Four years ago, I was an undergraduate student looking for a job to help lessen the mounting expenses of university. I applied for the position of behavioral therapist. The job description involved working as an in-home behavioral therapist with children diagnosed with autism. After a few rounds of student teaching and a bit of background as an educational assistant, I thought I was well equipped for the assignment. Besides, it was simply one child, as opposed to a classroom full of them. This was going to be a "piece of cake."

The job began with some initial training. I sat in a room and learned about behavioral theorists, antecedents, behaviors, and consequences, primarily topics that I had already covered in university while developing my classroom management plans. Following completion of the training days, my supervisor informed me of a "setup meeting" with a family. The purpose of the setup meeting was to gauge if I was going to be a good match for the child and the family.

I wasn't entirely sure what was meant by "he's predominantly nonverbal," but I was relatively optimistic as I headed up the driveway toward the family's home. At 22, I was very used to my quiet, self-involved life. I moved out of my parents' house at 17. Children, routines, and especially noise had not been a substantial part of my life for a solid 5 years. On entering the house, I was bombarded with children's shoes and toys, movies playing, and general noise. I sat down with the mother of the home to talk about her son (Brayden). We got settled in on a cleared spot on the basement floor and began the discussion. I learned from that conversation that two boys in the family were on the "autism spectrum," and that I'd be working with the younger boy, Brayden, who was 7 years old. She told me a bit about his likes and his dislikes and mentioned that he had a few words in his vocabulary that could get him through his day. We conversed about the long wait list for families to get in-home therapists and how great it was to be able to have another therapist there to help him again. At this point, he bolted out from one of the bedrooms and into the living room where we were now sitting.

He had a lot of energy, running around the room at speeds that seemed unnatural. Watching him, it appeared as though he had no awareness of anyone else in the room. Eventually his mother was able to direct his attention and introduce me as the lady who would be working with him next. To this introduction, he replied with an immediate, "Good-bye." His guard was up now though, and after a few more laps around the living room, he repeated to me, "Good-bye." He was now aware of me and I could tell he was not particularly pleased about the new lady who was going to be working with him. Perhaps this job was going to be somewhat more challenging than I had anticipated.

The first couple of months were indescribably difficult for me, and I'm assuming that they were just as difficult for Brayden. He did not approve of me being there in the least. Day after day, I'd open up the front door and, as soon as I was spotted, he'd start screaming and run in the opposite direction. Being 7, his screams were markedly louder than your typical infant screams. But we'd carry him to the room in which we'd start our Boardmaker schedule (software that creates easy-to-understand symbols to represent activities for special-needs individuals), and I'd wait while he kicked at the door and rolled around on the floor. I couldn't explain any of it—why he was acting like this or why I even bothered showing up every evening after a day spent teaching a classroom full of kids. Before my finger even touched the doorbell, I wanted to back away slowly and retreat in the direction I came from. Something kept me coming, and I proceeded to push on that doorbell and continue the journey inside.

I knew it was going to be a rough day when I'd show up just as he was starting to line up his hundreds of action figures in their rightful positions on the table, or when he had just pushed the "play" button on his favorite Dora movie and was expecting to watch it for the next hour. We were both a bit strung out after our days at school, but there was work to be done.

Our days would begin with the visual schedules we would use to explain our course of events for the evening. Goldfish crackers and Cheezies were always handy for positive reinforcement, and we didn't stray too far from home without them. I'd lay out his options, which were pictures that he could stick on a path, and he would place them into the order that he desired to complete the tasks or events. The pictures displayed activities such as going to the library, going to the grocery store, reading books, sharing toys, going to the park, eating supper, and so on. These may seem like enjoyable activities, but in his opinion, they really weren't. When he had completed all activities on the path, and we were back at home, I would start up a Dora the Explorer DVD for him and leave for the night.

The more I got to know Brayden, the more I learned of his peculiar ways. He was terrified of birds; it didn't take much more than the mere sight or sound of a bird and he'd be in hysterics. On the occasions that involved him coming into contact with other children, he just didn't understand why anyone would have to share his posses-sions, his spot on the swing, or why he even had to hang out with the other kids in the first place. He loved green. He wanted all the green items in the grocery store to take home with him, which he assumed were free. Waiting in line was also usually a disaster because he didn't understand that we, the general public, wait for our turn in the order that we arrived in line. Why would mom insist on eating anything for dinner besides green apples and peanut butter? If it wasn't what he wanted to eat, we could forget about him participating in that event.

What I found fascinating was that as I became more immune to the tantrums caused by these daily events, and persisted consistently with the routines and the rules, he slowly started accepting me into his "trust bubble," as I called it, which, in all probability, included a grand total of six adults if you counted parents, grandpar-ents, and his previous therapist. It was as if he was coming to the realization that no matter how profusely he tried to test my trustworthiness, I cared about him and we were going to stick with the plan.

After roughly 2 months, I rang the doorbell and entered the house as usual, but this time he didn't scream. Gradually, as time went on, he'd merrily say my name as I walked in. He still didn't appreciate me taking him away from his Dora DVDs or his toys. I suffered the occasional kick or hit while insisting he stay perfectly calm and peaceful during church. We must have experienced thousands of tantrums while a team, but at least we were getting to know each other.

Fridays were swimming days for Brayden and me. I had to find strategic methods of getting dressed while making sure he didn't run away from me. I found one way to convince him to wait was to sing his favorite song while getting dressed as rapidly as I possibly could, and he would hum along and wait for me. That was, "by a landslide," the easiest part. Brayden loved getting into the pool, but getting him out was another matter. Congruent with the rest of our relationship, the first couple of months at the pool were a challenge. Like other obstacles we had overcome, there was a gradual rise in cooperation as he came to realize that we were going to stick with the plan. I found that the technique he responded to the best was the "countdown." I'd let him know when there were 15 more minutes left of swimming, then five, then one, then one. This was an especially effective technique at the pool, and had good success in other situations as well.

The activity he struggled most with, right up until I left my position as his behavioral therapist, was our the trips to Wal-Mart. The objective was to successfully go shopping without tantrums or instances of running away. We didn't once achieve that goal, but my hope is that he is improving or has mastered it by now. Wal-Mart shopping was difficult for him for the same reasons that the local grocery store was challenging. There were confusing lines of people who glared at you when you tried to walk past them, and there were toys all over the place that you could only look at and not take with you. It was a mild form of torture for Brayden. He loved toys. What better place to go than Wal-Mart to have access to all the toys and DVDs that you could ever imagine? During these trips to Wal-Mart, I endured countless tantrums in the aisles and endless glares from the shoppers around us.

On one occasion at Wal-Mart, an infuriated Brayden darted into the men's washroom and hid. I asked a man nearby if he would please check on him for me. I was a bit taken aback as he retorted, "No. You wouldn't need me to if you were a better parent." My first instinct was to retaliate with, "I am not his mom; I am his therapist! He's autistic, and he thinks everything in here is free, and he's mad at me for not letting him take things home with him!" But then I thought, "I'm going to take this one for all of those parents out there that I have judged in the grocery store, those that we all judge when their children are crying in the aisle. We have no idea of the whole story."

Sometimes, if people can't hear us, we assume they can't see us talking about them. Sometimes, if people cannot speak, we assume that they can't hear us talking about them. Occasionally, I would ask Brayden, "Do you understand what I'm saying?" Generally, my question was met with a blank stare; however, I had a sort of uncanny feeling that he did understand what I was saying, despite the vacant gaze. Then there was the moment of truth when I realized he was fully aware of everything that was taking place around him. So aware, in fact, that he was trying to tell me, the only way he could, that I had made a mistake. It was unnerving how much information I had assumed he didn't comprehend.

One day, we met up with Brayden's family at a Dairy Queen for a quick lunch together. The family members decided they would prefer to eat at McDonald's instead of Dairy Queen, and briefly discussed which McDonald's they were going to eat at prior to leaving. Brayden had been nearby during this discussion, playing with some loose rock on the asphalt, appearing to be in his own world as usual. After all the details were sorted out and the rest of his family had left, Brayden and I got into my car. He was in a phenomenal mood. The entire family, including dad, was going to be joining us today at McDonald's. As we were driving toward McDonald's, he had one of the worst tantrums I'd ever seen from him. He was screaming and kicking as I struggled to yell over him, "What is your deal?! We are going to McDonalds! You love McDonalds!" But he didn't stop. I drove all the way to McDonald's while he continued kicking the dash of my car with as much force as possible. Eventually, we arrived at McDonald's but oddly, Brayden's family wasn't there. This puzzled me because they left earlier than we had. Then it hit me. We were at the wrong McDonald's. Brayden screamed the whole way there because we were going the wrong way. Although playing with rocks while we were at Dairy Queen, he was simultaneously picking up our conversation and, apparently, comprehended it better than I did. We got back into the car, buckled up his car seat, and started the engine. I looked over at him, and said, "We are at the wrong McDonald's, buddy. You were trying to tell me that weren't you?" He laughed and started humming, and as we drove away from that McDonald's and in the direction of the other one, a tear rolled down my face.

After approximately 6 months of working together daily, Brayden started to do something out of the ordinary, something that made me realize just how ordinary he was. Every so often, while I was verbalizing my thoughts to him, he'd hold onto my face with both of his hands and stare intently into my eyes. At first I was a bit startled, but when I looked into his eyes, I knew that it was his way of speaking to me. It was as if he was trying to say to me, "I wish I could speak to you, but I can't. I'm trapped in here, but I'm a normal human being. I'm just like you."

I've been a teacher now for 3 years in an inclusive classroom. It is a demanding as well as a gratifying job, but nothing compares to the challenges and rewards I experienced in my day-to-day personal relationship with Brayden. There were drastic ups and downs and extreme joys and challenges. The opportunity provided me with a respect for parents and caregivers, especially those caring for individuals with autism. It provided me with a better ability to empathize—to appreciate someone else's life and know better how that life is lived.

I am presently a graduate student of child psychology and attempting to more accurately understand children with disabilities. My aspiration is to assist and support others, as well as raise other's awareness of disabilities. Although we may perceive individuals with disabilities as substantially different from us, we actually have more similarities than differences.

Working Beyond the Diagnosis: Respecting the Bigger Picture

I am currently a psychologist working with children diagnosed with autism. My work now is primarily consultative; however, I began working as a frontline aide with children with autism almost two decades ago. When I first started, I did not know anything about autism (I had not even heard the term before), but I was immediately fascinated with the kids I worked with. They were so diverse, with so many different levels of skills, mannerisms, behaviors, and needs. I loved coming to work every day because it was so challenging, yet rewarding, and I had different experiences every day.

Immediately upon graduating from university, I got a job at a local agency in an early intervention program for children diagnosed with autism. After several years, I moved my way up to the psychology department as a consultant. I remember quickly becoming frustrated in my role because now I was to work directly with parents, rather than the child, and found that very few parents followed through with any of the recommendations I provided. For those few who at least attempted to implement suggestions, most were only able to stick with it for a day before complaining that "it didn't work." And then the complaints would roll in: "Why isn't my son toilet trained yet?" or "Why is my daughter still sleeping in my bed?" or "What are you going to do about these behaviors?" Parents wanted me to "fix" their child, but did not seem willing to put effort in themselves. I remember having numerous conversations with colleagues stating, "If they would just do what we tell them, their lives would be so much easier in the long run!"

The frustration I had with parents persisted until I had a daughter of my own. Now I appreciate the fact that parents have a lot on their plate. As wonderful as it would be to commit to your son or daughter and do everything "by the book" all the time, it is simply not realistic. Now I thoroughly understand how much easier it is said than done. Sometimes it truly is easier to put them in their training diapers than in underwear because it is a busy day and there's no time to clean up an accident. Sometimes it really is easier to let them sleep in our beds, instead of struggling to get them to sleep in their own, just so we can get some guaranteed sleep.

Once I had my daughter, I understood the struggles parents endure and the importance of considering the "big picture." I began to focus less on the child with autism, and more on the family as a whole, considering everything families deal with. Many of the parents I support have other children to take care of (who often also have diagnostic concerns), their own emotional difficulties to cope with, marital issues, and the need to find balance between home and work life. Many families, even years after a diagnosis, experience grief and struggle to come to terms with the diagnosis and their child's prognosis. This is true for almost every transition period in their child's life. Families have to deal with the multitudes of appointments, people coming through their home and disrupting any sense of calmness. Most families I have worked with are often feeling as though they are living in "survival mode," simply trying to create a sense of normality in the midst of chaos. Unfortunately, some service providers do not understand where the families are coming from, failing to look at the family context. Instead, they focus all of their attention on the one child with the diagnosis. There is also sometimes a tendency to stop services when parents do not follow through. Indeed, many of my colleagues insist that parents follow their recommendations and threaten to pull services if parents do not follow through. This approach does not take into account the pressures the parents face outside of their child with autism and creates a situation where the family will not benefit in the long term. Certainly, many funding agencies require parent involvement, and parent involvement is certainly necessary for long-term gains. However, parent involvement is not about imposing recommendations on the parents and demanding they be followed. Parent involvement includes collaborating on goals, strategizing ways parents can realistically meet those goals, and supporting parents along the way. Sometimes plans do not work initially, so there also needs to be some flexibility to maximize parental success.

After years of agency experience, it was not until I started running programs privately that I truly transitioned into my role as support for these families. Upon shifting my efforts from child-focused therapy to supporting families as a whole, my frustration with the families began to dissipate. I had developed a treatment philosophy closely linked with the agencies I had worked with, but now I could extend my professional mandate to be flexible and to support families in a way that best suited their situation. I could work with parents based on their needs and priorities to provide them with goals that they were ready and able to work on. Together, we would create behavior plans that would increase their chances of successful implementation, rather than imposing a rigid plan that did not consider their unique circumstances. Although I was ultimately responsible for knowing effective intervention strategies, it was through this collaborative approach that we were truly able to achieve success. This simple adjustment in philosophy resulted in parents following through on most recommendations and achieving goals they helped to create for the family. This all sounds like common sense now, but it took a journey to get here, a journey that some service providers still need to pursue.

Although this new method provided more focus on the family, I still needed to work directly with the individual child diagnosed with autism. When working with the child, I approached my work with the understanding that children with autism are more different than they are alike. There truly is no "cookie-cutter" approach, and yet so many programs try to fit the triangular child into their circular plan. Being on my own, I began to investigate what would work best for this child within this family with these needs. In doing so, I was better able to understand the family and child on a deeper level than I had the opportunity to do previously, which again facilitated success in our work together.

In retrospect, much of my learning was a result of the first family I worked with privately. A challenging family, I stepped in after their previous psychologist had left, indicating that she could not provide the support they needed. The mother, who I will call "Wendy," was intimidating to many people because of her thorough understanding of the policies and procedures of the funding agency. This coupled with her extensive knowledge of specialized services legislation, her close relationships with influential government officials, and her relentless ability to pursue a matter until she received the necessary supports could be off-putting to people, but fostered an intense respect in me. By not taking the time to uncover what motivated her, the need to be an advocate for her son, others were unable to understand her position and simply labeled her as "difficult."

Understanding the importance of the family context, and seeing Wendy's intensity, I decided that my first objective was not to immediately initiate programming for her son with autism, but to instead spend time getting to know Wendy and her family. By letting her tell her story, I was able to gain insight into the many complexities in her life. By having the opportunity to have someone listen to her, she was able to free herself from the feeling of having no allies and found someone she could trust to do what was in the best interest of her son. By taking a single afternoon to listen to Wendy, we developed a rapport that expedited our trust in one another and our ability to work effectively together.

Wendy's son, I'll call him "Max," was one of the oldest kids I had ever worked with. He was 10 years old and almost as big as me. Max had very little language and poor pronunciation, making it quite difficult to understand him, which proved to be frustrating for all parties. When I first started working with him, Max had a myriad of behaviors that impeded his ability to learn new skills. He had trouble sitting at a table or attending to tasks. He would often run away and needed to be watched constantly, which was exhausting. Indeed, when he did not have direct adult supervision, he would always find trouble. For example, he would climb on or knock over furniture, dump toothpaste or shampoo everywhere, pretty much anything he could do that caused his mother grief. Max would also run out of the house and down the street (often to the park or convenience store across a busy street). If the front door was locked, Max would go out the back door and climb over the fence into the neighbor's yard. He also loved throwing things over the fence, which caused a lot of trouble with the neighbors. This was especially true because it was usually chairs, bricks, or rain eaves that went over the fence. Max was also very aggressive and would often pinch, hit, push, or kick. I remember days when I would come to the house and Wendy's arms would be black and blue from Max. We were also concerned about her youngest son, age 5, as he was smaller than Max and a frequent target of Max's aggression. The older sister, age 12, was not in immediate physical danger, as she could easily defend herself now, but she had a lot of emotional difficulties associated with having a brother with autism.

As you can imagine, this family had bigger issues than just Max needing intervention. Upon getting to know the family better, I quickly realized how much the entire family unit needed support, not just Max. There were little things Wendy needed help with, like building a positive relationship with her angry neighbors. Together, Wendy and I wrote a letter that she delivered to everyone on her street to help increase their awareness about autism, Max, and his needs (and to keep their eye out for him in case he ran away from the house). As "small" as this was, I know this was hard for Wendy to do; I have worked with a lot of families who try to keep the autism a "secret," worried about the stigma and what people will think. There were also other issues, like helping Wendy engage in self-care because, as many mothers do, she was running herself ragged trying to run the family. Wendy has been the only mom I have known that knew exactly what was needed and, when finally given the permission and tools to do so, she took care of herself. This is one lesson I wish other mothers would take. Many mothers feel guilty about taking time for themselves, but, by taking care of herself, Wendy was much more effective in running the household and managing Max's needs, and was a better mom overall.

Max's siblings' well-being was also a concern, as mentioned previously. Most of my graduate research was on siblings of children with autism, so I knew that the siblings' needs are often greater than those of the child diagnosed with autism. As such, I began doing some one-on-one work with Max's brother and sister. Although they seemed to be coping okay, through our work together, significant social and emotional difficulties were revealed. Now, finally, *they* had someone they could go to for support, after all these years of Max getting all of the support.

While I could not help with all the family needs, such as marital conflict or money difficulties, I became sensitive to all of these challenges and demands the family struggled with. In doing so, I was more effective in providing services that best fit the family needs, which only facilitated their ability to function successfully. By addressing the family needs, we were better able to maximize Max's programming. Indeed, by focusing on family priorities and resources, Wendy had more willingness, energy, and patience to follow through with recommendations outside of therapy time, and the siblings were better able to interact positively, which reduced a lot of stress within the house. Behavioral programs were more effectively implemented, and Max's behaviors significantly reduced. With greater compliance and an increased ability to attend, Max quickly developed a solid foundation of skills across a number of areas. We focused a lot of attention on Max's strengths and interests, which helped increase his motivation and further facilitated his skill development. We realized Max loved to do domesticated activities and quickly had him doing chores around the house (e.g., vacuuming, laundry, and cooking). We were able to expand his repertoire of activities he could do independently, which immediately reduced the number of problematic behaviors he engaged in during his unstructured time. In turn, this further reduced stress within the family.

Although his overall behaviors reduced significantly, Max still displayed some aggression due to frustration. Unfortunately, despite his gains, there was so much focus on his maladaptive behaviors that his "hard-to-manage child" label stuck with him. It was easy to forget how severe and frequent his behaviors used to be and the significant progress he had made during those times when he did get frustrated and lashed out. Those rare occurrences usually occurred when he was ineffective in communicating his needs. I can only imagine the frustration he would feel. I know that whenever someone asks me to repeat myself two or three times, I often get annoyed and reply, "Never mind." Thinking of my own experience, I wonder how much Max feels the same way. He probably experiences even more anger because he is constantly trying to express himself. And, because he does not know what else to do, he hits the person to let them know he is upset. During these episodes of aggression, it was easy to forget that there was a need behind that aggression. Indeed, there is always a purpose as to why children behave the way they do. I believe that children are not "bad" and that they engage in maladaptive behaviors simply because they do not have the skills to engage in more appropriate behaviors. Many people forget this, label the child as "bad" or "difficult," and react or punish accordingly, though often inappropriately. For children with autism, who have such a difficult time expressing themselves in a meaningful way, it is easy to forget that behind that frustration and anger, there is a warm, loving, vulnerable child who wants to reach out but who does not know how. After a while, I wonder how many of these kids simply give up trying.

Within a few months of working with Max, we had a team meeting. As the adults were coming in and chatting, I could see Max becoming increasingly frustrated in the corner. He was pointing insistently at his train, but I had no idea what he wanted. Although he did not have the skills to gain anyone's attention, he did look in the general direction of the group of adults and then back to his train. It was evident that his frustration was elevating because his vocalization increased, he started flapping his hands more intensely, and he started to bounce on his toes. Any other time, Max would be redirected to his calming area as soon as he displayed these behaviors. As I watched now, I knew that was not an appropriate course of action. Max did not need to calm down as much as he needed to say something. In the past, we would redirect him rather than try to understand him. No wonder his frustration never truly subsided. I could clearly see that he wanted something and that removing him to his calming area would only upset him more.

I began to think that maybe he wanted to share something with us. I had never known him to want to share anything with anyone before. Now I realize that perhaps it was not because he did not want to share, but because we did not understand how he shared. Without the necessary skills to get anyone's attention, his attempts to engage someone were missed, resulting in his frustration. This frustration manifested in behaviors that would soon lead to aggression. When I finally went over to see what he was pointing at, I noticed that he had put a figurine inside the train station. Max was excited when I came over and began to say, "duck" over and over. I repeated "duck" back to him, and he got mad. Thinking harder about what he was trying to say, I finally said, "conductor," to which he immediately squealed, jumped up and down, and smiled from ear to ear. I had never seen him so happy. He left after that, and for the first time, he was able to occupy his time contently for the entire duration of our meeting. Shortly after that day, we were able to get Max an augmented communication device, which he excelled at. Finally, he had a voice. Although his frustration decreased, it did not abate entirely, as he still had difficulties expressing his needs and wants effectively. Max was limited in what he could say using the device, as the vocabulary was quite limited (i.e., to certain foods and requests). "Conductor" was not in the computer's vocabulary. To make matters worse, others did not consistently take the time to genuinely listen to him. Although Max was initially excited to use the device, after awhile, he stopped using it and no one ever encouraged him to keep using it.

I have seen a lot of children go "underground," internalizing their behaviors because we were unable to fully understand and meet their needs. We were unable to fully understand what it was they were trying to communicate to us through their behaviors. As early as newborns, children are always communicating with us. When they do not have language, it is often through behavior; babies cry when they are hungry. We are usually attentive to babies and quickly meet their needs, especially when they are screaming. As soon as those babies get a little older and start talking, our expectations change for them. What would happen to a preschooler who screams? We would expect him to calmly tell us what is wrong, when realistically, he has no more skills in doing so than an infant does. Expectations change for children with autism too, even if they do not have expressive language. All of a sudden we start implementing behavior programs, token systems, response–cost systems, time-out; there are even shows promoting "punishment." And yet we fail to take the time to find out why the child is misbehaving. What function does it serve? What do we do when a child engages in attention-seeking behaviors? The child is telling us exactly what he or she needs (i.e., attention), yet we do the opposite and ignore them. What we need to do is *give* attention. That is what the child is telling us he needs, and yet that is the very thing we take away. I agree that the best time to give attention is not in the heat of a tantrum. However, we need to be proactive rather than reactive (the latter of which is what we typically revert to), and give them enough positive attention for all the praiseworthy things they are doing to reduce their need for engaging in negative attention-seeking behaviors.

After my work with Max, he was eventually transferred to a residential program, but I maintained close contact with the family. When Max turned 14, I had the opportunity to spend a day observing his programming on a consultative basis. I was shocked at what I saw. Over an 8-h day, I saw Max play Pop-Up Pirate, sing "Row Your Boat," listen to a preschool book, color a picture from the story, and go for a walk in the community. During the walk, we stopped at a toy store, and Max spent most of his time lying on the floor looking up at the lights. When I queried his occupational therapist about this behavior, he commented that Max enjoyed lying on the ground and looking at the lights and that it was very calming for him. The rest of the time Max was either allowed to rock in the corner or watch television, again to help "keep him relaxed." This struck me as a waste of an 8-h day. Max was 14: Playing preschool games and lying on the floor in a public place did not seem appropriate to me. What Max needed to work on was adaptive functioning skills to increase his independence, such as shaving, showering, and learning vocational skills. We had begun working on these skills with Max when he was 9 and, had they been maintained, he would have had increased independence with each of these skills. Now, he still had others bathing him each night.

The unfortunate part is, I have consulted on numerous cases with school-aged children and I rarely see age-appropriate programming or the incorporation of future planning to prepare for the transition to adulthood. Many people argue, "But he's only 10; he still has a long time before we need to worry about adulthood," or "I know he's 12, but he loves *Barney*." We need to think proactively about the child's current skill set, age-appropriateness, and how their skills translate into adulthood. It may be acceptable for a 4-year-old boy to use the toilet with his pants around his ankles, but it is entirely unacceptable for a 14-year-old boy to do so. As service providers, we need to think ahead. I know it can be hard for parents to think of their children as adults, but as professionals, I believe it is our duty to help them understand the importance of building skills that will form the foundation of their independent functioning as adults.

Although my professional journey is in some respects still beginning, I have been fortunate to have acquired insights from my experiences working with families of children with special needs. I gained many of these insights from working with Max, and I try to share my experiences from working with him whenever possible. To conclude this narrative, I will outline the key factors that I believe are critical for anyone working with children with autism and their families.

First and foremost, to facilitate meaningful changes for a child with autism, services must be focused on the entire family instead of just one member. A crucial element of looking at the bigger picture for any family is to understand the parents' journey. We all know the power of the working relationship and how imperative collaboration is. Parents are powerful resources that we need to respect; after all, they are the experts of their child and their family. Understanding their priorities and including them in the development of goals, behavior plans, and strategies will empower them and may increase their receptivity to our recommendations.

Second, we need to demonstrate empathy and thoroughly understand the children we are working with. Kids are always telling us something, from the first day they enter this world. We need to observe them and understand their needs and respond accordingly, instead of simply grasping for a solution. If they need attention, we need to find proactive ways of providing positive attention. If they are unable to communicate with us conventionally, we need to understand how they communicate and adjust our method of understanding them.

Finally, no matter what the age of the child is, we need to think about transition planning for adulthood. As always, a team effort is required to create effective transition planning; as such, professionals, parents, and, if possible, the individual should be part of the planning process. I do not believe any age is too young to get started. Indeed, the sooner we get started, the greater chance of success the child will have, especially for individuals with more severe needs. When working on goals, it is important to ask whether a particular activity or skill will be necessary for success in the individual's adult life. Critical considerations include teaching kids how to use their unstructured time productively, to attend to and complete tasks independently, working within groups and social skill development. Age-appropriate programming is essential, and activities of daily living should be a key focus. Educational pursuits should be secondary until the child is able to demonstrate responsible behavior in the areas of safety, personal grooming, personal hygiene, respect for others, and respect for personal property. During secondary school, vocational skill development should be a major focus.

When thinking of transition planning, careful consideration and preparation of necessary supports and accommodations are critical in successful outcomes. No matter what the ability level of the individual, there are a number of elements that need to be addressed. These include, but are not limited to, the following:

1. Postsecondary education and training involves collaboration between agencies and family and the preparation of goals.
2. Employment, which involves focusing on the development of vocational skills. Building from his strengths and interests, we had Max developing vocational skills that would be transferable to the adult world (e.g., sorting recycling, stacking books, working with animals). No age is too young to get started.
3. Living arrangements involve planning ahead and recognizing those skills the individual will need to live independently, despite the level of supported living conditions that will be required.
4. Developing daily living skills such as personal care (showering), community skills (street safety), and domestic skills (making a simple meal) are all valuable programming goals.

5. Other things that should be considered are engagement in and access to recreational opportunities, exercise and nutrition, and adequate medical and behavioral health care, all of
which contribute to one's physical and social well-being.

When completing an appropriate transition plan, it is critical to complete a
detailed transition process with a variety of goals and timelines to meet those goals.
In creating the plan, it is necessary to consider the individual's current skill set, the
skills he or she will need as an adult, the necessary supports and resources the individual will need, and any associated training that will need to happen for others
who will be working with the individual. It is also crucial to remember that a plan is
only useful if it is consistently applied and maintained. Max was well on his way to
independent functioning in terms of personal care; unfortunately, skill development
in this area was not maintained and, as a result, he regressed considerably and will
likely require ongoing support to meet his basic personal care needs into adulthood.

Working with children diagnosed with autism has been an amazing experience,
and I have gained valuable insights that will benefit my work for the rest of my
career. After many years of working with individual children, I have learned the true
value of working empathically with the individual and the importance of working
within the entire family context. Understanding the children's individual needs, in
addition to their larger family unit's needs, will pave the way to developing meaningful interventions to facilitate long-lasting positive changes for all people involved.
My challenge to other service providers is to be flexible with your training and apply
your knowledge and expertise to have the greatest positive impact you can for children and their families alike.

Epilogue

From our perspective, some themes within and across many of the stories in this section
include the following.

- When children with autism have difficulty expressing themselves in an age-appropriate way, it's easy to forget or not understand that behind their display of frustration and anger is a child who is trying to reach out to have their needs met.
- Early intervention for children with autism typically focuses on attention, compliance, communication, social skills, maintenance of routine within the daily schedule, transfer of developing skills to new situations, and high family involvement.
- Parents have to be strong advocates for their autistic children.
- Knowledgeable, skillful, empathetic, and supportive teachers can make a huge difference in the quality and amount of success of an autistic child in school.
- Having a child with autism affects every aspect of family life.

Themes that parents and professionals have expressed numerous times among the narratives or thoughts over the years include the following.

- Children with autism are more different than they are alike and there is truly no one
  approach that will fit every child.
- Many families need to extinguish spontaneity if they want to have a "good" day
  with their child.

- As a consequence of the extreme and overt nature of their autistic children's behaviors, many parents feel that others judge them to be poor or inadequate parents.
- Many mothers and fathers experience grief and struggle to come to terms with their child's autistic diagnosis.
- To effectively work with children with autism and maximize their development requires an understanding of their world and their needs so treatment approaches and strategies can be adapted accordingly.

Part IV

Eating and Health-Related Disorders

9 Eating Disorders

Prologue

Children and youth with eating disorders have abnormal attitudes about their weight and shape, have severe disturbances in their eating behavior, and exert unhealthy and maladaptive ways to control their weight. The two most common types of eating disorders are anorexia nervosa (characterized by a refusal to maintain a normal body weight for their age, gender, and height, an extreme concern about their weight and shape, and sometimes excessive exercise to control their weight), and bulimia nervosa (characterized by repeated episodes of binge eating and inappropriate behaviors intended to control their body shape and weight such as fasting, misuse of laxatives, and self-induced vomiting). Typically, youth who have anorexia nervosa fear gaining weight or becoming fat, refuse to keep their body weight at a level that would be considered normal for their age and height, and have disturbed ideas about the importance of shape and weight. They also tend to be characterized as having perfectionism, as well as being rigid, obsessive, and sometimes impulsive. Typically, youth who have bulimia nervosa have a sense of self that is highly influenced by their body shape and weight and are often ashamed of and secretive about their eating problems. They also tend to be characterized as unrestrained and impulsive.

In this section are four stories: two personal stories, one parental story, and one professional story pertaining to eating disorders.

The personal story, entitled "When Gaining Is Losing," is written by a young woman who reflects on and shares her thoughts and feelings about when she had an eating disorder in her childhood. Throughout her story, she reveals her worries and struggles, eventual acceptance of her anorexia, and her continual battle for control and a better sense of self. The second personal story, entitled "Externalizing the Emptiness Within," is written by a young woman who talks about her struggles with anorexia throughout her childhood and youth and shares her journey toward seeing herself in a way that is more than what her eating disorder made her to be. The parental story, entitled "She Looks like Death," is written by a mother who presents a very vivid and moving account of her fears, struggles, torments, contemplations, and resolutions while "coming to grips" with her teenage daughter's eating disorder. The professional story, entitled "Finding Oneself Within Ambiguity," is written by a psychologist who reflects upon and shares her story about her working relationship with her first teenage client with anorexia and how it taught her to embrace uncertainty.

Exceptional Life Journeys. DOI: 10.1016/B978-0-12-385216-8.00009-6

Personal Stories

When Gaining Is Losing

I still don't trust men with facial hair, over 30 years later. My pediatrician had a mustache. My psychologist had a beard. I hated them both. No, *hate* is too gentle a word for what I felt. The loathing I felt toward those men was only exceeded by the loathing I felt toward my own body. If I hated them, I detested myself. Over my 3 months of hospitalization, and as I got "better," through the interventions of those men and my family, I hated myself more and more.

Getting "better" was measured, very simply, by the scale. I was weighed every morning in half-pound increments. If I gained weight, I could spend the day out of bed, wandering the hospital halls. If I stayed the same weight, I had to stay in bed for the whole day, but was allowed to get up to use the bathroom (where I would do jumping jacks, frantically). If I lost weight, I had to spend the day in bed and suffer the humiliation of using a bedpan and being taken to my therapy sessions in a wheelchair.

I saw every half pound of weight I gained during those 3 months of hospitalization as a sign of defeat—a sign that the doctors (and my out-of-control appetite) had "won," and I had lost. They beat me down, and I gave up. I gained enough weight to start sixth grade, just 2 weeks late. In my desperation to maintain my academic perfectionism (I was so scared of falling behind my peers in school), I had to let go of my starvation perfectionism. So I gained enough weight to get out of the hospital, and went back to school. But getting better, in my mind, meant getting fat. Getting better was a defeat that would haunt me for years.

I don't remember when I started restricting my food intake (I never saw it as dieting—just taking control of my eating), or why. When people ask me if I was a "fat" child, I always say yes, because in my memories I am always fat. But I don't know if I ever was. There are only a few pictures left of me as a young child, as I have destroyed most of them over the years—hating what I saw and the memories they evoked of self-loathing and loneliness. I know that by the middle of fifth grade, my parents were worried about my weight loss; they asked the teachers to keep an eye on my eating when we went away for the weeklong fifth-grade field trip to Camp Auburn. Years later, I found my diary from that field trip—filled with promises to "do better" and "eat less tomorrow."

By the end of fifth grade, I was in my doctor's office regularly, and I remember him telling my mother, "I just don't know what to do with her. I'm going to refer you to a psychologist." Our first meeting with the psychologist was at our house; he came to watch us eat dinner as a family. I did not speak, and I ate only cucumber pieces, carefully chopped and arranged on my plate. That man was the enemy, and I was going to do everything in my power to make sure he failed at his job—because for him to succeed, I would have to give up. At the end of that meal, my father screamed at me, "You're going to kill yourself!" I remember being shocked at these words. Kill myself? I couldn't even control the fat on my thighs, how could I possibly gain enough control over my body to kill myself?

I spent the first 2 weeks of the summer between fifth and sixth grades at home, getting as much exercise as possible and eating as little as possible. I would run the 2 miles through the woods to our swimming club, and then swim laps for hours on end. I remember eating green apples, cut into very thin slices, and ice cubes (I had read somewhere that chewing ice cubes actually consumes calories). There are a few pictures left from just before I was hospitalized that summer—huddled over and wearing a winter coat. I was all bones. I can see that now, but all I saw and felt then was fat, disgusting flesh. And I was tired, so tired, and scared. Being admitted to the hospital was almost a relief; maybe someone could stop those voices in my head?

I made myself at home in the hospital, moving in with most of my stuffed animals and lots of books. Visitors brought plants, cards, toys. I loved the nurses, who cared for me unconditionally. I got a new name band on my wrist every week or so, and we put my old name bands on my stuffed animals. We fed my plants with IV tubes (using colored water so that people knew the tubes weren't going in my arm!). The nurses were the ones who weighed me everyday, and who pronounced my verdict every morning (full bed rest, partial bed rest, or walking privileges). But somehow I knew they hadn't set the rules. I knew the doctors were to blame for that.

I met with the psychologist thrice a week for 3 months. I walked (or was wheeled) down the hallway to an ironically bright sunny room at the end of the ward. And then I sat, in silence, for an hour, and stared defiantly out the window. As far as I remember, I never once spoke a word during those sessions. And I worked very hard to always avoid eye contact with that man. Looking back as an adult, these memories evoke two emotions: pride and despair. I am proud of that little girl who was so strong that she did not let this man whom she did not trust in on her fears. Yet I despair of my parents and the medical system that did not see this match as a failure, and did not try to find someone else who I trusted to help me.

Years later, when I was going through therapy as an adult, I contacted that bearded psychologist by mail, and asked if I could see my records. I received a letter back from his office stating that they did not keep records that far back. I will never know whether my memories of this failed therapeutic relationship are true, and I will never know what it was about the psychologist that evoked such fear in me. I am sure that to him I was just another "defiant" patient—a "typical anorexic." At a professional eating disorder conference recently, I heard another psychologist refer to anorexics this way, as "notoriously difficult" to treat. Now I want to ask, "After 30 years of failure, perhaps it is the treatment that needs to change?"

I went through dozens of roommates that summer on the pediatric ward, some recovering from surgery, some suffering from terrible illnesses like leukemia. But one I will never forget. About a month into my hospital stay, a very thin girl was wheeled in and put in a bed across from mine. A nurse whispered to me, "She has anorexia, too." It was the first time I had ever heard the word *anorexia*. I remember being thrilled—I had a disease that had a name! It was a huge relief to me; if I had a disease, maybe someone could make me better? Maybe all of this wasn't my fault? I told all my visitors that week, "I have anorexia. It's a disease."

And yet, after the initial excitement of naming my illness, nothing changed. The voices inside my head continued to remind me how disgusting I was, how out of control, how lazy and fat. And the voices outside my head continued to blame me for not eating, for not gaining weight fast enough, for not getting better. Much to my dismay, I was actually gaining weight. Slowly, the numbers on the scale crept up. I started getting jealous of the girl in the bed across from me, who kept losing weight. She was put on a feeding tube, and I felt like a failure for never having been *that* sick. Eventually, she was moved from my room.

As the start of sixth grade approached, I realized I needed to gain just enough weight to get out of the hospital so that I would not fall behind at school. So I gave in, complied, and ate. In the early fall, we had one last family therapy meeting with the psychologist and another therapist in a nearby city. Not wanting to leave the hospital for this meeting, I refused to get dressed, and wore my nightgown under a big heavy duffle coat. My father had to drag me out of the hospital kicking and screaming. The hospital security stopped my father to ask him what he was doing. I remember him yelling, "She's my daughter. I can do what I want with her." At that point, I knew that he, and the doctors, had won. I was no longer in control, and my fat body was evidence of my defeat.

The day after that family therapy meeting (at which I did not say a word; I remember hiding my face in the hood of my duffle coat), I was discharged from the hospital. I went back to school the next week. While I was "better" physically, I was an emotional wreck. I would spend the next 20 years of my life dealing with the aftermath of having "lost" this epic battle.

I spent my teen years and most of my twenties hating my body. I agonized for hours and hours every single day over what to eat, what not to eat, how much to exercise, how to get rid of the calories I had already consumed. Although I was never rehospitalized, and never again classified as anorexic, I certainly lived within an anorexic mind-set. The ultimate irony of eating disorders is that they consume you. Anorexia consumed me for 20 years.

During those years, I was very successful in other areas of my life—academically, professionally, and even socially. But deep inside, I still felt that I had "failed" at the one thing that was most important—controlling my body and being thin. I often wonder, now, what I could have achieved had I been more productive with all those hours I spent obsessing over food, my body, and my weight. Surely, I would have been happier.

It was not until my late twenties, when my weight got dangerously low again and I developed a heart condition, that I sought professional psychiatric help. Finally, I was ready. Recovery was certainly not easy, or pleasant, and took many years. I had a minor mental breakdown along the way, and struggled to redefine myself as someone without an eating disorder. I will never forget, for example, trying to get my head around the idea that perfectionism was not a personality trait to be proud of, and that I needed to work *against* that tendency in myself.

Several years and many therapists later, and with the aid of selective serotonin reuptake inhibitors, I was healthy enough to get rid of my scales. I stopped weighing myself in my mid-thirties and have been happier ever since. There was a brief period during my pregnancy at age 39, when the doctors weighed me every week; that time was admittedly very difficult for me. I still struggle with body image, and doubt that my perception of my body will ever be very accurate. But now, in my early forties, I can proudly say that my perception of my mind, of my *self*, of who *I* am, is more accurate, and more accepting, than ever.

Externalizing the Emptiness Within

In my early childhood, I was a vibrant, happy, and friendly child who grew up in a loving household with married parents, an older brother, and a twin brother. My life started early as my mother gave birth prematurely to my brother and I at 26 weeks gestation, which followed by months of intensive care, attached to lifesaving equipment until we were both able to survive without medical help. Due to this situation, I was faced with retinopathy of prematurity, an eye disorder. At 6 weeks of age, I underwent laser surgery, which resulted in the prevention of blindness but having to wear glasses, and amblyopia (lazy eye) in my right eye. Apart from my first few years of frequent hospital visits, I lived like any other happy and healthy child. I loved attending school and learning, as well as the social aspect of friends. I was a high achiever and a perfectionist at everything I did, from school to sporting activities, but it was for my own self and not to please anyone. My mom was diagnosed with cancer when I was around 8 years old, and I do not have a clear memory of my mother being unwell from that time, as my parents and family friends made sure we were well cared for and continued to live as a child. I do understand that she was quite sick, but she has been in remission for many years. Like many children, I had a great upbringing and was taught the appropriate behaviors and manners. I loved playing sports and had many friends and not many enemies. I was always shy at first and would be perceived as rude, but once I got to know people, I would become more outgoing. I was happy to be friends with anybody. I accepted people for who they were or who they chose to be.

At 14, I began to be another person—a grumpy sad teenager. In the beginning, my parents thought it was just a phase, but I got progressively worse. A few things occurred that really disheartened my willingness to live and I didn't know how to cope or what to do. After showing continuous signs of low mood and self-harming behaviors and after making several suicide attempts, I was referred to an outpatient service by my school counselor. I started seeing a psychologist and was diagnosed with major depression and anxiety. She soon resigned from the service after a few months, so I started seeing someone else. The first time I saw a psychiatrist was in late 2004; this doctor immediately admitted me to a voluntary adolescent mental health unit.

I continued to eat and drink like a "normal" teenage girl does but at the time, I had low thoughts about food because of my low self-worth, my body image, and myself in general. After being admitted a few times, I started to feel more out of control because I couldn't do self-destructive behaviors. I just stopped eating, which developed into anorexia. It wasn't my intention to lose weight but once I started to lose weight, it became my only focus. I felt I was in control of my out-of-control life. I couldn't control my thoughts, but I could control what I put in my mouth. It also became a way of coping with my sadness. I wasn't ready to deal with my sadness so I just focused on losing weight. My weight loss soon became a concern for my treatment team. I became medically compromised and had to be placed in a medical ward. Initially, I accepted that I had to be rehydrated through a nasogastric tube and be put on a heart monitor to achieve physical stabilization. However, the eating disorder had already consumed my thoughts, feelings, and behaviors, and I became resistant to accept any further oral food intake or liquid nutritional supplements. My case team made the choice and sent me to an involuntary adolescent unit where treatment was forced upon me. I didn't have a choice in the treatment I was given nor did my parents. I was held down by nurses and force-fed via a nasogastric tube without sedation and received very little psychological intervention to help with my thoughts. I was not allowed to partake in many activities or allowed out the unit. My weight was increased to a healthy weight and, for me, that was unbearable. Weight gain made me feel out of control, and all I wanted to do was fade away to get away from my thoughts. I felt heavy inside my head, and having my body feel heavier only increased the depression and self-hatred. It was to save my life but it was torture being forced to do the opposite of what my thoughts were telling me. My mental state only further deteriorated.

After *nearly 3* months of being detained at the unit, the treatment team decided that I could go back to the children's hospital and be a part of the eating disorders program. The program involved both medical and psychological treatment. It was a better environment for me and had staff that had been trained to work with people with eating disorders. I had become more depressed and consumed by the eating disorder, and had little insight into what was happening. I was still being held down but sedated when fed via a nasogastric tube. I stopped school in tenth grade, refused to see friends and family, and didn't talk much apart from mumbles. I spent the majority of the next 2 years in the hospital, being fed via a nasogastric tube and spending very little time with my family and friends. My mother and father would come as often as they could, and then my dad started to avoid me as he felt hopeless that he couldn't help me. When I was discharged from the hospital and sent home, I would stay in my room, avoiding people and food. I would sleep or sedate myself so I was able to sleep. Within days of discharge, I would have an appointment for a medical checkup and would be admitted back to the hospital due to medical instability. At times, I resisted attending appointments, which resulted in my parents and brothers placing me in the car with the child lock on the doors so I was not able to escape. I was too frightened to eat, look after myself, and confront the feelings of shame, guilt, and self-hatred—it was easier to be force-fed and blame the docs for making me "fat." Back then, everything I felt and believed was "fat" because I couldn't and didn't know how to identify with my thoughts and feelings. *Fat* summed up everything. I eventually stopped resisting medical treatment and just accepted the tube, but I continued to deteriorate in my mental state. I was given medication to minimize the symptoms of anxiety and antidepressants for my depression, but they didn't improve my motivation to live. It wasn't until my eighteenth birthday that I started to see that I was more than what the eating disorder made me be. The day before my eighteenth birthday, I was still being fed via a nasogastric tube. I reflected on everything that made me become who I was at that current time. I was a sad dead soul in a just living body. I wasn't me. I was consumed by the eating disorder. From that moment, things started to change. I wanted my life to change. I missed out on my adolescence because of the eating disorder. I was meant to be directly transferred to an adult program, but I remember telling my case manager that I had to go home for my birthday. He arranged to have me admitted a few days after my birthday on the provision I would adhere to some negotiated requirements. I was not well but I went home for my birthday.

I was admitted to the adult program and was automatically given a nasogastric tube and then was told to eat or my target weight was going to be increased. I started to eat due to the fear of the eating disorder. After a month, I was discharged due to noncompliance and losing weight and I was not referred to any further services. I didn't know what to do. I was medically stable but was not a healthy weight by any means. I was stuck because I had been left to my own decisions, but from then on, I was determined to change. However, I learned how to manage my life with an eating disorder.

I began to experience my eating disorder differently. Although I began to eat again, I became rigid and very obsessive about how and what I ate. Any changes in my eating would create severe anxiety so I felt I had to adhere to all the rules so I could still live my life. It was not any easier, but I started university. I started to eat the bare minimum so I could study. I ate six meals a day. It was not a lot in terms of what "normal" people ate, but I made myself eat enough so that I was able to function and maintain a very low weight without being medically compromised. My body adapted. Living with an eating disorder was a daily struggle. Each day was a strict routine. I would wake up and weigh myself multiple times just to make sure I hadn't gained weight. The number on the scale defined how I felt about myself, my mood for the day, and whether I could go out and see people. I ate the same foods everyday and had to know the nutritional content of everything I consumed. I had to weigh and measure food 10 times before I could eat or drink anything. My lunch, dinner, and snacks were prepared by me at the beginning of the day. The way I felt about the way I looked and what I had to eat consumed my mind throughout the day. I felt extremely guilt-ridden when I ate because I did not feel good about myself. My thoughts were illogical. For example, I got scared and anxious if I even thought about not eating. I could not drink water by itself. The prospect of drinking pure water was extremely anxiety provoking. I felt undeserving to drink water, yet I was able to drink other liquids. This was one example of my many strict rules that I had to follow or I would feel my sense of control disappear and my thoughts would become erratic. It was really sad.

I lived this life with my daily rituals and routines right up until August 2009. To this day, I do not know how I managed to study and survive. However, I think it was my determination that I wanted to recover so that I could be able to help others. I tried to overcome my fears and severe anxieties surrounding food and my weight but changing my thoughts and behaviors would increase my anxiety, and I was only seeing a psychologist once a week to set goals about what I needed to change. I had actually moved out of my family home because of my eating disorder behaviors. They were too stressful for my family to watch. I felt so hopeless that I couldn't change. I was living on campus at university with little support from the friends I had because they simply didn't understand. I needed intensive support and encouragement. I was ready to recover but was so consumed by the eating disorder that recovery seemed impossible. I found seeing a psychologist helpful, but I also did not know how I was going to recover because I was not psychologically strong enough to complete the goals alone and I did not have the constant support to change. Change made me feel out of control. Doing something that I did not agree to do made me feel out of control. I knew within myself I needed to be supported and guided but I did not want forced treatment. I continued to look into services but none were suitable to what my needs were. The hospital programs primarily focused on weight restoration, not challenging behaviors and thoughts related to the eating disorder.

I looked internationally and, in August 2009, I traveled overseas to a residential program that allowed me to have insight into the thoughts, feelings, and behaviors of the eating disorder. The approach the program took was brilliant. The program's focus was not solely on weight restoration. It also provided the necessary guidance to help me make choices toward a healthy lifestyle. The way I lived was extremely ritualistic, obsessive, and quite sad to the point that I was not able to live with my family or friends. I had a lot of food issues and fears, including not being able to drink pure water. They helped me increase my food intake enough for me to be able to start challenging the eating disorder thoughts, change my eating behaviors, and cease my rituals. With their help, I started to focus on my self-esteem and body image issues. My recovery was facilitated by way of cognitive behavioral therapy. Being able to make recovery-orientated goals and choices allowed me to have a sense of control. Finally, I was more willing to achieve them because they were not being imposed and forced. Living in a small residential facility created a realistic "life" setting, and I learned to look after myself, cook and clean just like any other person would. The center had a small intake rate so I was able to work closely with the staff everyday. They didn't just treat the illness. They treated me as a person. I was a person suffering from mental illness. I was not the illness.

Feeling better did not come quickly. I relied on the positive approach, and enthusiasm of the staff. I didn't have my family to support me and I didn't make many friends at first. My emotions were erratic. I would isolate myself. I was not happy. Trying to change my behaviors and thinking patterns that could make me feel safe was very difficult for me. The sadness was very evident. But I knew I had to work through the emotions to recover. It was not until the day before Christmas that I realized I was actually getting better. Despite having changed so many behaviors and thoughts already, I couldn't acknowledge or see that I had, until then. It was a wonderful moment. I started to drink water and eat chocolate again too, two of the hardest things that I thought I would never be able to do because I felt I didn't deserve it. I can now happily drink water and eat chocolate without feeling guilty.

Restoring my body was very difficult to accept. I stayed a low weight for over 3 years, and I had gotten used to being that size. I believed I was already overweight, and increasing my weight seemed irrational. As I began to work on my thoughts surrounding my body image and the actual reality of my physical structure, I accepted that I needed to restore my weight to be healthier. It was quite overwhelming adjusting to a "normal" body, especially when the thoughts of feeling fat were still constantly barging in. I started to improve the way I would talk to and about myself, which I found beneficial for improving my body image. For example, I no longer said weight "gain" or "gaining" weight. Instead, I said and thought, "I am restoring my weight." I personally felt better when I was using the word *restoring* because I was open to the prospect of becoming a healthy weight to live a healthy life. I became more aware of what the eating disorder thoughts were and replaced them with ones that were rational. I learned to separate the eating disorder from myself. I began to think that recovery was possible.

In March 2010, I was ready to come home. It was a scary prospect at the time. I was healthier and happier. I no longer had a strict routine of what or how I ate. I could allow anyone to cook for me and enjoy eating and tasting food without fear. Food was never the issue. It became the issue. I had overcome many food fears, and improved how I felt about myself. The meaning of recovery differs for every sufferer. To me, recovery is being able to live without the eating disorder thoughts and behaviors consuming my life. I may have occasional negative thoughts but I am able to cope with them and continue with living my life. I believe I may always have some doubts about myself but I think I am strong enough to cope with them without resorting back to the eating disorder.

My experience with anorexia resulted in osteoporosis and low iron levels. However, I am a healthy young woman. I do not take any medication to help stabilize my mood, obsessive habits, or anxiety symptoms, and I rarely have panic attacks. I am able to cope in social situations without fearing that people are looking at me and thinking the worst of me. I take care of myself and dress like a female instead of covering up in oversized shirts and shorts. After 21 years, I have finally stopped biting my nails and have long and healthy nails that I confidently show off. I can wake up in the morning with a smile instead of dreading waking up to a new day. I think about my future but focus on living for today. I am proud of who I am and what I have accomplished. There are days and moments that still test my mental strength but they are less often now.

At the age of 22, I am living like any other 22-year-old, without the restriction of the eating disorder. I continue to learn about myself and who I want to become. Life is enjoyable, with the usual ups and downs of life. I am learning how to drive, had my twenty-second birthday, and enjoyed eating my delicious birthday cake with my beloved family and friends. Even though I had isolated myself from my family and friends and refused to be loved by them, they all have supported me toward recovery in one way or another. I now enjoy going out with friends and family and doing activities without being consumed by my eating disorder. I have been sharing my experiences at the children's hospital as a way to increase awareness and insight within others about eating disorders. I am about to start university again, with the intention of studying behavioral science and counseling. Recovery is possible. It just takes time.

Parental Story

She Looks like Death

It is April 2005, and I am sitting in my home office. I freeze and strain to listen. I can hear Melody downstairs through the heating vent—she is vomiting.

"NO. This isn't happening!"

I have heard her several times over the last few weeks. It is always the same pattern: huge meal consumed quickly, hurried exit from kitchen, bathroom door closes, water running in the sink, and emerges red-eyed and flushed.

In the beginning, I push the thought of an eating disorder far from my mind.

This is *not* happening to my daughter!

This is *not* happening to our family!

I am just imagining this. ...

She looks well. ...

She is not overly thin. ...

She seems happy. ...

I stare at the heating vent and listen, but my heart is pounding so loudly I can't hear anything. I can feel my breathing getting faster. I feel a lump forming in my throat. I can't swallow. My mouth is suddenly very dry. My hands are shaking. I don't want to hear this. I want to run downstairs and tell her to stop it, but I don't. I can't face it—*or her*—or this horrible reality. Instead, I just sit and listen and shake and fight back tears.

I hear myself murmur, *"Please God, make her stop this. Heal her body and mind. Help her love herself."*

I am rambling on and on in my head. I don't know what I am saying or if it makes sense. I do know that this disease kills young women, and now it has attacked my daughter. *Oh God, no!*

I feel alone and terrified and powerless.

God, please take care of this. I can't do it alone.

My mind goes back to Melody's birth in 1987. She is a beautiful, tiny, quiet baby. I love holding her. With blond hair and blue eyes, she looks just like her older sisters. As a toddler, she is very shy and quiet and clings to me when we are out. She does not seem to adjust well when her younger sister is born and constantly competes for my attention.

When my husband passes away in 1996, I am left with four children. My daughters are devastated, and I am emotionally unavailable to help them through their grief. I have to work overtime shifts as a nurse to pay the bills and therefore am rarely home. My daughters have to be self-reliant as I am very busy and tired. I know Melody struggles with this much more deeply than her sisters. She needs more of me than I have to give.

She becomes an anxious teenager and rarely steps out of her comfort zone to try new activities or projects. She is moody and yells at lot ... *no*, she screams a lot, and her sisters avoid her at all cost, and I end up being the peacemaker.

Melody is diagnosed with clinical depression at age 16. Her family doctor says she is sad about losing her dad and is afraid that I might die too. She becomes hypervigilant and hovers at my bedside whenever I am sick. I roll over in bed, and there is her anxious, worried face looking at me.

"Do you want some water mom?"

"You have to eat the sandwich I made or you won't get well."

"Do you need some pills mom?"

"Mom I think you should go to the doctor."

I quickly recognize that I cannot show any signs of illness, tiredness, or weakness as Melody becomes anxious, moody, and screams at everyone in the house. I plaster a permanent smile on my face to show that I am well.

As I sit at my desk reminiscing about Melody's childhood, I feel God nudging me: *"You have to deal with this, Cathy ... NOW! She is sick and needs you."*

I can't God. ...

I can't. ...

I don't know what to do. ...

I don't want to deal with this right now. ...

I don't know what to say. ...

I don't want to tell anyone. ...

I can't tell our family doctor as she is like a friend to me. She thinks I am amazing to have done so well since my husband passed away. How can I tell her? She will think badly of me and wonder how I let this happen. She will silently judge me when we leave her office. I can't tell my family as they have always told me to slow down. They have told me to spend more time with the girls, and they will throw my busy lifestyle in my face. I can't tell my church as they will judge me and gossip about me *in the name of prayer*. They will say, "If you have enough faith, your daughter will be healed." I can't tell my nursing colleagues as I will be one of those parents we have silently judged. I will be that controlling mother who caused her daughter to rebel against the control and starve herself. *And I can't take that right now!*

I walk slowly out to the kitchen and wait for Melody to come back upstairs.

I am sweating.

My heart is pounding so hard that my chest hurts.

I am shaking uncontrollably.

I feel sick to my stomach.

Oh God, here she comes. ...

Melody walks into the kitchen, and I murmur a silent prayer asking for the strength, courage, and wisdom to confront her.

I want to run away.

I want to leave it for another time when I have more courage.

I know that once the words are said, there is no turning back!

I am not strong enough for this.

Why do I have to do this alone?

Why do we have to go through this at all, God?

Why us?

We have been through enough.

I try to swallow past the lump in my throat. ... I take a deep breath and quickly blurt out:

"Melody, we need to talk. I don't want you to deny it, because I know it's true. You have been throwing up and have lost a lot of weight."

There I said it ... no turning back.

I take another deep breath and swallow, "I want to help you."

Melody stares at me for a moment, then bursts into tears.

"I have been trying to stop, but I can't. I'm scared, mom."

I am stunned!

I am expecting yelling, stomping out of the room, denial—not a vulnerable, scared, and crying child.

... Now what do I do?

Instead of my mother's instinct kicking in, my nursing experience takes over and we talk about the importance of eating and getting vitamins into her body. I am comfortable in my nursing role, and I need comfort right now.

Melody agrees to go to the doctor for a full checkup and have her weight recorded. She agrees to sit at the table with me for all meals and wait 30 min before going downstairs or into a bathroom. We hug and spend the afternoon shopping for foods she likes and supplements she needs.

Okay, good. I have control. ...

I can handle this alone. ...

I don't have to tell anyone except our family doctor, and maybe all this will be behind us before all the relatives arrive for Melody's high school graduation in 2 months. *Yes, it will work out! Everything will be okay. ...*

Melody continues to lose weight, despite my vigilance and control, and is referred to a psychologist. Extended family members arrive for her high school graduation and begin discussing Melody's thin body. I brush it off and say she is just stressed, and find myself changing the subject whenever her weight is brought up. When the graduation ceremonies are over and everyone goes home, I am relieved as the stress of lying and making excuses about Melody's weight becomes unbearable.

Melody's weight drops to 95 pounds, and she begins displaying bizarre behaviors such as opening up cans of food and hiding them in cupboards. I discover this when we develop an ant problem. I scream whenever I open a cupboard, as ants are swarming among the open food. The smell in the kitchen is disgusting as there is rotten and decaying food hidden behind dishes and cans. My other daughters are grossed out and start eating meals at their friend's homes. I spend hours crying while hunting for rotten food, cleaning the cupboards, and spraying the bugs. Melody won't talk about it. She says it is part of her eating disorder, and I have to "deal with it."

Is it possible to hate your own child?

I struggle with this thought daily as she becomes more belligerent in our conversations and continues to lose weight.

It is September, and Melody refuses to go to the doctor or psychologist anymore, and I am powerless to do anything as she is now 18 years old and able to make her own decisions. I am living with a *seething anger and an overwhelming fear* that we are losing the battle. Our home is a war zone and no one wants to be there, me included. I begin working more and more hours per week. I just can't face the horror at home. I sense I am throwing in the towel, and I can't even pray for help as I am worn out.

I stop crying.

I stop visiting friends.

I sink into a deep depression.

It is a cold November day, and Melody has just walked past me.

I gasp. *She is a human skeleton!*

I clamp my hand over my mouth to keep from screaming out loud from the shock. *She looks like death.*

"Oh God, please help her!"

I run into my room, fall on my knees beside my bed, and cry like I have never cried before ... wracking, wailing, sobbing that seems to last for hours. My daughter is going to die, and I can't go on without her. I begin to pray and bargain the day away. Let me die instead or let her live and I'll help others suffering with this disease.

Whatever you want me to do, God, I will do it; just *please God, don't take her from me ....*

It is at this time I finally realize I have to start being open and honest about her disease and look for the support we both need. If I don't, she *will* die. I take a deep breath, silently pray for courage, and phone my pastor. I am amazed at receiving compassion and encouragement instead of judgment. I admit to him that I am very scared and depressed. This is tough for me to admit, as I am independent and intensely private. It is then that my pastor shares with me that his child has suffered with an eating disorder as well, but is now completely recovered and doing well in life. *I can't believe it!*

That means there is hope for Melody too.

After we pray together, I begin to feel hope, faith, and the much-needed courage to fight this deadly disease.

Using my newfound courage, I phone a close friend and ask if we can go for a walk. I share with her about Melody's eating disorder, and she hugs me and says, *"You have been a great mom. This is not your fault."*

Relief rushes over me. She isn't condemning me or blaming me or admonishing me for working too much.

I am beginning to realize how important support is!

My family and friends become my lifeline and build me up whenever I get scared or worried. I am able to be around Melody and find joy in little things. I convince Melody to go back to the doctor, and am alarmed to discover that her weight has plummeted to 85 pounds. Her blood pressure is dangerously low, and her heartbeat is irregular and very slow. The doctor says she needs to be admitted to the hospital. Melody has not eaten anything in 3 days and has had no water for 24 h. She is crying but refusing to go to the hospital because she doesn't want to be force-fed or take medication.

I am powerless. ...

So Melody and I go home.

We lay on my bed, and I hold her in my arms. She is so weak and pale.

I am afraid to close my eyes and sleep in case she dies before I wake up.

I just lie there and stare at her.

Tears are spilling down my face.

I offer her a sip of water. No, I beg her to take a sip of water.

"I can't mom, I can't."

"Please Melody, just a little. *Please*, for me."

She takes a small sip and cries, "It hurts, mom, I can't."

My heart is breaking, and I am trying desperately to hold onto my faith. While Melody sleeps, I call my pastor and he sends an urgent prayer request throughout the church.

I cry and cry and pray and begin bargaining again with God.

"Please God, let her live. ..."

"I will do whatever you want me to do."

"I'm begging, please God!"

The next morning, while I am pacing in the living room and praying, Melody walks out of the bedroom and starts crying, "Mom, I need help. Help me."

I am shocked, elated, hopeful, scared. ...

Melody insists that she doesn't want to go to the hospital, but wants to go to a treatment center. She wants to go where they do not force-feed or drug with pills. She wants to go where they provide therapy to help her understand *why* she is doing this. Her nutrition counselor had previously mentioned an eating disorder treatment center that provides several hours of therapy per day and has an excellent recovery rate for both anorexia and bulimia. I phone immediately and, miraculously, they have a bed available. They say they usually have a wait list, but because it is close to Christmas, one girl has backed out until January.

They agree to take Melody. She is both excited and scared to go, but says she will stay until she is well. I cannot go with Melody as I have work obligations, with no replacement on short notice. Maybe subconsciously, I am too afraid to go and see other anorexic girls ... too real ... too close to home. I drive Melody to the airport on a cold, snowy morning at the end of November. Melody is frail-looking and very pale. She is so skinny that people stare at her as we check in at the airport and I want to yell, "Stop staring and mind your own business!"

As soon as she is checked in, I tell her I have to leave—it is too painful to prolong the good-bye. I hug Melody and then leave. Just before I go out the door, I turn and look at her. ...

Is this possibly the last time I will see her?

Is she going to die?

I don't want to leave her here alone.

Why am I sending my frail, sick daughter alone to get help?

Oh God, I should be going with her!

WHY DIDN'T I TAKE A LEAVE FROM WORK?

Suddenly, I hate myself.

I hate my job.

I hate leaving Melody.

I hate everyone and everything at that moment.

But I especially hate myself. ...

I push these thoughts from my mind and smile bravely at Melody, whisper that I love her, and turn to walk out of the airport. Before I reach my car, the tears start. ...

I cry so hard I can't see to drive. I just sit in the car and tears pour like water down my face and onto my winter coat. I can't pray ... I am too grief-stricken. *"What a terrible mother"* runs through my head over and over. ...

Suddenly, I hear God's voice in my ear and I immediately stop crying and feel a peace come over me. I sense God telling me Melody will be healed and to have faith.

By the time I reach the hospital to work, I am no longer sad or crying. I can't wait to phone Melody when I get home.

I phone the treatment center and discover that Melody has arrived safely and has eaten dinner. That was the best news I could have received, as she had been refusing food for days at home. *God is answering my prayer.*

With encouragement from friends and my doctor, I attend a support group for families while Melody is away and begin to cope and live again. Melody also responds well to the therapy and support at the treatment center. We talk frequently on the telephone, and I am encouraged by her commitment to get well. She arrives home in January 2006, after 6 weeks of treatment, with a gain of 11 pounds and a much healthier outlook on life. *Hallelujah.*

Melody agrees to see her family doctor and nutrition counselor regularly after arriving home, and with inspiring determination and only a few minor setbacks, she has recovered well over the following year.

It is October 2010, and Melody and I are sitting in a restaurant eating lunch. She is smiling mischievously as she swirls the creamy pasta on her fork and pops it into her mouth.

"Mmmm, this is my favorite dish."

I sit and enjoy watching her eat and think back to a time when I could never have imagined this scene. Our journey began when Melody was 17 years old, and she is now a healthy and happily married 23-year-old woman. I am so proud of the work she has done to get well.

"Do you want a bite mom? It is soooo good."

Oh, how I treasure our time together. Melody has gained over 30 pounds and has been able to maintain it. She works full-time and has a very supportive husband.

"Melody, I have been thinking about writing about your eating disorder and how I dealt with it. How would you feel about that?"

"Hmmm, what would you say?"

"I would say this has been an incredible journey as a mother of an anorexic daughter, and I have learned many things about myself and your disease."

"I would say how I have realized the importance of being your mom, not the professional, as this was detrimental to you when you just needed me, your mom, to love you and support you during this time."

"Yeah, I agree with that mom. I needed you to love me and not treat me like one of your patients."

"You should also talk about getting support because it really helped you."

"Yes, it sure did. I would say how dangerous it is for parents to isolate themselves and emphasize the importance of seeking support from family, friends, and professionals."

"I think you should go for it."

She pops another forkful of pasta in her mouth.

"I want it to be completely honest, so can I use your real name?"

"Yes, of course mom, no more secrets right?"

Melody smiles, licks her lips, and winks at me.

Oh how I love her.

"Mom, you have to tell them that I still struggle sometimes when I am stressed, but also tell them that the treatment center gave me some good tools to help me."

"I will."

"I love you, Melody."

"Love you too, mom."

Professional Story

Finding Oneself Within Ambiguity

Any statement about a professional experience reflects how we as "care providers" live our personal lives. I have specialized in the treatment and prevention of eating disorders for over 10 years and have been recovered from my own eating disorder for almost 18 years. I must first state my gratitude for the many lessons my clients (past and present) have taught me about myself and the art of living. Secondly, I must acknowledge, as I do with most of my clients in session, that my personal experience with my own eating issues strongly influences my work with youth with eating disorders. I have worked with many youth and women struggling with eating issues within a sociocultural context, which perpetuates the ideal of thinness and abhors fatness. I have learned important lessons about being a counselor and living an authentic life from each of my clients but my most important lesson was from one of my very first clients. I met "Jill" during my doctoral internship in applied psychology. I fondly remember Jill as she was one of my first clients with anorexia nervosa, and she taught me to embrace uncertainty.

Jill, an 18-year-old nursing student, arrived to counseling with her mother, Karen. Jill appeared tired, extremely thin, and wore baggy clothes. Jill's speech was slow and effortful. Karen appeared to be of normal weight and healthy. Jill described her presenting problem as some physiological difficulty with her digestion but was frustrated because years of tests and specialists revealed no digestive abnormalities. Medical conditions unrelated to digestion included a loss in bone density and amenorrhea for 3 years. Informal observation and initial assessment revealed extreme food restricting, perhaps anorexia nervosa. I did not know if Jill would come to our next session, but it was clear to me that she did not believe she had anorexia.

Despite my uncertainty, Jill and her mom did arrive at the next session. Initially, counseling was directed at helping Jill accept that there was no physiological cause for her difficulties. Next, the goal was for Jill to accept the possibility that the problem may be psychological and eventually entertain the idea that it might be somewhere on a continuum of disordered eating. It is not uncommon for clients to be uncomfortable with the language of eating disorders. In my experience, eating-disordered behaviors often start out as an adaptive response to a unhealthy culture, and it is difficult for people to see this when their behaviors have gone too far, particularly when their behaviors result in praise and attention. For example, "You look great, have you lost weight?" is a question heard in hallways across schools and workplaces. Similarly, exercise is seen as such a virtuous activity that it is difficult for people to accept that exercise in the extreme can be unhealthy. Regardless of what we name the "problem" for which Jill came to counseling, the first and fundamental step was to help Jill, through collaboration with her medical doctor, step out of medical danger. There were many sessions that ended with me not being sure if Jill would want to or be physically able to return.

A more formal eating disorder assessment indicated that Jill was struggling with anorexia nervosa (although this hypothesis was not shared with the client). This information was collected because 3 years of medical inquiry indicated no medical reason for her emaciated appearance. Further, Jill appeared to have the potential to be medically compromised due to her weight and was referred by her family physician for an eating disorder assessment. However, Jill was not convinced she had an eating disorder and was not comfortable initially even utilizing the term. In the spirit of engaging with this young woman and developing a trusting therapeutic alliance, the only agreed-upon contract between Jill and I was to work toward getting her out of medical danger. Specifically, this meant that Jill needed to gain weight. No contract beyond slow weight gain and an agreement to not lose weight could be established at the beginning of therapy.

I designed a formal contract to stand as an agreement between Jill, her doctor, and myself:

CONTRACT TO FIGHT AGAINST EATING DISORDER

I, Jill, agree to the following contract:

1. I agree that if my weight drops to 85 pounds, I will be hospitalized.
2. I agree not to attempt to kill myself.
3. I agree that if I am thinking about killing myself I will contact my family doctor, _____, my counselor, @ (phone number), or the Distress Center 24-Hour Crisis Line @ (phone number).
4. I agree not to use laxatives or any other substance to induce purging behaviors.
5. I agree to take the medications as prescribed by my doctor.

I understand this contract is necessary at this time to fight the influence of the eating disorder. When my weight drops to 85 pounds, I am medically compromised and could die; therefore, I accept this contract as a way to help me in my battle against the eating disorder.

This contract is an agreement between the following:

Client   Date

Counselor—   Date

Family Physician   Date

Once the medical condition improved, I felt Jill would have the cognitive capacity to choose further treatment or not. I recognize now that this was an attempt to deal with the ambiguity around the medical stability of my client. I guess I have certain limitations with discomfort and uncertainty, like not knowing whether my client will live or die. The fact that Jill signed this contract indicated to me and to her that she was choosing to live.

Of all the various approaches to the treatment of eating disorders, the most extensively studied and thus most empirically validated form of treatment for eating disorders is cognitive behavioral therapy. Cognitive behavioral therapy for eating disorders, adapted from treatments for depression and anxiety, targets behaviors and cognition that maintain the eating disorder. Short-term treatment goals, because they are more focused around the symptoms, tend to be based on cognitive behavioral therapy. Although cognitive behavioral therapy has limitations, it did provide the structure Jill needed at the beginning of her treatment. For example, a cognitive behavioral therapy approach includes self-monitoring of food intake. Daily food logs can be helpful, especially initially, to begin to identify eating patterns, trigger foods, how restriction leads to bingeing, physiological versus psychological binge–purge episodes, and so on. A food log was an important intervention and evaluation tool at the outset of therapy with Jill. Jill kept a food log daily and brought her food log into every session. The food log I use looks something like this:

Date	Time	Location	Food Eaten	Binge	Purge	Feelings and Thoughts		
						Before	During	After

Food logs provide information about the amount, type, and frequency of eating behaviors. They also provide a means to evaluate progress in terms of quantity of food eaten in a day. Food logs also allow insight into the thoughts and feelings that Jill had about eating. Finally, my intention was to use these logs as a way of separating the two "voices" or "influences" in her mind. The feelings and thoughts column provided commentary that Jill and I could use to determine if messages were eating-disordered messages or healthy messages.

Conceptualizing the eating disorder as the problem and not locating the problem within the person seems to offer freedom to clients as it provides a choice about rejecting or accepting the influence of the problem in their lives. It also allows a recognition of times when the person has resisted the influence of the eating disorder and opens up real possibility of that continuing in the future. Externalizing involves speaking about and conceptualizing the problem as separate, as existing outside any person or family member. The externalizing helps to stop the blaming and allows enough distance for the family members to fight against the problem rather than each other. It frees up space for the client to see the choice of submitting to or fighting against the problem's influence.

Instead of fighting and blaming each other, I hoped externalizing the eating disorder would help Karen and Jill team up together to fight the influence of the eating disorder. I also hoped that externalization would allow Jill to gain weight without feeling she was defeating herself; rather, she was defeating the eating disorder. Although Jill did not want to be aware of her weight gain, every 2 weeks she was weighed (backward on the scale) at her doctor's office. Her weight was tracked carefully and used as an evaluation of the success of treatment. Weight gain was my benchmark of success while Jill's level of energy on a 10-point scale was her benchmark. Jill gained about 12 pounds in the first 2 months of treatment and was no longer as medically compromised. She also consistently reported feeling increased energy and did seem livelier during our sessions.

Karen and Jill reported that Jill also began to take responsibility for her own food intake rather than relying on Karen to shop, cook, and feed her. Both described this as a step in the direction of freeing their family from the grip of the eating disorder. Further, Jill continued therapy even after her mom was no longer driving her and encouraging her to come. Jill made her own way to therapy sessions on the bus.

Jill was uncertain at this point about whether or not she would like to choose "to stay where she is and just survive or agree to more change and potentially flourish." Our work together moved from a focus on food intake and medical issues to deeper, more meaningful, philosophical exploration. For example, one intervention that Jill reported impacted her decision was the "empty chair" technique, where the eating disorder is invited into the room and given a chair to sit in. Jill talked to the eating disorder chair and described what it was like to have its influence in her life. It is important to note that the discussion here was not all negative; Jill perceived certain advantages to having the eating disorder in her life. She then switched chairs and became the eating disorder and talked about what it had been like to gain control over Jill. Finally, Jill switched back to the original chair and talked to the eating disorder chair about the ways she has influence and power over the eating disorder. This exercise was a big risk for Jill, and she became very emotional during the intervention. She was able to see the eating disorder as something she might be able to give up, but at the same time, this hope was also very scary.

To work with Jill around her hope for next steps, a letter was co-constructed between Jill and I for distribution to people Jill chose to involve.

Dear Friend of Jill:

Hello, I am a psychologist who is working alongside Jill in her battle to be free of anorexia. As you may already know, anorexia has been pushing Jill and her family around for close to 3 years now. What you may not know is that anorexia tells Jill "she has no reason to live" and "no one likes her because she is too fat."

Did you also know that anorexia pushes Jill into "despair" and "rituals" and sometimes convinces her that she is "better off dead"? Anorexia has waged its best efforts to kill Jill, but somehow she remains alive and wants to "find her way to truth and freedom."

Jill believes that the time has come for her to reach out and ask her family and friends to write her a letter of support that may counteract anorexia's claims against her. Is it possible that you write Jill a note that outlines a story of (1) what you remember most about her, (2) who she is as a person outside of anorexia's grip on her, (3) what you think is in store for her future once she goes anorexia-free, and (4) what you think your relationships will be like with her once she gives anorexia the slip.

We want to thank you in advance and let you know Jill intends to return all correspondence with her return letter of acknowledgment. Thanks!

Yours anti-anorexically,

Jill and I

This letter clearly speaks in language that separates Jill and anorexia. Further, it uses a metaphor of a battle or war to indicate that Jill does have power over the eating disorder. Additionally, it expands Jill's support group outside of her family, which proved to be helpful for both Jill and her family. Jill cherished the letters of support Jill received from family and friends. She would arrive at sessions with a letter to share and light up as she read them aloud. Through expanding her social support network, Jill was able to embrace the hope that life without the eating disorder may have some appeal.

It is important to note that my work with Jill lasted for over a year, and sessions especially at the beginning were often twice weekly. Much of the work toward the end of our therapeutic relationship was successful because we were able to move beyond the medical issues at a pace that Jill chose. There is a tricky balance between nonnegotiables (contract about medical stability) and negotiables (the pacing of the counseling work), focus on behaviors (food log) and meaning (empty chair), hope and fear (choosing life). In the end, Jill taught me that my success as a counselor is in large part a function of my ability to be comfortable with ambiguity. There are so many things to be uncertain about when working with clients with eating disorders. Often in the initial stages of treatment, there is doubt about whether the client will live or die. There is also the ever-present ambiguity, expressed directly or indirectly, about if the client really wants to be in treatment; part of the client wants to get better and the other part holds fast to the eating disorder. At times during the course of treatment, the issue seems to be all about food and at other times, it really is not about food at all. Although all of these uncertainties may appear contradictory, it is the counselor's job to see the "both/and" and not the "either/or" of these ambiguities.

If I were to write a letter to Jill now, it would read:

Dear Jill,

You may not know that you were indeed a very good teacher to a beginning counselor. Your journey paralleled my journey in development as a counselor in many ways. You taught me to endure the discomfort of "not knowing" and allowed me to live out the complexities of giving up "a problem" that has been a companion and maybe even a friend while growing up. I realize now that, because of the practice you gave me, I am better able to embrace the sense of ambiguity that all of my clients present to me. I thank you for not knowing, for thinking carefully about your decision to choose life over the eating disorder, and for taking that leap of faith with my younger, less experienced counseling self. It was a privilege to learn with you.

Thank you for teaching me that if I can understand and accept the ambiguity that my clients present to me, the clarity will follow. Because of you, in many areas of my life, I am able to see the "both/and," not the "either/or."

In hope and wellness and with fondness.

Epilogue

From our perspective, some themes within and across many of the stories in this section include the following.

- Caregivers sometimes have an overwhelming fear that they are powerless to help and that losing the "help battle" means losing a life.
- Individuals with eating disorders often have a very negative body image that, at times, seems impossible for others to influence. In this regard, they perceive themselves as overweight, fat, or obese, whereas others report to them that they are too slim or skinny.
- Individuals with an eating disorder often have an extraordinary need for control of some aspect of their life and have arrived at the conclusion they can exercise almost complete control over their eating.
- Parents sometimes express that they are incredibly frustrated by their inability to identify or understand what triggered the eating disorder in their child.

Specific thoughts parents expressed in the stories and that we have heard expressed on more than one occasion over the years include the following.

- It is a very sad, scary, and helpless feeling to watch your child starve herself and continually have her say how dissatisfied she is about how she looks.
- Seeing your daughter in a hospital bed being force-fed is traumatic, and to think that she might die because of her despair about her sense of self results in overwhelming feelings of parental inadequacy and guilt.

Specific thoughts some youth expressed in the stories and that we have heard expressed on more than one occasion over the years include the following.

- I just keep on saying to myself how much I hate my body.
- Much of my life has been about being in and out of control of what I think and do.

Specific thoughts professionals expressed in the stories and that we have heard expressed on more than one occasion over the years include the following.

- The key to treating eating disorders is much about weight restoration and stability but also much about adjusting the thoughts and behaviors of the individual related to the eating disorder and facilitating a long-term healthy lifestyle.

10 Somatoform Disorders

Prologue

The primary feature of somatoform disorders (e.g., conversion disorder, hypochondriasis, and pain disorder) is persistent physical complaints or symptoms (e.g., nausea and dizziness, stomachaches, and headaches), for which there is no organic cause. Although the etiology of these disorders is unclear, psychological factors are considered to be associated with the complaints and symptoms.

One of the more common somatoform disorders in children and youth is conversion disorder, which is defined by the presence of unexplained symptoms that affect motor and/or sensory functions (e.g., inability to perform particular movements, numbness and tingling feeling, body weakness, difficulty with swallowing, feelings of a lump or tightness in the throat, hearing and/or visual problems, and pain sensation) and that is indicative of some medical (physiological/organic) condition but not proven to have a physiological or neurological cause by way of medical evaluation. For the diagnosis of conversion disorder to be made, the individual's problem is considered clinically significant due to personal distress (e.g., psychosocial stressors such as peer rejection, family conflict, perceived inadequacies) and impairment relative to academic, social, or vocational areas of functioning. Conversion symptoms are sometimes thought to be the result of anxiety associated with some unconscious conflict, a form of physical communication of an emotional idea or feeling when one is unable to verbally express the problem or conflict due to personal or social issues.

In this section are three stories: one personal story, one parental story, and one professional story pertaining to conversion disorder.

The personal story, entitled "Believing in Myself and Going from Bad to Good," is written by a young girl who shares her story about her ongoing and awakening experience with respect to her conversion disorder and how she got control of her situation. The "parental" story, entitled "Ups and Downs and Turnarounds," is written by a young girl's aunt who shares her story about her experience with and support of her niece while her niece went through conversion problems in her childhood that seemed to be not well understood. The professional story, entitled "A Matter of Control," is written by a clinical psychologist who shares his story about a young girl with conversion disorder, with respect to the problems she was experiencing and his approach in treating her.

Exceptional Life Journeys. DOI: 10.1016/B978-0-12-385216-8.00010-2

Personal Story

Believing in Myself and Going from Bad to Good

The difference between each of us is extraordinary. It's what we do with that difference that makes us who we are. Each person tends to recognize what's not right with a person when they come face to face, and never to focus in on the good. Having a disease or disorder makes a person no different than those without, but we always seem to forget that and treat others unfairly. Although children go through life with fewer situations than adults, some children have situations that are unforgettable.

As a young child, I always thought life was easy. There was nothing for me to worry about, because my parents took care of everything. My mom died when I was only 9 years old, after fighting a 6-year battle with breast cancer. After this horrifying event, my opinion on life started to change. I noticed more negative than positive. I wondered how it could be possible that my mom was taken away from me, but I had to learn to live with it.

After my mom died, I knew nothing would ever be the same, but life carries on and I had to as well. It was only a year later that I started having gruesome migraine headaches once or twice every few weeks. Theses headaches were so bad I could hardly move. I would spend the day sitting in the same spot, crying out in pain. No amount of medicine made me feel better; the medicine actually made me more ill. After 6 months, I decided enough was enough. I was referred to the neuroscience clinic at the Children's Hospital.

After the referral, I knew the situation was bad. After being sent by a neurologist to undergo a CT scan followed by an MRI, I began to get really scared. Only hours after the two scans, I was brought into a room were my whole family was waiting with a doctor. The doctor told me I had a very large brain tumor overpowering my brain, which was the source of my headaches. I was to have surgery just days later.

No 10-year-old ever wants to be told they have a tumor growing inside their brain; most children don't even know what a tumor is, or think that such a bad thing could exist. That was my position almost 4 years ago. All I knew is that the situation was at the extreme point. As I prepared for surgery, I became more scared. I hardly knew what was going on. All I knew was my head was going to be operated on for 6 h.

After a successful surgery, with a week stay at the hospital, life continued on. I had a few effects from the surgery: no sensation on the left side of my body and a loss of balance due to my tumor's overpowering depth and mass. After a year of occupational and physical therapy, I regained most of my balance, but I had to learn to live with having no sensation on the left side of my body. I continued at school; it was like nothing had changed. School never became an issue after the surgery; no one treated me differently or unfairly, and I still had all my friends. Clearly, I thought nothing else could go wrong.

Unfortunately, something did go wrong, just as I was completing seventh grade, only 2 years after surgery. In May 2009, I arrived at school, when I suddenly felt this feeling of a "lump" in my throat that made me feel very nauseous and dizzy. The lump felt something like a grape that was stuck at the back of my throat. This lump of mine was with me every moment of every day. It became such a part of me that it scared me. It made me so scared of going to school because I was afraid of getting sick in front of my classmates. Certain things such as water and mint-flavored gum seemed to make the lump smaller, or maybe that's what I wished for.

No matter how many assignments and tests I was missing, or the thought of how badly my marks would go down, I was still scared of going to school with this lump. Everyone close to me had ideas that I was afraid of going to school because of a person bullying me. Or I was trying to skip school because I didn't like it. But both of those opinions weren't true. The days I did go to school, I got scared when I got there and phoned someone to come take me home. This went on for the rest of May and all of June. I knew that, even if no one believed what I said, the lump was there and I had to do something about it.

One of my biggest problems with the lump was taking the bus to and from school. I was asked many times why I was avoiding the bus. Many people thought again that someone was bothering me on the bus. However, whether or not people were bothering me was the last thing on my mind. Something about the bus made the lump stronger. Months before the lump, I had gotten sick in front of my peers on the bus, which I think contributed to my fear of the lump and bus at the same time.

Over the months, I began to forget about all the people trying to tell me to stop "making this story up." I found ways to help myself. I decided that I would ignore everyone who told me I was making this lump story up. Air conditioning also seemed to make the lump shrink, but not everywhere has air conditioning. I worked it out with an aunt of mine that she would pick me up from school so I would only have to take the bus one way.

After going back to see my neurosurgeon, he sent me for an EEG, to search for the possibility of epilepsy. I was actually hoping that the test came back saying I did have epilepsy. At least that way I would know what was going on. The EEG came back fine, and the search continued. After many blood tests, and searches for diabetes and thyroid problems, again everyone began to question me, accusing me of lying about the lump actually being there, but I knew it was, and it had to be taken care of.

My aunt, one of the few people who believed me about the lump, came up with the idea of sending me to see a psychologist. I agreed. The idea of seeing a psychologist made me feel better. I felt that, even if this lump was caused by my previous experiences, a psychologist could give me hope that I could overcome this issue. My aunt sent a letter to my neurologist who had agreed, and booked an appointment for me to meet with a psychologist.

My first appointment with the psychologist made me very nervous. I came up with ideas that he wouldn't believe my story or have any way to help me. I began to feel hopeless. When I explained my situation, I immediately knew that I was wrong before. The man clearly had many ideas and ways to help me. I decided it would be good for me to give this solution to my problems a positive start.

My first few psychology sessions were very basic. I discussed more about my life and what I thought of myself with the psychologist. He had many ideas from the start. It didn't take me long to realize that I made the lump worse by worrying about it. As the weeks went on, I noticed more and more how much my thoughts impacted me, how much I worry, and how powerful my mind was. As the sessions continued, I learned more about myself. I learned a lot that I never thought about before. These sessions began to change my life. They changed how I viewed life and situations I was put into in the outside world. After a few months of the sessions, I learned so much about myself that I was able to work with the lump. I began to become in control over the lump and was able to make it go away for periods of time.

After a few months of working to gain more control over the lump, I got used to it. The lump was no longer my problem. Instead, my psychology sessions were based on how I was feeling. Trying to recognize my thoughts, deal with uncertainty, and balance negative and positive in my life. As time went on, I began to feel proud of myself. I was overcoming something that I thought was going to ruin my life.

Psychology was the one time during the week that I knew I could say anything without being looked at as different. I felt safe, welcomed, and valued. My psychologist and I had a trusting specialist–patient relationship. The more the time passed, the more I could trust him and the more I knew he could help me. My psychologist began to feel like a best friend to me. He taught me to live in the moment and be grateful for all the good things in life.

An important thing I learned in this experience was to always take what you believe and develop it yourself. I learned to not let other people tell me I was different or wrong. Inside I knew that my situation was not common, but it gave me an inspiration. The inspiration was to believe in myself. I stopped letting other people direct my life and made life my responsibility.

Nothing is ever perfect. It's how we develop situations that make them what they are. The lump showed me that there is more to life than we ever know. It made me believe that I am who I am and no one can change that. It taught me to keep the important things to me close and never forget them. I'm not afraid to tell my friends or family I love them, because the lump taught me that having these people around me shapes who I am. It caused me to believe that I have the power to direct my life, and that I can do anything with it. I just have to believe in myself.

The brain tumor gave me an inspiration to help people. I plan to go to university and major in neuroscience. I owe it to myself to show people and me that I had some rough times but I made it through okay. I currently consider myself to be doing very good. Before, I thought the lump would destroy my whole life, and now it's become my helper. Any bad situation can turn into something good. You have to take it and change it to the way you want it.

Some advice I have for others is to not let anything get in the way of your life. There are so many experiences that you will go through in your life, good and bad, but they will all make you a stronger person and teach you to believe in yourself and to love yourself. Life will teach you how valued you are and how you should value time. Live your life to the fullest and don't let anything stop you from achieving your goals.

Four years ago at the start of my brain tumor experience, I thought life was bad. Most people would think I am insane to say this, but I am grateful for the tumor. If not for the tumor, I wouldn't have met so many amazing people that helped shape my life. I wouldn't be as strong as I am today. My life was saved by a neurologist, a neurosurgeon, and a psychologist. We can all save each other. If we stop violence, we will become stronger people. Save yourself from negative, turn it to positive and live life the way you want it.

Today, I am currently in ninth grade, preparing to go off to high school. The lump is still a good friend of mine. When it shows up, I know that I am worrying too much about something and need to stop. I still live with no sensation on the left side of my body, but I think it makes me very unique. I try my hardest every day to succeed and to help others succeed. I know that my life has had some bad moments, but I work on trying to make them good. I still see my psychologist; we focus on helping me to live in the moment, but my sessions have become more like checkups. I have made much progress in psychology. My life might not be the happiest one, but I love it just the way it is.

Parental Story

Ups and Downs and Turnarounds

When you first meet Katherine, you see a quiet and respectful person. She is independent and responsible for many aspects of her life and has been for many years. She loves children and has a strong sense of herself. She reminds me of the analogy of the duck floating on the water. Quiet, calm, majestic is what we see on the top of the water, but under the water, where we do not see, she is paddling like crazy to stay afloat.

The paddling part of Katherine is what takes time to understand. After all that has happened to her, I think she struggles between the responsible person she has been forced to become because of her life events and being a normal teenager, with all the challenges that go with that. Some skills that she has developed are not complete, because she has been making it up as she goes along and maybe is not at the maturity level needed to complete these skills.

Katherine is my niece. I married late into the family, when Katherine was 2 years old. Our initial connection was mostly at family functions and through her mother. I am a nurse at the Children's Hospital. My three children, 7-year-old twins and a 5-year-old are the youngest of the 17 grandchildren on Katherine's mother's side of the family.

Katherine holds so many characteristics of her mother, so to understand Katherine, you need to start with understanding her bond with her mom. Katherine's mother, Virginia (Gini), was a very courageous, energetic person, whom you couldn't help but love. She welcomed me into the family and, as my children arrived, she became my friend and confidante.

As I watched Katherine grow, her attachment to her mom was so strong. Her mother courageously fought through her cancer (for 6 years), and their bond grew. As Gini became sicker, Katherine seemed to take on even more responsibility, helping her mom with tasks, brushing her hair when she had it, and sleeping in the same room with her. It is hard to know if, as a child, she knew how truly sick her mom was, because her mom would not give up. She felt that every hour she lived, it was one more hour with her children. Despite how she felt, she kept up a great front for everyone, and only a few of us knew how much she was really suffering.

For Katherine's ninth birthday, her mom wanted it to be so special for her. She struggled to breathe as she made the preparations. As I arrived, she quietly asked me to drain her lungs, through the two existing tubes, so she could breathe better for the party. No big drama; we just quietly went upstairs and then the party began.

Gini coped with laughter and making things fun for everyone, and she was the queen of special touches and knowing you needed something before you did. She was a strong advocate for right and wrong and let her opinion be known if she felt things were not right.

In her short 14 years of life, Katherine has coped and learned from her mother's long illness and the increased responsibility that came with her death, the acceptance of her brother's special needs, "his special brain," and the knowledge that the load would never be shared equally between them. She has been trying to adapt to life without her mom and accepting her father who is trying to be both parents, yet struggles along the way. She would be part-time parented by grandparents who were still living in an old culture from Italy and, although they were doing the best they could, it conflicted with what Katherine believed. Her other grandparents, who had been a big part of her life, were failing in health, her grandmother now lost in Alzheimer's, so they could no longer help.

Katherine then developed headaches and vomiting. This went on for months. We thought maybe the stress, maybe her hormonal changes, and maybe being the stoic one that she is caused this, but we underestimated what she was going through. Finally, she was brought to her doctor who started the path to her brain tumor diagnosis. She was so very strong as she went through her diagnosis, her surgery, and her ICU stay, and finally she went home. Then she spent many hours in rehab to learn to cope with her residual weaknesses. The left side of her body is weak and she still cannot feel parts.

Just as her world was starting to become "normal," her uncle was diagnosed with lung cancer and, 72 days later, he too had passed away. A year later, her maternal grandfather was diagnosed with two types of cancer and his chemotherapy started.

You would think that, after all this, she might have become a very bitter young lady, but no, in despite of this and with tremendous courage, she carried on with life. She laughed with her cousins, went to school, and kept up her marks. Finally, we thought, she could start to regain some definition of normalcy in her life!

In May 2009, she became unwell. She started having dizzy spells, lumps in her throat, and feeling she may throw up on the bus ride. As a nurse, warning bells started to go through my head. I hoped it was just a flu. But this flu did not go away, and my thoughts went to the very worst. As time went by, physician opinions were explored; she went on multiple courses of antibiotics, new medications, more MRIs, and more tests. One by one, they came back negative, to our relief. But what could it be that made her feel so awful?

I was starting to get increased calls from school, seeing if I could pick her up, as she felt unwell. She was missing so much school. We discussed the stress of what she had been through and if the obvious events common to teenagers were an underlining cause. We spoke about bullying, about self-image, and about her friends, and I think she was being truthful about those things not being the cause. She was adamant though that the bus ride could illicit nausea, and her fear of throwing up on the bus was so strong; it made her unwell just to think about the embarrassment of throwing up.

There was a short-term easy solution: If I promised to pick her up at the end of the day, she would ride the bus there, and knowing she did not have to face the bus ride home gave her enough comfort to last the day. And so it went. Cousins and uncles helped on the days I could not.

It was a difficult time of wanting some answers for what she was feeling, balanced with hoping there was nothing wrong. We spoke about how anxiety and stress could make a small symptom grow and how we had the power, not always to control the symptom, but to control its intensity.

Many of the family members felt she was not telling us everything and perhaps this was her way of crying out. I could feel her frustration as she answered one more time she was not making it up and she was not being bullied. We thought that because Katherine, like her mother, carried on through her discomfort and appeared to be able to function, maybe she was not as sick as she was saying, although I suspect it was taking all she had to keep it together.

I was told I was feeding into her dependency and making it worse. I should let her try to ride the bus. I struggled with the guilt of not picking up the signs of her initial tumor earlier, and of not being able to give her a reason for her spells. I enjoyed our 15 min of speaking together after school on the ride home. My kids loved seeing her every day as we picked them up, before we dropped her off. Was I also getting secondary gains from this? And as I struggled with this, I spoke to a friend who had lost his mother as a child. He said, "What I would have given to have a mother figure/friend for even 15 min a day Don't stop. Work this through with her." That gave me so much strength and helped me realize that I was doing the right thing.

As the tests finally came back as all normal and I felt that she had been assessed by the best, I began to explore a different area. Maybe all her life struggles, so well contained in her façade, were coming out in physical symptoms. I acknowledged that I knew her symptoms were real to her. We just didn't know where they were coming from.

I work with some incredible psychologists at the Children's Hospital and started to speak to them about Katherine. They felt that, even if it still could be something physical, she might benefit from someone to help her sort out what she was feeling. Perhaps they could help her develop some new coping skills and help her feel that she is in control and be confident to cope with all she was facing.

I asked Katherine if she would be willing to speak to someone who had spoken to many children about the challenges she was facing. She surprised me and said she knew there was nothing else wrong with her brain, but she might have some work to do with her mind! Yes, she was willing to speak to someone.

How proud I was that she was so self-aware and would consider this. I know how difficult it was for her to agree to share her thoughts with a stranger, and I prayed she would make a connection with her psychologist. I wrote a letter to her neurologist with her and her father's approval, explaining why I felt Katherine would benefit from a psychologist and, with the help of the clinic nurse, it became reality.

Katherine connected with her psychologist and expressed she could finally speak without being judged, and she looks forward to her appointments. Her ability to discuss things at home, for now, is limited with her father and brother, and this gives her such a great outlet to express her concerns and test out guided solutions without fear of judgment. I have seen her grow so much.

I think we have also developed over time. I feel confident that I can ask her questions, and she knows it is out of concern, not judgment. She fears being lied to, and I think she knows she will get a honest response from me. I may not tell her all the details but the ones that are important for her to know. We work through some issues together; she has some great cousins and aunts to whom she also goes. We all have our roles, and she has sorted out to whom she approaches with different issues.

As we continue along this journey, I am sure there will be more hills and valleys. She has grown a lot, but she is still a teenager and life has many more doors for her to open. Through this process, I think she has developed some new coping skills. She does not wallow in self-pity, but can now recognize it, or will let herself be confronted when she allows it to overwhelm her. I know she has a network of people who love her and she knows to whom she can share her concerns and her joys. I too am a richer person for sharing in her journey, and I hope that I will always be part of her life. How lucky am I.

Professional Story

A Matter of Control

This narrative explores the work of a clinical psychologist in relation to an adolescent conversion disorder. By case study example, an overview of symptoms, background, and development of psychosocial stressors leading to the expression of psychological distress through bodily symptoms are described. Various successes and challenges are encountered during this young girl's psychotherapy to resolve her alarming and impairing symptoms. Significant issues relating to assessment and treatment of children and youth with conversion disorder are described. Clinician and patient experiences of somatoform symptoms are also described prior, during, and nearing the conclusion of treatment.

Imagine having physical symptoms that are disturbing to you: unexplained nausea and dizziness that comes in waves, the increasing sensation of a tightening in the throat that does not go away. Next imagine making repeated visits to the emergency department for these symptoms because, aside from their disturbing nature, you had a brain tumor removed only 2 years ago. Your mother and recently your uncle have died from cancer, which does nothing to calm your fears about what may be happening. Now imagine that the medications given by the doctors do not seem to be helping and that no one seems to be able to provide any good explanation for what you are experiencing. Then you are asked to see a psychologist.

This was the exact situation Katherine found herself in, over a year ago today. It is also important to tell you that, despite all she had been through, Katherine was only 13 years old at the time we first met. Her story begins long before she met me. When referred by her child neurologist, Katherine's symptomatic "picture" was not clear. In addition to the unexplained symptoms already described, she had residual left-sided weaknesses in her arm and hand as a result of the partial removal of her right temporal lobe. The dizziness and throat sensation had persisted for about 4 months, and she was taking an antiseizure medication, as there was a very small, but plausible explanation for her symptoms: seizure activity arising from the area of the previous surgery.

After some time however, her neurologist began to suspect a different explanation for her symptoms, that these were more likely related to psychological phenomena. There was no change in her symptoms with an antiseizure medication. All investigations, including computerized images of the brain and EEGs, had returned normal. During an ambulatory EEG taken over several days, Katherine reported several of her episodes involving the dizziness, nausea, and the throat sensation; however, no associated seizure activity was recorded. Katherine's problem was a complex one indeed.

As with most referrals, my initial time with Katherine was spent in assessment. She struck me as an engaging and pleasant young girl who did well in school and enjoyed spending time with her family and friends. Unlike most children her age, however, she had been through a lot, beginning with the loss of her mother at an early age. She later began having severe headaches, which led to her own treatment of a tumor. Though surgery was successful, the residual deficits left her with motor weaknesses. More recently, she had lost an uncle to cancer and a grandfather was now being diagnosed as well. Although she described herself as generally happy, with careful questioning, it became clear she intensely wished to reduce her fear over the current symptoms but also to overcome feelings of fear and anxiety in many other areas of her life. These included discomfort around peers at school and performing well academically.

The results of psychological tests Katherine and her father completed were telling: an extremely high number of physical symptoms for which there had been no explanation including complaints of headache, stomach and chest pain, shortness of breath, and fears of becoming sick. Unusually high levels of anxiety, worry, and hopelessness with emerging symptoms of depression were also indicated. Finally, there was strong evidence that Katherine had a tendency to devalue her own self-worth and experienced high levels of insecurity around peers. She also responded very similar to youth dealing with stressors by way of inhibiting their responses and becoming submissive in the face of others. It was clear that Katherine had significant psychological distress impacting her functioning on a daily basis. *What was perhaps less clear was whether these were contributing to her physical symptoms, or being caused by them.*

I am a medical psychologist in a neuroscience department. It may be surprising to many that, in addition to seeing children with neurological disorders, I am also commonly asked to see children where there is no medical or structural reason for their symptoms, often termed a "functional neurological condition." A large portion of these children will ultimately receive a psychosomatic explanation for their symptoms, sometimes with an accompanying diagnosis identified in the current version of the DSM, under somatoform conditions. Conversion reaction is one such disorder thought to involve expression of psychic distress through the central nervous system by way of motor or sensory pathways, effectively causing impairing sensations or disabling of normal movements such as an arm or leg.

Although historically a controversial diagnosis of exclusion, conversion disorder has become much more well understood in the recent decade, largely as a result of imaging studies that clearly show how the brain functions differently in conversion disorder, as if the responsible areas had been "shut off" while other areas of the brain become overactivated in an effort to compensate for the localized inhibitions. There is no gold standard for treatment of a conversion reaction, though many neurology clinics have developed their own practices, which usually start with education and reassurances to the patient and family. Regardless of the approach, treatment goals are clear: a return of function.

Katherine's psychological profile was very similar to most children and youth I had seen previously with a conversion reaction: evidence of one or more significant psychosocial stressors with an inhibited and submissive or conforming coping response, with resulting anxiety not immediately apparent to the child. These children are seen by neurologists I know well and trust, and they receive thorough investigations to rule out the possibility of less obvious causes for the symptoms.

There may be telling signs of a conversion reaction, but not always. When these occur, they usually fall into one of two categories: a lack of consistency in symptoms with cognitive distraction and/or lack of conformity to known anatomic pathways, sometimes to the point of physiologic impossibility. When these signs accompany the impairment, they make diagnosis much easier to determine for both clinician and parent. Real certainty about the diagnosis, however, is often best achieved with time and can be elusive in the early stages of treatment. Katherine's case was complicated by her known neurologic history and lack of a clear diagnostic sign for a conversion reaction. What was known was her significant psychological distress and predisposition to psychosomatic symptoms. So, we proceeded.

Katherine was determined to be a good candidate for individual therapy, which commenced immediately with weekly sessions. We began to track her symptoms and cognitions using cognitive behavior therapy methods such as daily recording of events. By having her rate the strength of her symptoms at various intervals, we were able to determine subtle patterns and changes over time. Over the course of sessions, Katherine began to notice her sensory symptoms reduced with absorption in a movie or other cognitive task. She observed that they worsened when in anxiety-provoking situations. With time, she observed that when experiencing high levels of stress, a lump in her throat developed, she became dizzy, nauseous, and even at times developed a headache. This awareness was not enough in itself to resolve the symptoms, but it did form a good basis from which to institute behavior changes.

Katherine's hypnotic ability was assessed early on with good response, indicating her symptoms may respond to suggestions while under trance. Given the long-standing acceptance of conversion reaction as related to psychological dissociation, and in turn dissociation with hypnotic phenomena, this seemed a logical place to start. Katherine underwent multiple sessions during which suggestions were made for increasing her ability to cope with stressors and regain control over her body. Hypnosis was used to alter her symptoms, initially by showing her the power of the mind to stop sensing body parts, and then "reconnect" them in an integrated fashion. With time, hypnotic suggestions were even used to increase and decrease the disturbing sensations in her throat. She was taught self-hypnosis and provided a recording of an induction for practice in between sessions.

It should be mentioned that, although I was quite excited by Katherine's hypnotic ability to alter physical sensations, I believe she was only marginally impressed by these experiences, which she sometimes described with trepidation. For example, despite my attempts to make her comfortable, she described the gradual dissociation of body parts while under trance as rather unpleasant. I suspect this related to heightened fear and sensitivity in response to new bodily sensations, which caused these to be interpreted as signals of a new threat to her well-being. Nonetheless we persisted, and I believe hypnosis was a powerful and useful tool that aided in Katherine's recovery.

With persistent cognitive behavioral therapy work, Katherine was able to identify cognitions responsible for increasing her anxiety. This led to further identification of deeper core beliefs she had held for many years as a result of her past experiences. These included such beliefs as "I'm unlucky," "the world is not fair," and "trying to be positive make things worse." Given Katherine's experiences in life, these discoveries were far from surprising; however, they were just that, discoveries, and identification paved the way for an understanding of how her past was causing her to experience fear in the present. Also important to discover was that these beliefs guided her actions and emotions in the present, especially at times of significant psychosocial stress.

Continuing, we acknowledged Katherine's fear of death, in an attempt to perhaps tame the impact this had on her daily living. Doing so had surprisingly beneficial consequences. Katherine was provided homework assignments to complete in order to confront the irrational core beliefs she held about herself and others. I recall one such assignment: to remain negative about her life situation for an entire week, which she believed would be entirely possible, as this was likely to occur anyway. Surprisingly, she found this a very difficult task, as there were at least several instances throughout each day during which she was very positive about living! Gradually, in this manner, she was able to alter some of her irrational beliefs, making them more consistent with what actually occurred in her day-to-day experience.

After many sessions of therapy, Katherine's physical symptoms gradually began to resolve. She continued to complete behavioral experiments, and we were able to develop organized flowcharts that described the cascade of trigger cognitions, resulting fears/emotions, and bodily sensations that occurred in various day-to-day situations. The picture of what was going on in her body and mind became clearer, and with that, became easier to control.

After Katherine's physical symptoms resolved completely, we began to consider therapy termination. This would be gradual over time, in a transition typical of winding down a longer-term therapy case. What I failed to predict, however, was that once her physical symptoms had resolved, Katherine's worries and fears continued to impact her life. Only this time, the effects on her behavior were much more obvious and no longer concealed within bodily manifestations. Initially, these presented as new anxious behaviors and fears such as difficulty resisting the urge to repeatedly recheck door locks at night prior to sleep. At times, it actually seemed that things were getting worse for Katherine. With time, however, it became obvious to us both that the anxiety was simply more obvious and open, and once this occurred, we were able to deal much more directly with ways of increasing her tolerance for the unknown in daily life. We continued to plan and complete behavioral experiments, including graduated exposures to feared situations.

I don't wish to suggest that any of Katherine's successes were necessarily easy to accomplish. Gains, when made, were often subtle and successive over several weeks and months. Her symptoms did not immediately resolve and, in fact, she developed new concerns including abdominal pain, prompting new medical investigation. Although these fortunately turned out negative, Katherine continued to experience frustration and renewed feelings of helplessness in the face of her body's symptoms. These events, when they occurred, provided a constant stream of opportunity to apply new strategies and insights she had learned, as well as develop new ones.

At one point, when many of her anxious behaviors began to resolve, I recall Katherine disclosing repeating thoughts of "not being around anymore." This was greatly concerning such that we devoted an entire session to determining her safety and the extent that she may act upon such thoughts. Fortunately, this was determined to be low. What followed, however, was what I now consider to have been a "breakthrough" in her therapy. Through the attention directed at her thoughts and feelings, Katherine was able to reveal a deep core belief: a fear of being left alone, identifiable at least since her mother had died. She went on to view her symptoms, both past and present, as attempts at control of this fear. I recall her describing at a later point the way in which her fears had been expressed through the body: "You know, underneath, I did not really want to lose the lump in my throat." Katherine went on to understand that complete control over such events as life and death was not possible, but ultimately could be accepted at some level of being human.

In the weeks that followed, Katherine realized she was still needing to grieve the loss of her mother. We continued to work on this as much as she was comfortable. We gradually began to increase the time between sessions from weekly to biweekly, and she found continued benefit in these sessions as check-ins, to help her determine solutions to new problems with my support. Recently, Katherine summed up her progress in therapy by stating the following: "Something is different, better. I feel more positive, mostly happiness. I have so many connections to others in my life. The days go by faster. Things are starting to make more sense."

I continue to see Katherine in therapy. At some point shortly, I expect it will be time for us to part. Her (and my) experience over the course of her treatment has certainly been a winding road, but with many pleasant surprises along the way. Throughout Katherine's therapy and even today, there remains a doubt for her when new sensations arise. This is part of being human. Her fears remain of a new disease or return of a brain tumor at some point. This fear, however, is more balanced, the biggest change being a prevailing fear at the start of therapy that is now only very small and miniscule. I know she will continue to question the sensations her body produces. Through such experiences, however, she is certain to develop a new trust in herself, as well as her mind, body, and being.

Epilogue

From our perspective, some themes within and across many of the stories in this section include the following.

- Conversion disorder within children and youth is a very difficult disorder to understand and diagnose.
- Conversion disorder within children and youth is a very challenging and complex disorder to treat.
- Conversion disorder in children and youth is very difficult to comprehend by parents.

Specific thoughts parents and professionals expressed in the stories that we thought were insightful and that we have heard expressed in similar ways over the years include the following.

- Interventions used to help children deal with their conversion disorder include education, cognitive behavioral therapy, physical therapy, insight training, and coping methods training.
- Interventions need to not only focus on the child but also on the parents and significant others in the child's life to promote a better understanding of the child's situation and provide ways that can best support the child within the home and school.

Conclusion

From our perspective, all of the contributors in this book have written stories that are enlightening, insightful, engaging, informative, and inspirational as well as being heartfelt, riveting, and truthful accounts of their experiences with a childhood disorder. All of the story contributors have indicated that they have experienced child suffering, particularly with respect to abnormal thinking, feeling, and behavior. However, most importantly, they have also experienced a newfound understanding of themselves as individuals with a childhood disorder, as parents of children with a childhood disorder, or as professionals working with children and youth with disorders.

The stories in this book are from individuals who are presently dealing with their childhood disorder or are from individuals who have reflected on their childhood disorder and the impact it had on their lives. The stories also come from parents who are still dealing with their children who have a childhood disorder or from parents who have reflected on their past experiences with their children who had a disorder. Lastly, the stories have come from professionals who reflected on a particular child or youth with a disorder whom they are working with currently or have worked with in their past. In every one of these stories, we are reminded of the impact of developmental processes (cognitive, behavioral, social, and emotional), the importance of context (home and school), and the influence of multiple dynamic and intersecting events (e.g., genetic influences, parenting influences, or social–environmental influences) that shape the development of childhood disorders.

The stories in this book represent individual experiences and do not reflect all children's experiences with disorders, all parents' experiences with childhood disorder, or all professionals' experiences with childhood disorder. Nevertheless, all of the stories underscore some general understanding of childhood disorder. In this regard, the stories present common thoughts, feelings, and behaviors of children with various disorders; common challenges and struggles of many children with disorders and of the parents of these children as they go through various stages of development; common highs and lows experienced and felt by many children and their parents; common approaches and strategies utilized by professionals (such as psychologists, psychiatrists, teachers, clinicians, therapists) that have helped children cope with their disorder; common co-occurring conditions of children with disorders (e.g., comorbid disorders, i.e., learning disabilities and ADHD); and common attitudes of children, parents, and professionals relative to their experiences with childhood disorder.

All of the stories have highlighted particular styles and approaches parents and professionals utilized that have influenced the pattern of behaviors in these children and youth. In this regard, the stories revealed different intervention approaches parents used and different treatment models and approaches professionals used in their interactions with children and youth to help them with respect to their learning, behavioral, and emotional problems (e.g., psychodynamic models, behavioral models, cognitive behavioral models, affective models, and family system models).

These stories bring more consensus to the view that child psychopathology needs to be considered in light of the frequency, intensity, and chronicity of the maladaptive behaviors of children and youth as well as the antecedents and consequences, severity, cultural context, and gender qualities of these behaviors. Some of the personal and parental stories reveal to us that sometimes there is great resistance on the part of children and youth to change how they feel and act, which requires greater understanding, patience, persistence, flexibility, adaptability, and skill from parents and professionals. The stories also remind us that children and youth with disorders are not disordered in all areas; they have both strengths and weaknesses across a variety of personal, academic, and social areas, and they can fluctuate in and out of healthy or unhealthy states throughout their development. Moreover, it is further understood in some of these stories that, although some childhood disorders can continue into adulthood, some others can be effectively treated while in childhood and adolescence and not persist into adulthood as a result of such things as coping ability, intellectual ability, social ability, communicability, and the presence of support systems (e.g., home and school) associated with the child or youth.

The stories in this book also highlight the complexity of childhood disorders. The identification and treatment of children and youth with disorders need to be undertaken by well-trained, knowledgeable, skillful, and wise professionals. These professionals also need to be, among other things, personable, empathetic, and adaptable. Most importantly, there needs to be a very good fit between professional and child, professional and family, and treatment approach and child.

We end this book with a quote we used at the beginning of the book because we believe that the personal, parental, and professional stories captured the following understanding relative to having a childhood disorder, parenting a child with a disorder, and treating a child with a disorder:

> *"The keys for success are within us; once we find them, all we need to know is when and where to use them."*
>
> *Andrews, 2011*